Current Status of Low-Grade Minerals and Mine Wastes Recovery: Reaction Mechanism, Mass Transfer, and Process Control

Current Status of Low-Grade Minerals and Mine Wastes Recovery: Reaction Mechanism, Mass Transfer, and Process Control

Editors

Shenghua Yin
Leiming Wang

MDPI • Basel • Beijing • Wuhan • Barcelona • Belgrade • Manchester • Tokyo • Cluj • Tianjin

Editors
Shenghua Yin
School of Civil and Resource Engineering
University of Science and Technology Beijing
Beijing
China

Leiming Wang
School of Civil and Resource Engineering
University of Science and Technology Beijing
Beijing
China

Editorial Office
MDPI
St. Alban-Anlage 66
4052 Basel, Switzerland

This is a reprint of articles from the Special Issue published online in the open access journal *Minerals* (ISSN 2075-163X) (available at: www.mdpi.com/journal/minerals/special_issues/WastesRecovery_Minerals).

For citation purposes, cite each article independently as indicated on the article page online and as indicated below:

LastName, A.A.; LastName, B.B.; LastName, C.C. Article Title. *Journal Name* **Year**, *Volume Number*, Page Range.

ISBN 978-3-0365-7971-9 (Hbk)
ISBN 978-3-0365-7970-2 (PDF)

Contents

Preface to "Current Status of Low-Grade Minerals and Mine Wastes Recovery: Reaction Mechanism, Mass Transfer, and Process Control"

This reprint focuses on chemical/bio-leaching and utilization of low-grade minerals, solid wastes, and tailings in metal mines. It will provide a good reference for researchers in the fields of mining engineering, mineral processing engineering, hydrometallurgy, etc. Thank you very much to the MDPI editors and reviewers for their tremendous support and assistance to guest editors and authors in this editing process. Last but not least, the author's academic ability is limited, and we sincerely hope that readers can provide criticism and valuable suggestions.

Shenghua Yin and Leiming Wang
Editors

minerals

MDPI

Editorial

Editorial for Special Issue: "Current Status of Low-Grade Minerals and Mine Wastes Recovery: Reaction Mechanism, Mass Transfer, and Process Control"

Shenghua Yin [1,2] and Leiming Wang [1,2,*]

1 Key Laboratory of Ministry of Education for High-Efficient Mining and Safety of Metal, University of Science and Technology Beijing, Beijing 100083, China
2 School of Civil and Environment Engineering, University of Science and Technology Beijing, Beijing 100083, China
* Correspondence: ustb_wlm@126.com

Mineral resources provide an important material foundation for industrial construction and are important for the economy, thus being closely integrated into our daily lives and near future [1,2]. However, mineral resources still possess limitations, such as low mineral grades and serious secondary/concomitant deposits [3]. This undesirable situation causes great difficulty in traditional underground/surface mining, increasing operational costs. In this regard, solution mining (heap leaching, in situ leaching, etc.) offers an environmentally friendly, low-cost, and efficient method for extracting these minerals, especially copper sulfides (chalcopyrite, chalcocite, etc.), sandstone uranium, and sandstone gold deposits. After years of development, solution mining has made great progress, but it has also encountered some technical bottlenecks.

To better understand the current status of low-grade mineral and mine waste recovery, we collected 13 contributions in this Special Issue, whose topics can be divided into the following three main aspects:

- **Section 1**: Reaction mechanisms of chemical/bio-leaching—includes the leaching kinetics of copper sulfides, assisted leaching (chloride acidic leaching, iodide-assisted leaching, etc.), in situ leaching of uranium and salt deposits, etc.
- **Section 2**: Process detection, characterization, and visualization—includes the detection of reaction products, the visualization of fluid flow and mass transfer, the pore structure characterization of leaching systems (ore-packed beds, etc.), microbial successions of leaching bacteria, etc.
- **Section 3**: Recovery, recycling, and reuse of mine wastes—includes cleaner leaching, disposal and production methods (dump leaching, etc.), the recovery of mine waste (waste rock, tailings, etc.), the assessment of operations problems, etc.

Section 1 includes one review paper and three research papers. Nkuma et al. reviewed the metals accessed from low-grade ores and the related environmental impact considerations and then carefully examined the benefits of conventional versus bioleaching strategies [4]. Dushyantha et al. presented a case of mine waste management related to the recovery potential of rare earth elements (REEs) from the gem mining waste of Sri Lanka [5]. Rakishev et al. sought to improve the efficiency of downhole uranium production using oxygen as an oxidizer [6]. Han et al. revealed the flocculation and settlement characteristics of ultrafine tailings and microscopic characteristics of flocs [7]. Section 2 includes four research papers. Alfonso et al. detected the process mineralogy of tailings from Llallagua and discussed the recovery potential of tin reserves [8]; Chen et al. studied the mineralogical characteristics of pegmatite tailings and performed a beneficiation assessment of pollucite in the recovery of cesium [9]; Wang L. et al. discussed hydrodynamic hysteresis and solute transport in agglomerated heaps under irrigation, stacking, and bioleaching

Citation: Yin, S.; Wang, L. Editorial for Special Issue: "Current Status of Low-Grade Minerals and Mine Wastes Recovery: Reaction Mechanism, Mass Transfer, and Process Control". *Minerals* **2023**, *13*, 778. https://doi.org/10.3390/min13060778

Received: 31 May 2023
Accepted: 6 June 2023
Published: 7 June 2023

control [10]. Stepwise irrigation also plays a key role in liquid retention. Thus, Wang L. et al. deepened our understanding of the effect of stepwise irrigation on the liquid holdup and hysteresis behaviors of unsaturated ore heaps [11]. In Section 3, there are two review paper and three research papers. Yang et al. reviewed the comprehensive utilization of magnesium slag and development prospects of preparing backfilling materials [12]. Similarly, Chen et al. reviewed the art of roof contacting in cemented waste backfill technology in a metal mine [13]. Wang J. et al. focused on mine waste disposal and backfilling technology, discussing the energy dissipation, AE characteristics, and microcrack evolution of rock–backfill composite materials (RBCM) [14], and the energy dissipation mechanism and damage constitutive model of roof–CPB–floor (RCF) layered composite materials were investigated [15]. Wang L. et al. evaluated the stope stability and displacement in a subsidence area using 3Dmine–Rhino3D–FLAC3D coupling [16].

The authors who contributed to this Special Issue represent more than 10 different institutions (including Monash University Malaysia, the University of South Africa, University of Moratuwa, General Research Institute of Nonferrous Metals, Henan Polytechnic University, Central South University, University of Science and Technology Beijing, etc.) in seven different countries (China, Kazakhstan, Malaysia, Spain, Bolivia, South Africa, Sri Lanka). We thank them for their academic contributions to this Special Issue.

Last but not least, we thank the *Minerals* editorial staff for their cooperation and the invited reviewers for their valuable input.

Funding: This research was funded by the Program of the National Natural Science Foundation of China (52204124, 52034001); Beijing Natural Science Foundation (2232080); China National Postdoctoral Program for Innovative Talents (BX20220036), China Postdoctoral Science Foundation (2022M710356), Fundamental Research Funds for the Central Universities (FRF-IDRY-21-010, QNXM20220001); Open Project of the State Key Laboratory of Coal Mine Resources and Safety Mining (SKLCRSM22KF006); Open Project of State Key Laboratory of Safety and Health for Metal Mines (2021-JSKSSYS-01), Open Project of Key Laboratory of Green Chemical Engineering Process of Ministry of Education (GCP202108); and Fundamental Research Funds for the Central Universities (FRF-IDRY-21-010; QNXM20220001).

Conflicts of Interest: The authors declare no conflict of interest.

References

1. Wang, L.; Yin, S.; Deng, B.; Wu, A. Copper Sulfides Leaching Assisted by Acidic Seawater-Based Media: Ionic Strength and Mechanism. *Miner. Eng.* **2022**, *175*, 107286. [CrossRef]
2. Johnson, D.B. The Evolution, Current Status, and Future Prospects of Using Biotechnologies in the Mineral Extraction and Metal Recovery Sectors. *Minerals* **2018**, *8*, 343. [CrossRef]
3. Wang, L.; Zhang, X.; Yin, S.; Zhang, X.; Liu, P.; Ilankoon, I.M.S.K. Three-Dimensional Characterisation of Pore Networks and Fluid Flow in Segregated Heaps in the Presence of Crushed Ore and Agglomerates. *Hydrometallurgy* **2023**, *219*, 106082. [CrossRef]
4. Nkuna, R.; Ijoma, G.N.; Matambo, T.S.; Chimwani, N. Accessing Metals from Low-Grade Ores and the Environmental Impact Considerations: A Review of the Perspectives of Conventional versus Bioleaching Strategies. *Minerals* **2022**, *12*, 506. [CrossRef]
5. Dushyantha, N.; Ilankoon, I.M.S.K.; Ratnayake, N.P.; Premasiri, H.M.R.; Dharmaratne, P.G.R.; Abeysinghe, A.M.K.B.; Rohitha, L.P.S.; Chandrajith, R.; Ratnayake, A.S.; Dissanayake, D.M.D.O.K.; et al. Recovery Potential of Rare Earth Elements (REEs) from the Gem Mining Waste of Sri Lanka: A Case Study for Mine Waste Management. *Minerals* **2022**, *12*, 1411. [CrossRef]
6. Rakishev, B.; Kenzhetaev, Z.; Mataev, M.; Togizov, K. Improving the Efficiency of Downhole Uranium Production Using Oxygen as an Oxidizer. *Minerals* **2022**, *12*, 1005. [CrossRef]
7. Han, C.; Tan, Y.; Chu, L.; Song, W.; Yu, X. Flocculation and Settlement Characteristics of Ultrafine Tailings and Microscopic Characteristics of Flocs. *Minerals* **2022**, *12*, 221. [CrossRef]
8. Alfonso, P.; Ruiz, M.; Zambrana, R.N.; Sendrós, M.; Garcia-Valles, M.; Anticoi, H.; Sidki-Rius, N.; Salas, A. Process Mineralogy of the Tailings from Llallagua: Towards a Sustainable Activity. *Minerals* **2022**, *12*, 214. [CrossRef]
9. Chen, Y.; Wen, J.; Song, Y.; Li, W.; Liu, S.; Liu, Y. Mineralogical Characteristics of Pegmatite Tailings and Beneficiation Assessment of Pollucite in Recovering Cesium. *Minerals* **2022**, *12*, 541. [CrossRef]
10. Wang, L.; Yin, S.; Zhang, X.; Yan, Z.; Liao, W. Hydrodynamic Hysteresis and Solute Transport in Agglomerated Heaps under Irrigation, Stacking, and Bioleaching Controlling. *Minerals* **2022**, *12*, 1623. [CrossRef]
11. Wang, L.; Yin, S.; Deng, B. Understanding the Effect of Stepwise Irrigation on Liquid Holdup and Hysteresis Behavior of Unsaturated Ore Heap. *Minerals* **2021**, *11*, 1180. [CrossRef]

12. Yang, X.; Dong, F.; Zhang, X.; Li, C.; Gao, Q. Review on Comprehensive Utilization of Magnesium Slag and Development Prospect of Preparing Backfilling Materials. *Minerals* **2022**, *12*, 1415. [CrossRef]
13. Chen, F.; Liu, J.; Zhang, X.; Wang, J.; Jiao, H.; Yu, J. Review on the Art of Roof Contacting in Cemented Waste Backfill Technology in a Metal Mine. *Minerals* **2022**, *12*, 721. [CrossRef]
14. Wang, J.; Zhang, C.; Song, W.; Zhang, Y. The Energy Dissipation, AE Characteristics, and Microcrack Evolution of Rock–Backfill Composite Materials (RBCM). *Minerals* **2022**, *12*, 482. [CrossRef]
15. Wang, J.; Zhang, C.; Fu, J.; Song, W.; Zhang, Y. The Energy Dissipation Mechanism and Damage Constitutive Model of Roof–CPB–Floor (RCF) Layered Composite Materials. *Minerals* **2022**, *12*, 419. [CrossRef]
16. Wang, L.; Zhang, X.; Yin, S.; Zhang, X.; Jia, Y.; Kong, H. Evaluation of Stope Stability and Displacement in a Subsidence Area Using 3Dmine–Rhino3D–FLAC3D Coupling. *Minerals* **2022**, *12*, 1202. [CrossRef]

minerals

MDPI

Article

Recovery Potential of Rare Earth Elements (REEs) from the Gem Mining Waste of Sri Lanka: A Case Study for Mine Waste Management

Nimila Dushyantha [1,2,*], I. M. Saman K. Ilankoon [3,*], N. P. Ratnayake [1], H. M. R. Premasiri [1], P. G. R. Dharmaratne [1], A. M. K. B. Abeysinghe [1], L. P. S. Rohitha [1], Rohana Chandrajith [4], A. S. Ratnayake [2], D. M. D. O. K. Dissanayake [1] and N. M. Batapola [1]

1 Department of Earth Resources Engineering, University of Moratuwa, Moratuwa 10400, Sri Lanka
2 Department of Applied Earth Sciences, Faculty of Applied Sciences, Uva Wellassa University, Passaara Road, Badulla 90000, Sri Lanka
3 Discipline of Chemical Engineering, School of Engineering, Monash University Malaysia, Jalan Lagoon Selatan, Bandar Sunway, Subang Jaya 47500, Malaysia
4 Department of Geology, Faculty of Science, University of Peradeniya, Peradeniya 20400, Sri Lanka
* Correspondence: nimila.dush@gmail.com (N.D.); saman.ilankoon@monash.edu (I.M.S.K.I.)

Citation: Dushyantha, N.; Ilankoon, I.M.S.K.; Ratnayake, N.P.; Premasiri, H.M.R.; Dharmaratne, P.G.R.; Abeysinghe, A.M.K.B.; Rohitha, L.P.S.; Chandrajith, R.; Ratnayake, A.S.; Dissanayake, D.M.D.O.K.; et al. Recovery Potential of Rare Earth Elements (REEs) from the Gem Mining Waste of Sri Lanka: A Case Study for Mine Waste Management. *Minerals* **2022**, *12*, 1411. https://doi.org/10.3390/min12111411

Academic Editors: Eric D. van Hullebusch, Shenghua Yin and Leiming Wang

Received: 26 September 2022
Accepted: 5 November 2022
Published: 7 November 2022

Publisher's Note: MDPI stays neutral with regard to jurisdictional claims in published maps and institutional affiliations.

Abstract: Sri Lanka is one of the wealthiest countries in terms of gems. Therefore, gem mining is extensively carried out in many areas of Sri Lanka, including districts such as Ratnapura, Monaragala, Matale, and Kalutara. During the mining process, only valuable gemstones are collected, and the remaining gravel fraction with many heavy minerals is discarded. Therefore, the gem mining industry produces a large amount of waste that is mainly used only for backfilling. To sustainably manage this waste stream, gem mining waste collected from a gem pit at Wagawatta in the Kalutara district in Sri Lanka was investigated for value recovery, specifically for rare-earth elements (REEs). The gem-bearing alluvial layer contained 0.3% rare-earth oxide (REO) that could easily be upgraded up to 2.8% (LREEs = 94%) with wet sieving and subsequent density separation via a shaking table. Therefore, the concentrates of gem mine tailings with REE-bearing minerals have the potential to be a secondary source for LREEs. The organic-rich clay layer underlying the gem-bearing alluvial layer contained 0.6% REO with 49% HREEs, including Gd, Dy, Er, Yb, and especially Y enrichments. Detailed explorations are thus necessary to assess the REE potential in Sri Lanka's gem mining waste, and value recovery flowsheets should be subsequently developed to economically extract REEs. In addition, the presence of high U concentrations (800 mg/kg) in the concentrated samples could be alarming when considering the health and safety of the people engaged in gem mining. This aspect also requires detailed research studies.

Keywords: gem mining waste; rare-earth elements; Sri Lankan gems; tailing management; uranium

1. Introduction

Since the mining industry produces large volumes of tailings, mine tailings of past and present mining activities have become a potential secondary source for many critical raw materials, including rare earth elements (REEs). Several authors presented mine waste management and value recoveries for different mine waste materials, such as copper [1,2], vanadium [1], gold [3], and REEs [4].

REEs are a set of 17 elements comprising 15 lanthanides, Y, and Sc. They are further categorised into two subgroups, namely, light rare earth elements (LREEs) (from La to Eu) and heavy rare earth elements (HREEs) (from Gd to Lu, including Y). Despite the crustal abundance, these elements are typically not found in nature in high concentrations, and similar chemical properties complicate the separation processes [5]. Therefore, the mining and processing of REEs are complex and expensive [6]. These elements have become indispensable ingredients in the global industrial revolution due to their unique properties such as high thermal stability, strong magnetism, and high electric conductivity. As a result, they are used in various applications in high-tech, green energy, military, and aerospace industries [5,7]. Therefore, the demand for REEs is continuously increasing, especially considering the developments of electric vehicles and wind turbines. However, the Chinese rare earth industry mainly fulfils the current global rare earth consumption, including value-added products such as rare earth magnets [5]. These factors have attracted significant attention to the exploration and extraction of REEs from new and secondary sources [8] to diversify global rare earth supply chains [5].

The enrichments of REEs were reported in uranium, iron-ore, and titanium mine tailings in Australia, Kazakhstan, and Sweden [9,10], and the authors in [11] investigated the economic viability of extracting REEs from iron-ore mine tailings in the Kiruna mine. In this context, the authors in [12,13] reported that gem-bearing sediments are enriched in rare earth-bearing minerals. However, in Sri Lanka, it has not been attempted to utilise gem mining waste to produce REE concentrates through physical separation techniques. This article thus addresses that research gap, considering the significance of the gem mining industry to the gross domestic product (GDP) in Sri Lanka. It is of paramount importance to design economic mineral processing techniques for waste materials such as gem mining waste in Sri Lanka while considering local socio-techno-economic factors.

Since 250 BCE, Sri Lanka has been famous for its wide variety of gems, gem industry, and international gem trading. Some gemstones have adorned crowns and thrones in the royal coronations and religious functions of ancient rulers [14]. The export value of precious and semiprecious stones was USD 145 million and 141 million in 2017 and 2018, respectively. This is equivalent to 1.3% and 1.2% of the total exports in 2017 and 2018, respectively [15]. This implies the requirement for the sustainable management and development of the gem mining industry in Sri Lanka, including the generated mining waste. Mitigating the damage to fragile ecosystems during gem mining is imperative, and artisanal gem mining activities in Madagascar [16] and Myanmar [17] have caused adverse environmental effects.

Secondary gem deposits are primarily found in Sri Lanka as sedimentary rocks and placers. During gem mining, gem-bearing sediments are piled up near the pit head and mechanically washed or manually panned using rattan baskets (i.e., storage vessels produced from rattan palm fibres) to separate the heavy fractions of sediments [14,18]. A fraction of the generated gem mining waste is only used for backfilling. Even though mining activities are regulated by the National Gem and Jewellery Authority, gem mining is a traditional industry in Sri Lanka [14]. Thus, a significant volume of gem mining waste has accumulated in gem pits. In addition, current gem mining activities in gem mining areas (Figure 1) produce additional volumes of gem mining waste. However, the waste volumes have not been estimated. Existing volumes typically result in environmental issues such as land degradation, river-bank and soil erosion, sedimentation, landslides, groundwater pollution and reduced irrigation efficiency. In addition, the gem pits with accumulated water become breeding grounds for mosquitoes; thus, this affects the health and safety of the people living around gem mining areas. These research gaps highlight the requirements of sustainable gem mining activities in Sri Lanka, including gem mining waste management and repurposing strategies. However, these are not explicitly addressed in the gem mining industry in Sri Lanka and other countries, such as Madagascar and Myanmar. This study was thus designed to address the identified research gaps. In this context, gem mining waste could also be a potential source for the economic extraction of REEs. Therefore, this study assesses the REE potential of gem mining waste in Sri Lanka by performing economic value concentration processes. Thus, the physical separation methods of wet sieving and density separation are investigated.

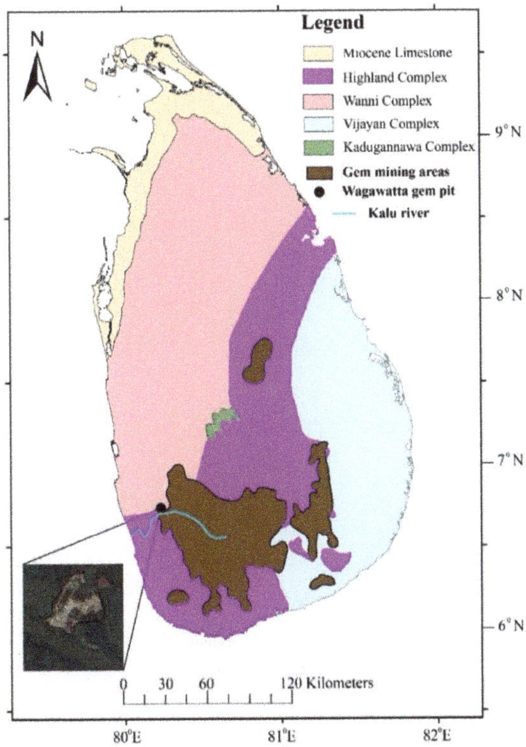

Figure 1. Location of the gem pit in Wagawatta on a simplified geological map of Sri Lanka (source: reproduced from [19] with permission from Precambrian Research) with the main gem mining areas [20].

2. Study Area

Geologically, the Precambrian basement rocks of Sri Lanka are divided into three main lithotectonic divisions: the Highland, Vijayan, and Wanni Complexes [19]. In addition, the island's northern and northwestern coastal belts are underlain by Miocene limestones (Figure 1). Among these complexes, gem deposits are abundantly found in the southern part of the Highland Complex (Figure 1). In addition, this area consists of REE-bearing mineral sources such as granites and pegmatites [14].

Kalu Ganga River has the second largest river basin in Sri Lanka, with a 2766 km^2 drainage area originating from the central hills [21,22]. Since Sri Lanka is a tropical country, a high yield of sediments is evident due to the extreme weathering conditions [23–25]. In this context, after the weathering and decomposition of the gem-bearing source rocks in the catchment area, the Kalu Ganga River transports these sediments and redeposits them as alluvial layers, especially in old river channels and flood plains where the gemstones are found [14]. A gem pit in Wagawatta in the Kalutara district was thus selected for this study. The gem layer is deposited in an old river channel with an organic-rich clay layer underneath the alluvial gem-bearing gravel layer (Figure 2).

In Sri Lankan gem mining, the alluvial layer is subjected to wet sieving or/and panning to find gemstones. Miners typically discard the remaining fractions, including gem gravel and fine tailing fractions containing heavy minerals, as waste. In this study, the authors employed such a fine tailing fraction with heavy minerals as the feed of the shaking table (Figure 3). In addition, the organic layer was mined to find the gem-bearing alluvial layers underneath. In this process, the mined organic layer is thus stored on site. Lastly, all the waste materials, including gem gravel, fine tailings with heavy minerals, and the organic layer, are used for backfilling the gem mining pit [26]. The authors thus focused on all these waste materials in this work.

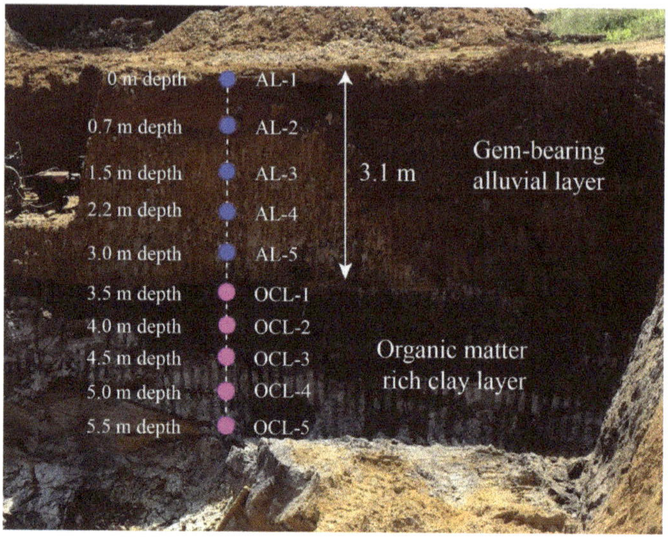

Figure 2. Sample locations with an illustration of the gem-bearing alluvial layer and organic-rich clay layers of the gem pit.

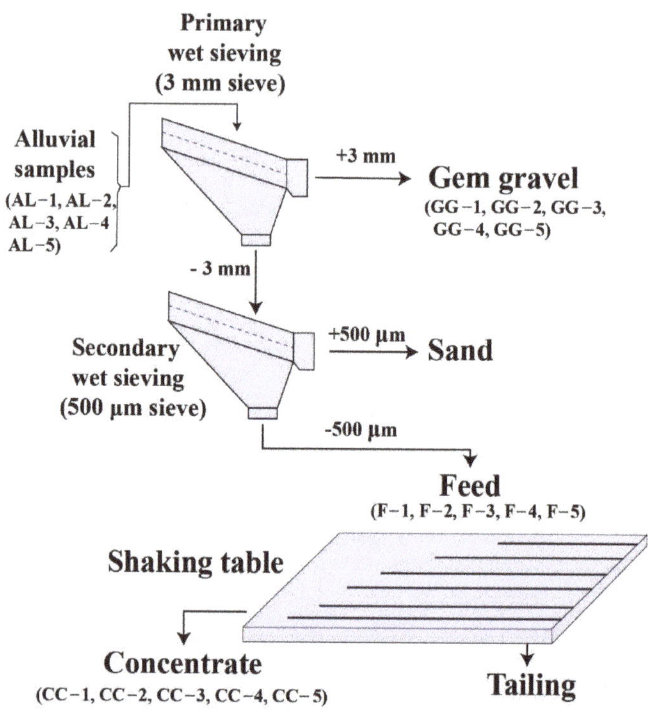

Figure 3. Physical separation flowsheet of gem-bearing alluvial samples.

3. Materials and Methods

3.1. Sample Collection

Five samples from the gem-bearing alluvial layer (AL-1 to AL-5) and five samples from the organic-rich clay layer (OCL-1 to OCL-5) were collected at different heights of the gem pit (Figure 2).

3.2. Physical Separation

Gem-bearing sediment samples in the alluvial layer (AL-1 to AL-5) were subjected to primary wet sieving using a 3 mm sieve. The oversized fraction (>3 mm) was obtained as the gem gravel samples (GG-1 to GG-5), and the undersized fraction (<3 mm) was further subjected to secondary wet sieving using a 500 μm sieve (Figure 3). After the secondary wet sieving, the oversized fraction (>500 μm) was obtained as the sand fraction. The undersized fraction (<500 μm or Samples F-1 to F-5) was further subjected to a density separation process via a shaking table (Figure 3). After the separation, the heavy fraction was considered to be the concentrate (CC-1 to CC-5). The lighter fraction was taken as the tailing (T-1 to T-5) (Figure 3). The density separation was conducted using a Wilfley shaking table employed at a deck angle of 8°, stoke amplitude of 10 mm, and wash water flow rate of 3 L/min for 1 kg of feed (pulp density 25% *w*/*w*). Optimal deck angle and stroke amplitude values were employed on the basis of preliminary experiments.

In addition, particle size distributions (PSDs) of the feed (F-1 to F-5), concentrate (CC-1 to CC-5), and tailings (T-1 to T-5) were ascertained using a laser particle size analyser (AimSizer HMK CD-02). The PSDs are shown in Figure 4. The D_{50} values for the feed, concentrate, and tailings were 90.2, 35.9 and 65.5 μm, respectively. In addition, D_{10} values were 20.3, 10.6 and 10.3 μm for the feed, concentrate, and tailings, respectively. Furthermore, the determined D_{90} values of the feed, concentrate, and tailings were 324.1, 152.1 and 327.6 μm, respectively.

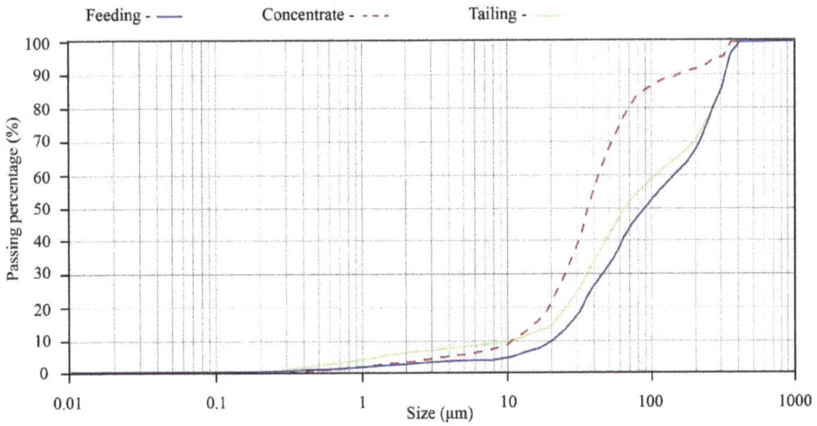

Figure 4. Particle size distributions of the feed (F-1 to F-5), concentrate (CC-1 to CC-5), and tailing (T1-T5) samples resulted from the physical separation process employed in Figure 3.

3.3. Sample Analysis

3.3.1. REE and U Analysis

All the samples were oven-dried at 105 °C, and the dried samples were then powdered using a laboratory Tema mill. The powdered samples were sieved with a 63 μm sieve, and the representative samples were selected via coning and quartering for further analysis.

Subsamples (0.2 g) were digested with an HNO_3 and HCl mixture (1 mL:3 mL) and 1 mL of H_2O_2 using a Mars-6 microwave digester (CEM; Mathews, NC) equipped with EasyPrep Plus high-pressure vessels. The digested samples were diluted with deionized water and then analysed with an inductively coupled plasma mass spectrometer (ICP-MS) (ICapQ-Thermo Fisher, Bremen, Germany). The instrument was calibrated according to multielemental ICP-MS standards (Sigma-Aldrich, Germany). Quality control in the analysis was maintained using certified international reference samples (San Joaquin NIST SRM 2709a from Sigma-Aldrich, Germany), replicate analysis, and blanks at every tenth sample.

3.3.2. Mineralogical Analysis of Concentrated Samples

Mineralogical compositions of samples were determined with powder X-ray diffraction (XRD) using a BRUKER D8 advance eco X-ray diffractometer with normal operating conditions. Phase analysis was carried out using the International Centre for Diffraction database (ICDD). In addition, scanning electron microscopic (SEM) and energy-dispersive spectrometric (EDS) analyses were carried out using a Carl Zeiss EVO-18 instrument with an EDS element on the SEM.

4. Results and Discussion

4.1. REE Concentrations of Alluvial and Gem Gravel Samples

The average total rare earth element (TREE) concentrations of alluvial (AL) and gem gravel (GG) samples were 2188 and 819 mg/kg, respectively (Table 1). These results reveal that REEs are associated with the finer fraction of the alluvial layer (<3 mm), since the gem gravels contained a relatively low content of REEs. Both alluvial and gem gravel samples showed a higher fraction of LREEs (85% and 80%, respectively) over HREEs (12% and 16%, respectively). The rare earth oxide (REO) grades of alluvial (AL) and gem gravel (GG) samples were 0.3% and 0.1%, respectively (Table 1).

4.2. REE Concentrations of Organic-Rich Clay Samples

The average TREE concentration of the organic-rich clay layer (OCL) was 5348 mg/kg, whereas HREEs such as Gd, Dy, Er, Yb and Y showed higher average concentrations (191, 219, 167, 135 and 1766 mg/kg, respectively) compared to other HREEs. As a result, HREEs were 49%, almost the same as the LREE percentage of 48% (the remaining 3% was Sc) (Table 1). In addition, the REO percentage of this clay layer was 0.6. LREEs and HREEs were thus enriched in the organic-rich clay layer (OCL) compared to the alluvial layer (Figure 5).

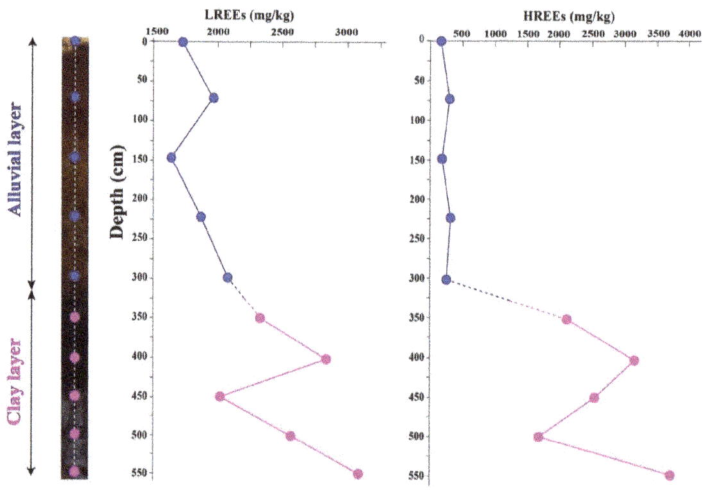

Figure 5. Distribution of LREEs and HREEs in the vertical cross-section, including alluvial (AL) and organic-rich clay (OCL) layers.

Table 1. REE and U concentrations (mg/kg) of alluvial (AL), gem gravel (GG), and organic-rich clay later (OCL) samples (bdl—below the detection limit, AVG—average, SD—standard deviation).

Sample ID	La	Ce	Pr	Nd	Sm	Eu	Gd	Tb	Dy	Ho	Er	Tm	Yb	Lu	Y	Sc	LREE	HREE	TREE	LREE%	HREE%	Sc%	REO%	U
AL-1	413	875	93	308	48	11	46	13	24	12	17	11	15	bdl	98	51	1746	237	2034	86	12	2	-	8.3
AL-2	483	955	121	343	50	12	58	14	27	16	20	12	17	bdl	122	74	1963	285	2322	85	12	3	-	9.1
AL-3	370	833	80	308	59	13	55	15	25	14	18	11	17	bdl	92	34	1662	246	1941	86	13	1	-	8.6
AL-4	440	913	108	343	61	14	67	16	27	17	21	12	19	bdl	117	57	1878	294	2229	84	13	3	-	9.4
AL-5	525	998	134	343	39	10	49	12	26	14	19	12	16	bdl	128	92	2047	275	2414	85	11	4	-	8.8
AL-AVG	446	915	107	329	51	12	55	14	26	14	19	12	17	bdl	111	62	1859	267	2188	85	12	3	0.3	8.8
STD	60	65	21	19	9.1	1.6	8	1.6	1.2	1.7	1.2	0.5	1.4	-	16	22	156	25	197	-	-	-	-	0.4
GG-1	133	268	35	108	23	10	21	10	11	10	8.5	9	8.3	bdl	30	31	576	108	715	81	15	4	-	6.3
GG-2	171	333	47	145	30	11	28	12	12	12	12	11	10	bdl	47	44	736	143	923	80	15	5	-	8.1
GG-3	109	217	35	92	18	11	18	11	11	12	10	11	10	bdl	19	29	481	101	611	79	17	4	-	8.7
GG-4	147	282	47	129	25	12	25	12	12	14	13	13	12	bdl	36	42	641	137	819	78	17	5	-	10.5
GG-5	195	383	47	162	35	10	31	11	12	10	11	9	8	bdl	58	47	831	150	1027	82	14	4	-	5.7
GG-AVG	151	296	42	127	26	11	24	11	12	11	11	11	10	bdl	38	39	653	128	819	80	16	4	0.1	7.9
STD	33	64	6.7	28	6.4	0.7	5.2	0.8	0.4	1.6	1.7	1.7	1.7	-	15	8	136	22	165	-	-	-	-	1.9
OCL-1	553	973	135	505	115	25	172	32	179	46	129	24	107	13	1473	113	2304	2174	4591	50	47	3	-	7.6
OCL-2	658	1205	178	624	149	33	212	57	248	79	207	48	162	18	2105	150	2846	3135	6131	46	51	3	-	8.9
OCL-3	493	850	123	450	93	31	147	30	171	39	86	11	85	15	1043	92	2039	1627	3758	54	43	3	-	8.4
OCL-4	598	1083	166	569	127	39	186	55	239	72	165	36	140	21	1675	129	2581	2588	5298	50	47	3	-	9.7
OCL-5	718	1328	190	679	171	27	238	58	256	85	250	61	183	16	2535	171	3111	3682	6963	42	55	3	-	8.1
OCL-AVG	604	1088	158	565	131	31	191	46	219	64	167	36	135	17	1766	131	2576	2641	5348	48	49	3	0.6	8.5
STD	88	188	29	91	30	6	35	14	40	20	64	20	40	3	575	31	425	803	1258	-	-	-	-	0.8

Economic concentrations of HREEs are abundantly found in ion-adsorption clay (IAC) deposits in South China and Madagascar in economic extractions. The REO percentage of IAC deposits in China varies from 0.05 to 0.2 [27], whereas the HREE concentrations of IAC deposits in Madagascar goes up to 1570 mg/kg [28]. Gd, Dy, Er, Yb, and especially Y in both deposits mainly contribute to the enriched HREE content. In this context, the organic-rich clay layer (OCL) in this gem deposit showed a similar variation, with a high content of Y, and possessed a potential source of HREEs, especially for Y. Therefore, the development of a suitable extraction method is necessary to extract HREEs from the organic-rich clay layer in this geochemical formation.

4.3. REE Concentrations of Concentrated Samples

The average REE concentration and REO grade of the concentrated samples (CC1-CC5) were 24,670 mg/kg and 2.8%, respectively (Table 2). This clearly showed that the wet sieving and subsequent density separation process via a shaking table significantly enriched the REE content in the concentrated samples, which upgraded the REO percentage by almost 10-fold compared to the original alluvial samples. The LREEs percentage in the concentrated samples accounted for about 94% of the TREE contents (Table 2). Therefore, these concentrated samples could be a secondary source for REEs, particularly for LREEs. In addition, the wet sieving and density separation process via a shaking table is a promising physical separation method to upgrade the REO grade of gem mining waste before further physical beneficiations and chemical leaching.

4.4. REE Patterns of Collected Samples

Figure 6 illustrates the chondrite-normalised REE patterns of the studied samples. Significant enrichments of LREEs in the concentrated samples (CC) and HREEs in the organic-rich clay (OCL) samples were evident in these normalised patterns. All samples showed positive Ce anomalies and significantly negative Eu anomalies. Comparing the chondrite-normalised patterns of IAC deposits in China, the organic-rich clay layer shows that HREEs such as Gd, Dy, Er and Yb were enriched in the organic-rich clay layer.

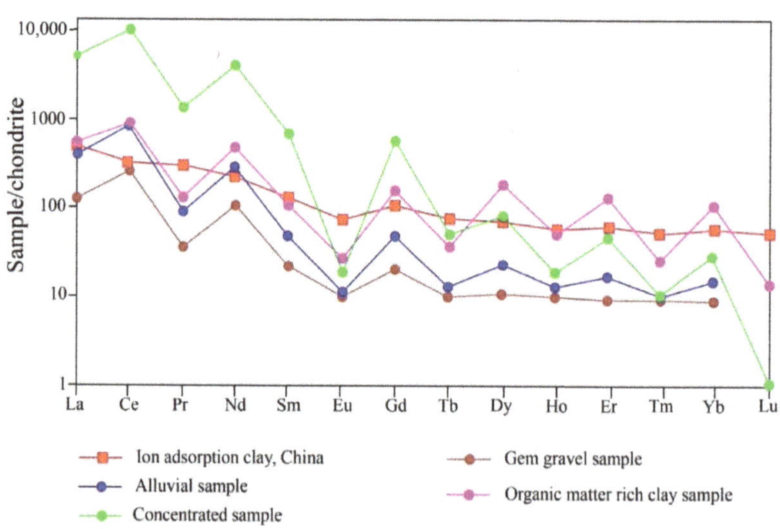

Figure 6. Chondrite-normalised REE patterns of the studied samples. Since Lu concentrations were below the detection limit, Lu values in alluvial and gem gravel samples are not plotted in this figure. Published REE values of IAC deposits in China and chondrite-normalised factors were adapted from [29,30], respectively.

Table 2. REE and U concentrations (mg/kg) of feed (F) and concentrated (CC) samples. (bdl—below the detection limit, AVG—average, SD—standard deviation).

Sample ID	La	Ce	Pr	Nd	Sm	Eu	Gd	Tb	Dy	Ho	Er	Tm	Yb	Lu	Y	Sc	LREE	HREE	TREE	LREE%	HREE%	Sc%	REO%	U
F-1	2854	6116	656	2238	392	11	271	27	43	11	25	7	14	bdl	123	66	12,266	521	12,854	95	5	0	-	457
F-2	2671	6012	682	2101	329	10	281	23	45	13	29	9	17	bdl	144	83	11,805	521	12,449	95	5	0		511
F-3	2816	5928	671	2865	389	12	278	21	39	12	27	5	18	bdl	137	74	12,681	561	13,292	95	5	0		472
F-4	2743	6231	599	2003	401	14	269	29	46	25	21	7	15	bdl	116	90	11,991	537	12,609	95	5	0		443
F-5	2903	6193	612	2362	381	11	288	30	44	10	29	9	11	bdl	126	87	12,462	528	13,096	95	5	0		470
F-AVG	2797	6096	644	2314	378	12	277	26	43	14	26	7	15	bdl	129	80	12,241	534	12,860	95	5	0	1.4	471
STD	92	126	37	337	29	2	8	4	3	6	3	2	3	-	11	10	352	17	344	0	0	0	-	25
CC-1	6138	12,455	1435	4953	718	25	635	52	103	23	52	14	34	1.5	293	77	25,722	1206	27,005	95	5	1	-	794
CC-2	5198	10,478	1768	3465	880	17	738	68	77	18	72	8.5	22	0.8	360	110	21,804	1364	23,279	94	3	1	-	806
CC-3	6020	11,868	1235	5513	685	22	615	47	97	24	41	14	43	1.5	234	52	25,342	1116	26,510	96	5	1	-	717
CC-4	5080	9890	1568	4025	848	14	718	62	71	19	62	8.3	32	0.8	302	86	21,424	1274	22,783	94	6	1	-	871
CC-5	5315	11,065	1968	2905	913	20	758	73	84	17	83	8.8	13	0.8	419	135	22,185	1455	23,774	93	5	1	-	814
CC-AVG	5550	11,151	1595	4172	809	19	693	60	86	20	62	11	29	1.1	321	92	23,295	1283	24,670	94	5	1	2.8	800
STD	492	1033	285	1065	101	4	64	11	13	3	16	3	12	0.4	70	32	2064	132	1945	-	2	-	-	-

4.5. REE Yield, Grade, and Recovery of Separation Experiments

During the shaking-table tests, the average yield of the concentrate was 37%, whereas tailing accounted for 63% of the yield (Table 3). The TREO grades of feed, concentrate, and tailings were 1.4, 2.8 and 0.6, respectively. The recovery of TREO via the shaking table was 74%. The recoveries of LREO (light rare earth oxide) and HREO (heavy rare earth oxide) via the shaking table with respect to the feed were 74% (La_2O_3 = 76, Ce_2O_3 = 71, Pr_2O_3 = 87, Nd_2O_3 = 74, Sm_2O_3 = 73 and Eu_2O_3 = 75) and 90% (Gd_2O_3 = 91, Tb_2O_3 = 79, Dy_2O_3 = 81, Ho_2O_3 = 72, Er_2O_3 = 87, Tm_2O_3 = 65, Yb_2O_3 = 79 and Y_2O_3 = 95), respectively. The results reveal that the shaking-table separation could remove major gangue minerals in the feed and enrich REEs in the concentrate. Due to the high recoveries, grade, and enrichment factors of REEs, the concentrate (CC) possessed a high potential of being a secondary source for REEs, particularly for LREEs.

Table 3. Results of physical separation through Wilfley shaking table.

Product	Yield %	LERO		HREO		TREO	
		Grade %	Recovery %	Grade %	Recovery %	Grade %	Recovery %
Feed	100	1.33	100	0.06	100	1.4	100
Concentrate	37	2.65	74	0.14	90	2.8	74
Tailing	63	0.54	26	0.01	10	0.6	20

4.6. REE-Bearing Minerals in the Concentrate

REE-bearing minerals such as monazite, zircon, and xenotime were identified in the concentrate (Figure 7). In addition, the SEM/EDX spot analysis indicates the presence of LREEs, Y and phosphorus in the surface sediment samples (Figure 8). The presence of monazite and zircon justified the high concentrations of LREEs. In contrast, xenotime acted as the main HREE-bearing mineral, particularly for Y.

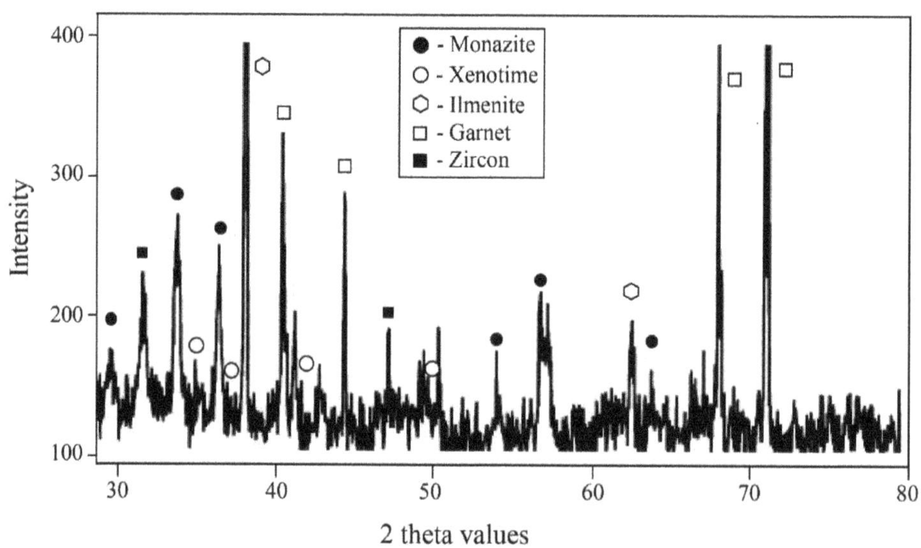

Figure 7. Concentrate's XRD patterns indicating REE-bearing and other heavy minerals.

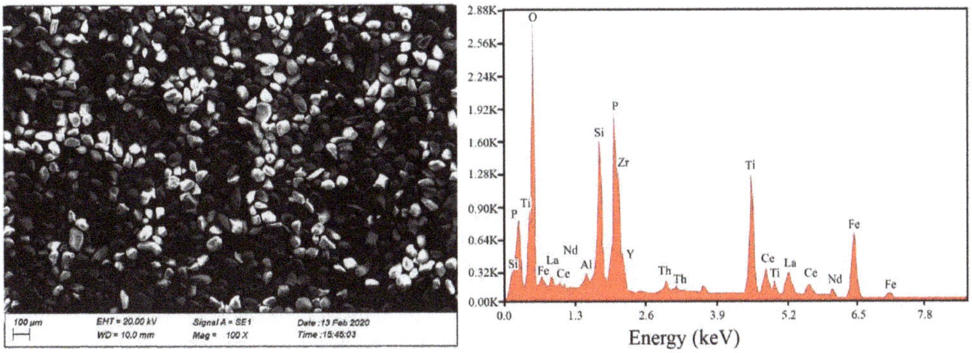

Figure 8. SEM/EDX spot analysis of the concentrate.

In addition, heavy minerals such as ilmenite and garnet were enriched in the concentrate. The organic-rich clay layer showed a lower abundance of REE-bearing minerals. However, the REO was as high as 0.6% compared to the alluvial layer. Therefore, that REE minerals may have been dissolved into the groundwater in the form of REE ions due to the weathering and decomposition of the alluvial layer (Figure 2). Once groundwater penetrates through the alluvial layer, dissolved components can be migrated into the clay layer, and REE ions could be adsorbed onto the surface of the clay minerals. A similar formation could be found in IAC rare earth deposits in China [31]. Despite the low REO content, such geochemical formations render REE extraction easy and economically viable.

4.7. Mine Waste, Value Recovery and Presence of U

Considering recent trends in mine tailing management and critical metals [1,4], the investigation of REE potential in gem mining waste in Sri Lanka is crucial on two fronts, namely, value recovery from mine waste streams, and sustainable and cleaner mine waste management. In addition, significant levels of U were found in the studied samples that need to be addressed, considering the waste management aspects of gem mining in Sri Lanka. U concentrations ranged from 8.8 to 800 mg/kg in alluvial and concentrated samples, respectively (Tables 1 and 2). Since zircon, monazite and xenotime were the REEs' primary carrier minerals, U may also be enriched by these minerals in the concentrated samples. According to the soil quality guidelines (SQGs) recommended by the Canadian Council of Ministers of the Environment (CCME), the permissible U concentration for industrial land use is 300 mg/kg [32]. Therefore, the concentrated samples in this study exceeded the permissible U level.

In some gem mines in Sri Lanka, gem-bearing sediments are subjected to a density separation process using a shaking table to recover any available gold nuggets. The concentrates are thus piled up in gem mining sites without proper safe disposal procedures. Therefore, workers may be exposed to a high level of radiation, and their health could be at increased risk [33]. Detailed studies are recommended to investigate this aspect in the gem mining industry in Sri Lanka to ensure that the industry is cleaner and more sustainable.

5. Conclusions

Gem mining in Sri Lanka is intensively carried out and annually generates tonnes of mine tailings. The quantification of REE content and REO grade is essential to sustainably manage this waste stream and to assess the REE resource potential in the gem waste of Sri Lanka. In the Wagawatta gem pit, the alluvial layer contained 0.3% REO. A simple physical separation method consisting of wet sieving and subsequent density separation via a shaking table was employed to upgrade the REO up to 2.8% with a 94% LREE. Therefore, the concentrated samples are a high potential source for LREEs. The organic-rich clay layer formed underneath the alluvial layer contained 0.6% REO and 49% HREEs. Therefore, the organic-rich clay layer could also be considered a potential HREE source.

Since this study was limited to a particular gem mine, a substantial amount of REEs could be extracted from the waste tailings in the gem mining industry. Therefore, it is necessary to assess the REE potential in gem mine waste in the entire country (this can be performed by estimating the gem mining waste generation at each mine for a given period). Subsequently, sustainable process flowsheets can be developed to extract REEs economically. In addition, it is recommended to investigate the background radiation levels in gem mining sites. Furthermore, adverse radioactive impacts must also be mitigated to maintain the sustainability of the REE extraction from the gem mining waste in Sri Lanka, fostering the United Nations Sustainable Development Goals (SDGs) such as responsible production and consumption (SDG 12), and life on land (SDG 15).

Author Contributions: Conceptualisation, N.D., I.M.S.K.I., N.P.R. and H.M.R.P.; data curation, N.D. and N.M.B.; formal analysis, N.D., I.M.S.K.I., N.M.B., R.C. and A.S.R.; funding acquisition, N.P.R., H.M.R.P., P.G.R.D., A.M.K.B.A., L.P.S.R. and D.M.D.O.K.D.; investigation, N.D., I.M.S.K.I. and N.M.B.; methodology, N.D. and N.M.B.; project administration, N.D., N.P.R., H.M.R.P., P.G.R.D., A.M.K.B.A., L.P.S.R., R.C., A.S.R. and D.M.D.O.K.D.; resources, N.D., N.P.R., H.M.R.P., P.G.R.D., A.M.K.B.A., L.P.S.R., R.C., A.S.R. and D.M.D.O.K.D.; supervision, N.P.R., H.M.R.P., P.G.R.D., A.M.K.B.A., L.P.S.R., R.C., A.S.R. and D.M.D.O.K.D.; validation, N.D., I.M.S.K.I., N.P.R., H.M.R.P., L.P.S.R.; writing—original draft, N.D. and N.M.B.; writing—review and editing, I.M.S.K.I., N.D., N.P.R., H.M.R.P., P.G.R.D., A.M.K.B.A., L.P.S.R., R.C., A.S.R. and D.M.D.O.K.D. All authors have read and agreed to the published version of the manuscript.

Funding: The authors wish to acknowledge the financial and other support provided by the Accelerating Higher Education and Development (AHEAD) Operation of the Ministry of Higher Education of Sri Lanka, funded by the World Bank (AHEAD/DOR/6026-LK/8743-LK), to carry out this study. Additionally, this work was supported by the Engineering Research Node of "Advanced Waste Engineering and Management (A-WEM)" of the Faculty of Engineering, Monash University.

Data Availability Statement: Data available on request from I.M.S.K.I. (saman.ilankoon@monash.edu) and N.D. (nimila.dush@gmail.com).

Conflicts of Interest: The authors declare that they have no known competing financial interests or personal relationships that could have appeared to influence the work reported in this paper.

References

1. Araya, N.; Kraslawski, A.; Cisternas, L.A. Towards Mine Tailings Valorization: Recovery of Critical Materials from Chilean Mine Tailings. *J. Clean. Prod.* **2020**, *263*, 121555. [CrossRef]
2. Rosado, S.; Gullón, L.; Martínez, L.F.M.; Llamas Borrajo, J.F. Potential Uses of Copper Wastes in the Building Sector: Inertization and Added Value Solutions. *Mater. Proc.* **2021**, *5*, 25.
3. Nwaila, G.T.; Ghorbani, Y.; Zhang, S.E.; Frimmel, H.E.; Tolmay, L.C.K.; Rose, D.H.; Nwaila, P.C.; Bourdeau, J.E. Valorisation of Mine Waste-Part I: Characteristics of, and Sampling Methodology for, Consolidated Mineralised Tailings by Using Witwatersrand Gold Mines (South Africa) as an Example. *J. Environ. Manag.* **2021**, *295*, 113013. [CrossRef] [PubMed]
4. Tunsu, C.; Menard, Y.; Eriksen, D.Ø.; Ekberg, C.; Petranikova, M. Recovery of Critical Materials from Mine Tailings: A Comparative Study of the Solvent Extraction of Rare Earths Using Acidic, Solvating and Mixed Extractant Systems. *J. Clean. Prod.* **2019**, *218*, 425–437. [CrossRef]
5. Ilankoon, I.M.S.K.; Dushyantha, N.P.; Mancheri, N.; Edirisinghe, P.M.; Neethling, S.J.; Ratnayake, N.P.; Rohitha, L.P.S.; Dissanayake, D.; Premasiri, H.M.R.; Abeysinghe, A.M.K.B.; et al. Constraints to Rare Earth Elements Supply Diversification: Evidence from an Industry Survey. *J. Clean. Prod.* **2022**, *331*, 129932. [CrossRef]
6. Balaram, V. Rare Earth Elements: A Review of Applications, Occurrence, Exploration, Analysis, Recycling, and Environmental Impact. *Geosci. Front.* **2019**, *10*, 1285–1303. [CrossRef]
7. Dushyantha, N.; Batapola, N.; Ilankoon, I.; Rohitha, S.; Premasiri, R.; Abeysinghe, B.; Ratnayake, N.; Dissanayake, K. The Story of Rare Earth Elements (REEs): Occurrences, Global Distribution, Genesis, Geology, Mineralogy and Global Production. *Ore Geol. Rev.* **2020**, *122*, 103521. [CrossRef]
8. Dushyantha, N.P.; Ratnayake, N.P.; Premasiri, H.M.R.; Ilankoon, I.M.S.K.; Hemalal, P.V.A.; Jayawardena, C.L.; Chandrajith, R.; Rohitha, L.P.S.; Abeysinghe, A.; Dissanayake, D. Leaching of Rare Earth Elements (REEs) from Lake Sediments around Eppawala Phosphate Deposit, Sri Lanka: A Secondary Source for REEs. *Hydrometallurgy* **2021**, *205*, 105751. [CrossRef]

9. Jha, A.; Lahiri, A.; Kumari, E.J. Beneficiation of Titaniferous Ores by Selective Separation of Iron Oxide, Impurities and Rare Earth Oxides for the Production of High Grade Synthetic Rutile. *Miner. Process. Extr. Metall.* **2008**, *117*, 157–165. [CrossRef]
10. Peelman, S.; Sun, Z.H.I.; Sietsma, J.; Yang, Y. Hydrometallurgical Extraction of Rare Earth Elements from Low Grade Mine Tailings. In *Rare Metal Technology 2016*; Springer: Cham, Switzerland, 2016; pp. 17–29.
11. Peelman, S.; Sun, Z.H.I.; Sietsma, J.; Yang, Y. Leaching of Rare Earth Elements: Review of Past and Present Technologies. In *Rare Earths Industry*; Elsevier: Amsterdam, The Netherlands, 2016; pp. 319–334.
12. Rupasinghe, M.S.; Dissanayake, C.B. The Rare-Earth Element Abundance in the Sedimentary Gem Deposits of Sri Lanka. *Lithos* **1984**, *17*, 329–342. [CrossRef]
13. Dissanayake, C.B.; Chandrajith, R.; Tobschall, H.J. The Geology, Mineralogy and Rare Element Geochemistry of the Gem Deposits of Sri Lanka. *Bull. Soc. Finl.* **2000**, *72*, 5–20. [CrossRef]
14. Dharmaratne, P.G.R. *Gem Industry of Sri Lanka*, 2nd ed.; The Geological Survey & Mines Bureau, Sri Lanka: Colombo, Sri Lanka, 2016; ISBN 9559370010.
15. Moon, J.W. The Mineral Industry of Sri Lanka. Available online: https://www.usgs.gov/media/files/mineral-industry-sri-lanka-2017-18-pdf (accessed on 20 October 2022).
16. Cartier, L.E. Environmental Stewardship in Gemstone Mining: Quo Vadis? Available online: https://www.uvm.edu/~{}shali/egems.pdf (accessed on 20 October 2022).
17. Thu, K. Gem Mining and Sustainability in Myanmar. Available online: https://sgiinstitute.com/wp-content/uploads/2019/10/Gem-Mining-and-Sustainability-in-Myanmar-presentation.pdf (accessed on 20 October 2022).
18. Dharmaratne, P.G.R.; Premasiri, H.M.; Dillimuni, D. Sapphires from Thammannawa, Kataragama Area, Sri Lanka. *Gems Gemol.* **2012**, *48*, 98–107. [CrossRef]
19. Cooray, P.G. The Precambrian of Sri Lanka: A Historical Review. *Precambrian Res.* **1994**, *66*, 3–18. [CrossRef]
20. Gunaratne, H.S.; Dissanayake, C.B. *Gems and Gem Deposits of Sri Lanka*; National Gem and Jewellery Authority of Sri Lanka: Colombo, Sri Lanka, 1995; ISBN 9559371002.
21. Ampitiyawatta, A.D.; Guo, S. Precipitation Trends in the Kalu Ganga Basin in Sri Lanka. *J. Agric. Sci.* **2009**, *4*, 10–18. [CrossRef]
22. Kumara, S.; Premsiri, R.; Hewawasam, T. Morphological Features and Depositional Patterns of Gem-Bearing Sediments in the Kalu Ganga Basin, Sri Lanka. *J. Geol. Soc. Sri Lanka* **2015**, *17*, 125–138.
23. Dushyantha, N.P.; Hemalal, P.V.A.; JAayawardena, C.L.; Ratnayake, A.S.; Premasiri, H.M.R.; Ratnayake, N.P. Nutrient Characteristics of Lake Sediments Around Eppawala Phosphate Deposit, Sri Lanka. *J. Geol. Soc. Sri Lanka* **2017**, *18*, 33–42.
24. Dushyantha, N.P.; Hemalal, P.V.A.; Jayawardena, C.L.; Ratnayake, A.S.; Ratnayake, N.P. Application of Geochemical Techniques for Prospecting Unconventional Phosphate Sources: A Case Study of the Lake Sediments in Eppawala Area Sri Lanka. *J. Geochem. Explor.* **2019**, *201*, 113–124. [CrossRef]
25. Ratnayake, A.S.; Dushyantha, N.; De Silva, N.; Somasiri, H.P.; Jayasekara, N.N.; Weththasinghe, S.M.; Samaradivakara, G.V.I.; Vijitha, A.V.P.; Ratnayake, N.P. Sediment and Physicochemical Characteristics in Madu-Ganga Estuary, Southwest Sri Lanka. *J. Geol. Soc. Sri Lanka* **2017**, *18*, 43–52.
26. Subasinghe, C.S.; Ratnayake, A.S.; Roser, B.; Sudesh, M.; Wijewardhana, D.U.; Attanayake, N.; Pitawala, J. Global Distribution, Genesis, Exploitation, Applications, Production, and Demand of Industrial Heavy Minerals. *Arab. J. Geosci.* **2022**, *15*, 1–28. [CrossRef]
27. Moldoveanu, G.A.; Papangelakis, V.G. Recovery of Rare Earth Elements Adsorbed on Clay Minerals: I. Desorption Mechanism. *Hydrometallurgy* **2012**, *117–118*, 71–78. [CrossRef]
28. Ram, R.; Becker, M.; Brugger, J.; Etschmann, B.; Burcher-Jones, C.; Howard, D.; Kooyman, P.J.; Petersen, J. Characterisation of a Rare Earth Element-and Zirconium-Bearing Ion-Adsorption Clay Deposit in Madagascar. *Chem. Geol.* **2019**, *522*, 93–107. [CrossRef]
29. Bao, Z.; Zhao, Z. Geochemistry of Mineralization with Exchangeable REY in the Weathering Crusts of Granitic Rocks in South China. *Ore Geol. Rev.* **2008**, *33*, 519–535. [CrossRef]
30. Sun, S.-S.; McDonough, W.F. Chemical and Isotopic Systematics of Oceanic Basalts: Implications for Mantle Composition and Processes. *Geol. Soc. Lond. Spec. Publ.* **1989**, *42*, 313–345. [CrossRef]
31. Ilankoon, I.M.S.K.; Tang, Y.; Ghorbani, Y.; Northey, S.; Yellishetty, M.; Deng, X.; Mcbride, D. The Current State and Future Directions of Percolation Leaching in the Chinese Mining Industry: Challenges and Opportunities. *Miner. Eng.* **2018**, *125*, 206–222. [CrossRef]
32. Canadian Council of Ministers of the Environment. *Canadian Soil Quality Guidelines for Uranium: Environmental and Human Health—Scientific Supporting Document*; Canadian Council of Ministers of the Environment: Winnipeg, MB, Canada, 2007; ISBN 9781896997643.
33. Subasinghe, H.C.S.; Ratnayake, A.S.; Sameera, K.A.G. State-of-the-Art and Perspectives in the Heavy Mineral Industry of Sri Lanka. *Miner. Econ.* **2021**, *34*, 427–439. [CrossRef]

minerals

MDPI

Article

Hydrodynamic Hysteresis and Solute Transport in Agglomerated Heaps under Irrigation, Stacking, and Bioleaching Controlling

Leiming Wang [1,2,3,4,5], Shenghua Yin [1,2,*], Xuelan Zhang [6], Zepeng Yan [1,2] and Wensheng Liao [7]

1. Key Laboratory of Ministry of Education for High-Efficient Mining and Safety of Metal, University of Science and Technology Beijing, Beijing 100083, China
2. School of Civil and Environment Engineering, University of Science and Technology Beijing, Beijing 100083, China
3. State Key Laboratory of Coal Mine Resources and Safety Mining, China University of Mining and Technology, Xuzhou 221116, China
4. State Key Laboratory Safety and Health for Metal Mines, Sinosteel Maanshan General Institute of Mining Research Co., Ltd., Maanshan 243000, China
5. Key Laboratory of Green Chemical Engineering Process of Ministry of Education, Wuhan University of Technological, Wuhan 430205, China
6. School of Mathematics and Physics, University of Science and Technology Beijing, Beijing 100083, China
7. Beijing Research Institute of Chemical Engineering Metallurgy, Beijing 101149, China

* Correspondence: ustb_ysh@163.com; Tel.: +86-138-1166-8481

Abstract: Hydrodynamic hysteresis exists widely in agglomerated heaps with well-developed intrapores, and it directly affects solute transports and bioleaching reaction. In this paper, the dynamic liquid retention behavior under different heap porosity and irrigation condition is quantified via a novel real-time, in-situ liquid retention characterizing system (RILRCS), and the potential effects of initial liquid retention on solute transport and leaching reaction are carefully discussed. The results show that the immobile liquid is dominant in agglomerated heaps. The ratio of immobile and mobile liquid (η) dynamically changes due to mineral dissolution and new flow path appearances. The η normally increases and mobile liquid occupies a higher proportion due to acidic leaching reactions, especially at a smaller R_g (10.32 mm) and a larger u (0.10 mm/s). The dynamic liquid retention is more sensitive to the diameter of packed feeds (R_g) and superficial flow rate (u) instead of leaching reactions. This might be because the damage of leaching reaction on minerals pores/voids is limited and cannot extensively change the potential pore channels or fluid flow paths. Based on pulse tracing and conductivity tests, we reveal that the solute resides longer under a slower u and smaller packed R_g condition, which corresponds well with desirable copper leaching efficiency. Specifically, the liquid hysteresis behavior is more obvious at a lower u (0.01 mm/s) and smaller R_g (10.32 mm). This paper gives a good reference to ascertain the liquid retention and hydrodynamic hysteresis and promote mineral leaching performance.

Keywords: agglomerated heap; liquid retention; hysteresis behavior; solute transport; fluid flow

Citation: Wang, L.; Yin, S.; Zhang, X.; Yan, Z.; Liao, W. Hydrodynamic Hysteresis and Solute Transport in Agglomerated Heaps under Irrigation, Stacking, and Bioleaching Controlling. *Minerals* **2022**, *12*, 1623. https://doi.org/10.3390/min12121623

Academic Editor: Luis A. Cisternas

Received: 8 October 2022
Accepted: 10 December 2022
Published: 16 December 2022

1. Introduction

Low-grade minerals, waste ores, and even e-wastes are diffusely deposited in the earth [1,2]. These secondary recyclable resources require low-cost, environmentally-friendly, and efficient mining methods. Since 1980s, heap leaching, which crushes ores and then packs them in heaps [3,4], has been successfully industrialized in the United States, Canada, Chile, South Africa, China, and other countries. The agglomeration, as an effective pretreatment method of crushed ores, is widely utilized in the heap leaching of copper oxides, copper sulfides, laterite nickels, and uranium minerals.

The experimental agglomeration refers to the procedure of using chemical binders to adsorb and bond crushed ores and fine powders to form well-shaped agglomerations (WAs), which could obviously ameliorate pore structure, fluid flow, and solute dispersion [5].

Similar with crushed ore heaps, both the unsaturated and saturated condition exist in agglomerated heaps simultaneously [6,7]. The liquid hysteresis behavior in unsaturated beds respectively packed by solid glass beads, crushed ores, and agglomerations are comparatively discussed. Recently, some novel detecting methods such as computer tomography (CT), magnetic resonance image (MRI), and particle tracking velocimetry (PIV) have been introduced to describe the liquid retention behavior. Although these undisturbed methods are mostly short-term and high cost, studies have indicated that developing differences of intra-aggregate and inter-aggregate porosity is an essential reason for differences in liquid hysteresis. Due to well-developed intra-particle pores and voids [8], the liquid retention and hysteresis appeared more easily than for the crushed ore heaps, which is regarded as one of the essential factors leading to desirable leaching efficiency. The fluid flow is closely related to the formation of flow paths and fluid dispersion in unsaturated heaps [9,10]. In addition to the superficial flow rate, the fluid flow pattern is also related to the flow pipe diameter, fluid viscosity, and density. The flow pattern in unsaturated crushed ore/agglomerated heap is commonly shown as laminar flow where the flow rate is higher in the central part of flow tubes.

It is impossible for the solute (including valuable metallic/non-metallic ions, soluble oxygen, etc.) to be independent of the leaching system and its transport process must rely on liquid medium. A consensus has been developing that the solute consistently existed in the reactive fluid inside unsaturated heaps, and solute transport is closely related to solution flow. The efficiency of solute transports represents the ion exchange rate at the reaction interface and limits the mineral dissolution to some extent. Specifically, the solute transport areas in the unsaturated heaps are closely related to mobile liquid (mainly shown as preferential flow), while the solute transport efficiency is closely related to immobile liquid (mainly shown as stagnant flow). The existing research shows that there are two key star-like inter-particle pores and intra-particle pores in the ore heap, and solute transport widely exists in the inter-particle pores with solution as the key medium[11,12]. These solute/mass (including solute oxygen, metallic ions, etc.) transports could be affected by channeling tortuosity [13,14], contact angle at the solid-fluid interface [15], and the diameter of pore throat [16,17]. In the heap heterogeneous, the segregation and stratification phenomenon appeared widely [18]. This could also be quantitively characterized by detecting liquid holdup difference from macro-scope [19]. Caused by the complex hydraulic conductivity and hysteresis in unsaturated heaps, it is hard to quantitively characterized solute transport [20,21]. To describe the solute transport and liquid retention behavior of different unsaturated porous systems, previous studies have mainly focused on numerical simulation to characterize the solute transport process under the effect of different porosity and irrigation condition. Except for the mathematical modeling studies mentioned above, hydrodynamics features in fluid flow and solute transport in unsaturated porous heaps are explored via column system. Some detection methods including calibration solute tracing and overflow conductivity measuring are utilized to quantify the solute transport behavior, and discuss the potential effect of liquid hysteresis on mass transfer in the minerals interface [22]. In addition, Neethling's team of Imperial College London innovatively used the suspension force measurement method to monitor the liquid holding behavior of the static ore particle pile system. Some computer tomography, nuclear magnetic resonance, and other techniques have also been applied to the liquid holding characteristics of unsaturated porous media stacking heap [23–25], but the above research findings are more static and transient without considering mineral leaching. Despite this, there are still few studies that both consider liquid hysteresis and solute transport in agglomerated leaching condition, leading to a research gap regarding a deep understanding of how agglomerated heaps enhance mineral leaching and solute transport intensification [26–28].

In this paper, using a novel developed in-situ, real-time liquid retention characterizing experimental system, the fluid flow, dynamic liquid retention, and hysteresis behavior in unsaturated agglomerated heaps are effectively discussed. To better show the potential connection between liquid retention and mass transfer of agglomerated heap leaching system, the dynamic liquid retention is quantitative by the liquid holdup value θ, the residual liquid holdup value $\theta_{residual}$, and the ratio of immobile and mobile liquid η; the dynamic solute transport in agglomerated heaps is quantitatively described by fluid parameters, such as overflow the overflow conductivity, and residual time distribution (RTD). The research measures and findings in this paper could promote the understanding of the dynamic liquid retention and hysteresis behavior in agglomerated heaps under different inter-aggregate porosity (comparing diameter of WAs), intra-particle porosity (comparing feeds with same diameters), and superficial flow rate condition.

2. Materials and Methods

2.1. Ore Samples and Its Agglomeration Condition

The ore feeds are sampled from copper sulfides mines in China. The valuable copper minerals mainly are copper sulfides (chalcocite and chalcopyrite, etc.) and copper oxides (malachite, etc.). The total mass fraction of Cu^{2+} is 0.70% (free copper oxides is 5.71%, primary/secondary copper sulfides occupied 92.86%), the gangue minerals is mainly quartz. The agglomerated condition is set as: rolling speed is 80 rpm, inclined angle is 30°, curing time is 7 days, and chemical binder is 9.3% sulfuric acid solution. The diameter of 50% crushed ore feeds is less than 2 mm, and 35% is less than 1 mm. The diameter of 80% WAs is larger than 10 mm. The difference of particle bonding collision in agglomeration procedure is not considered in this paper.

2.2. Real-Time, In-Situ Liquid Retention Characterizing System (RILRCS)

To quantitively describe the dynamic liquid retention and solute transports in agglomerated heaps, a novel real-time, in-situ liquid retention characterizing system (RILRCS) is well developed (Figure 1). Taking the functionality as evaluation factors, the structure of RILRCS is divided into (Figure 1a): liquid irrigating (9, 10, 11, 13, 14, 15, 16, 17, 18), tension measuring (5, 6), data visualizing (8), and supporting (1, 2, 3, 4, 7, 12) systems.

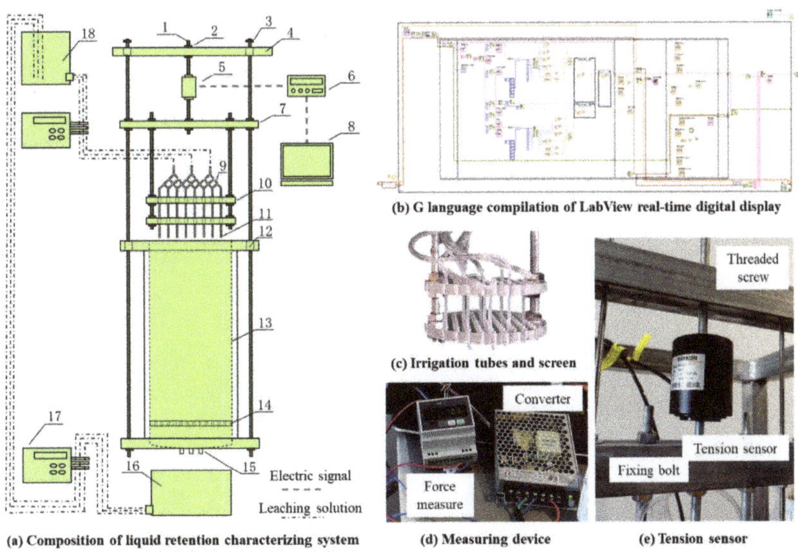

Legends			
	1 - Threaded screw	7 - Connecting board	13-Acid-resistant plexiglass column
	2 - fixing bolt	8 - LabView real-time digital display	14-Porous Screen
	3 - Anti-falling safety nut	9 - Single-way/Multi-way connector	15-Drain outlet
	4 - fixed bracket	10-Double-layer deflector	16-Liquid collection tank
	5 - Tension sensor	11-Diversion rubber hose	17-Peristaltic pump
	6 - force measurer	12-Catheter	18-Liquid storage tank

Figure 1. Scheme of novel Real-time, In-situ Liquid Retention Characterizing System (RILRCS).

The RILRCS idea is to use a tension sensor (Figure 1e) to monitor and converted dynamical liquid retention signal into electrical signal by force measurer and converter (Figure 1d). The effective measuring range of liquid retention is 30 kg, test accuracy is 0.001 kg. The A/D conversion rate is 200 times/s and counting step is 15 s. The electrical signals are compiled by LABVIEW and graphical language (Figure 1b), eventually obtaining the liquid holdup (θ), residual liquid holdup ($\theta_{residual}$), and ratio of immobile and mobile liquid (η). The dripping irrigation uses the uniform vertical hoses and glass porous board (Figure 1c). The pulse conductivity of overflow in liquid collection tank (Figure 1a) is measured regularly.

2.3. Experimental Scheme and Design

This paper utilizes column leaching experiment (**CLE**) and pulse tracing experiment (**PTE**) to study the dynamic liquid retention behavior, solute transport, and its influencing factors in agglomerated heaps. The detailed experimental scheme and key parameters are shown in Table 1. The dynamic liquid retention in leaching process is detected via RILRCS. The key parameters of overflow, namely Cu^{2+} concentration, pH/Eh, and bacterial concentration, are measured. The influencing factors are described: (1) packed feed types (solid glass beads, crushed ore, and Was), (2) geometric mean diameter of agglomerations (10.32, 16.02, and 24.36 mm), and (3) superficial flow rate (0.01, 0.02, 0.05, and 0.10 mm/s), respectively. Different WAs feeds and irrigation conditions were set up.

Table 1. Experimental scheme of pulse tracing and column leaching in this paper.

Experiment	Factors	Packed Feed Type	Geometric Mean Diameter (mm)	Superficial Flow Rate (mm/s)
Pulse Tracing Experiment (PTE)	Packed feed types	Solid glass beads	16.02	0.10
		Crushed ore	16.02	0.10
		Well-shaped agglomerations (WAs)	16.02	0.10
	Geometric mean diameter of packed feeds	WAs	10.32	0.10
		WAs	16.02	0.10
		WAs	24.36	0.10
	Superficial flow rate (Irrigation rate)	WAs	16.02	0.01
		WAs	16.02	0.02
		WAs	16.02	0.05
		WAs	16.02	0.10
Column Leaching Experiment (CLE)	Geometric mean diameter of packed feeds	WAs	10.32	0.10
		WAs	16.02	0.10
		WAs	24.36	0.10
	Superficial flow rate (Irrigation rate)	WAs	16.02	0.01
		WAs	16.02	0.02
		WAs	16.02	0.05
		WAs	16.02	0.10

In the **CLE**, the WAs are artificially dumped and packed the agglomerated heaps. Then the acidic solution is irrigated from the top of packed agglomerated heaps, the bacterial concentration, pH/Eh value, and cupric ions concentration (copper extraction rate) of overflow are detected regularly. The basic liquid medium is no-ferrous 9K solution containing 3.0 g $(NH_4)_2SO_4$, 0.5 g K_2HPO_4, 0.1 g KCI, 0.5 g $MgSO_4{\cdot}7H_2O$, and 0.01 g $Ca(NO_3)_2$ per liter. The initial pH value is set as 2.00. The functional leaching micrograms are injected and domesticated in leaching medium. The dominant bacteria species are

Acidthiobacillus ferrooxidans (*A.f*) and *Acidophilus thiooxidans* (*A.t*). The initial total bacterial concentration is roughly 2.0 × 10^6 cell/mL, the environmental temperature is 28 ± 2°.

In the **PTE**, to detect the solute transport under different initial liquid retention conditions, the tracer ion is thought as solute ions of valuable minerals reserved in the ore samples. The conductivity value of overflow contained tracer ions is continuously detected by conductivity sensor. The 4.00 mol/dm^3 NaCl solution is used as the experimental tracer, and this NaCl tracer was injected once in a pulse (shown in Figure 2), with an injection volume of 0.05 mL. This pulse injection adopts a single dipper head. The overflow conductivity is measured by the inserted sensors, and the residual time distribution (RTD) is calculated according to previous similar studies. The residence, spreading, and hysteresis behavior of solute (tracer ions) in agglomerated heaps are carefully discussed.

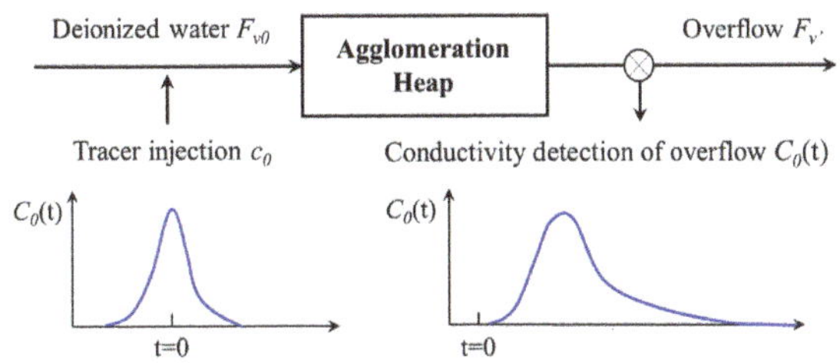

Figure 2. Experimental scheme of conductivity pulse tracer method using NaCl solution.

2.4. *Key Parameters of Liquid Retention and Solute Transports*

2.4.1. Liquid Holdup (θ) and Residual Liquid Holdup ($\theta_{residual}$)

In irrigation and drainage process, the agglomerated heaps reach static, and residual steady-state of liquid retention, which is corresponding with liquid holdup (θ, %) and residual liquid holdup ($\theta_{residual}$, %), respectively. As described in [29], two key parameters are calculated by Equations (1) and (2).

$$\theta = \frac{\int (v_{in} - v_{out})\, \mathrm{dt}}{V} = \frac{v_{in}t - m_{out}/\rho}{V} \tag{1}$$

$$\theta_{residual} = \frac{V_{steady} - \int (v_{out})\mathrm{dt}}{V} = \frac{\theta V - m_{out}/\rho}{V} = 1 - \frac{m_{out}}{V\rho} \tag{2}$$

where v_{in} is flow rate in, v_{out} is flow rate out, m_{out} is mass of liquid flow out, and ρ is liquid density. The immobile and mobile liquid co-existed in flow paths and stagnant areas of agglomerated heap. To evaluate existing status of liquid (leaching solution) reserved in the agglomerated heaps, the ratio of immobile liquid and mobile liquid (η) is utilized, which could be calculated by Equation (3).

$$\eta = \frac{\theta_{residual}}{\theta - \theta_{residual}} \tag{3}$$

2.4.2. Solute Transport and Resident Parameters

The solute transport in agglomerated heap relies on liquid dispersion, which is heavily affected by heap porosity and liquid retention. The hydrodynamic diffusion coefficient D is the sum of effective molecular diffusion D_m and mechanical diffusion D_h (Equation (4)). The two diffusion processes cannot exist independently.

$$D = D_m + D_h = D_m + \lambda v^{n_0} = D_w \tau_l + \lambda v^{n_0} \tag{4}$$

wherein the D_h mechanical diffusion may be expressed as λv^{n0}. The λ is the scattering coefficient, n_0 is an empirical constant, v is the pore water velocity, θ is liquid holdup. The relationship of tracer conductivity and spray time curve $C(t)$ can be obtained by **PTE**.

The conductivity distribution of the overflow water containing the tracer at the outlet is integrated and normalized, as shown in Equation (5).

$$E(t) = \frac{C(t)}{\int_0^\infty C(t)\,dt} \tag{5}$$

The residence time distribution (RTD) ranges from 0 to 1, representing the fraction of salt tracers whose residence time is less than a specific value. To better compare the solute transports under different liquid holdup conditions, take mean residence time (t_R) as an indicator, as shown in Equation (6).

$$t_R = \int_0^\infty t\,E(t)dt = \int_0^\infty t\,\frac{C(t)}{\int_0^\infty C(t)\,dt}\,dt \tag{6}$$

3. Results and Discussion

3.1. Key Parameters under Geometric Mean Diameter (R_g) and Superficial Flow Rate (u)

(1) **Cupric ionic concentration/copper extraction rate**

The liquid (leaching solution), which contains types of metallic ions, continuously and disorderly spreads in the irrigation or drainage process of heap leaching operations. The leaching medium and feed diameter influence leaching kinetics of minerals dissolution. Thus, the column leaching conducted here is aimed at discussing the effect of initial liquid retention on leaching reaction. Figure 3 shows the relationship between leaching time and copper extraction rate under different geometric mean diameter (R_g, mm) and superficial flow rate (u, mm/s).

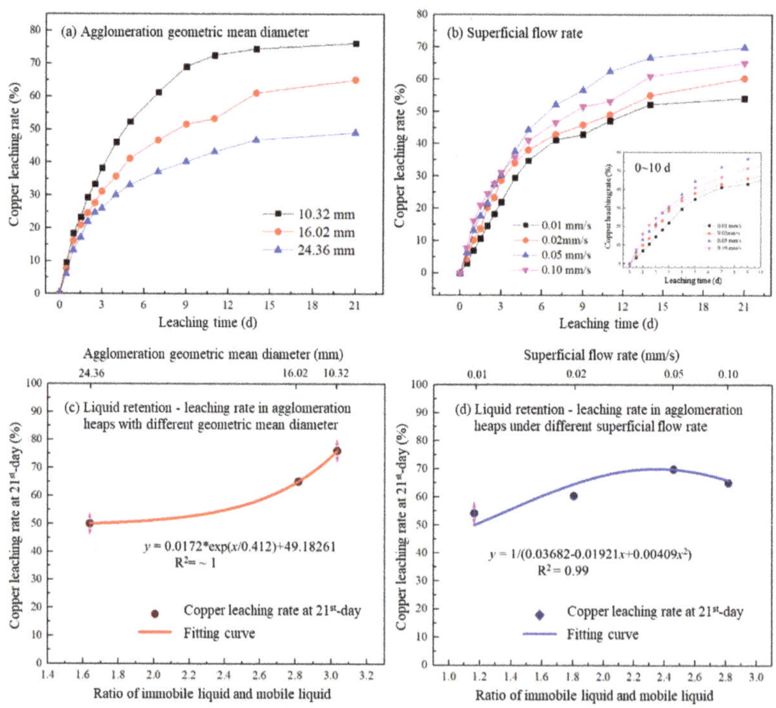

Figure 3. Copper leaching rate with leaching time under different packed and irrigation condition.

Due to a larger specific surface area, a better leaching reaction interface is provided in fine-grained (10.32 mm) agglomerated heaps where the peak copper extraction rate is

76.2%, which is much higher than 49.2% of coarse-grained (24.36 mm) agglomerated heaps (Figure 3a). For the copper sulfides leaching, it is known that its leaching kinetics could be roughly divided into two stages: the first stage is Fe^{3+} diffusion through the product layers controlled by solute concentration and temperature, which could correspond with the first 5 leaching days in this paper; the second stage is mainly controlled by mineral decomposition and Fe^{2+} reduction, which could corresponded with the 5–12 leaching days. In leaching solution, the reserved ferric ion (Fe^{3+}) is reduced to ferrous ions (Fe^{2+}) participated with oxygen and *A.f* bacteria, the S^{2-} is oxides to sulfate ions by *A.t* bacteria. To deeply understand leaching reaction under different effects of influencing factors, the chemical leaching reaction of the majority copper sulfide mineral compositions (including of chalcocite, covellite, and chalcopyrite) is simply shown in Equations (7)–(9), respectively.

$$Cu_2S + 2Fe^{3+} \rightarrow Cu^{2+} + 2Fe^{2+} + CuS \tag{7}$$

$$CuS + 2Fe^{3+} \rightarrow Cu^{2+} + 2Fe^{2+} + S^0 \tag{8}$$

$$CuFeS_2 + 4H^+ + O_2 \rightarrow Cu^{2+} + 2Fe^{2+} + S^0 + H_2O \tag{9}$$

Figure 3b shows that the cupric ions are released to liquid in leaching process, and the mineral dissolution is desirable under 0.05 mm/s irrigation condition. When the agglomerated leaching is operated for 21 days, the peak value of copper extraction rate is 70.0% (at 0.05 mm/s), which is much higher than 54.2% (0.01 mm/s). As Figure 3c shows, accompanied with the increase of peak copper extraction rate under different R_g condition, the η tends to increase as well. In contrast, the peak value of copper extraction rate is obtained at 0.05 mm/s instead of 0.10 mm/s (Figure 3d). This means that for continuously increasing u, the solute diffusion and minerals dissolution in agglomerated heaps cannot be linearly improved. This fact might be caused by the recognition that most liquid and solute transport is conducted via preferential flow paths under an excessive high superficial flow rate [30]. The detailed liquid retention will be discussed below.

(2) **Bacterial concentration**

The bacterial concentration in liquid medium keeps changing dynamically, and closely corresponds with sulfide mineral dissolution and reactions. Figure 4 shows the relationship between leaching time and bacterial concentration under the different initial liquid retention condition. Figure 4a,b infer that the initial liquid retention, which is designed by different packed and irrigation condition, could clearly affect the leaching efficiency. The bacterial concentration obviously increased from 2.0×10^6 cells/mL to around 1.6×10^8 cells/mL in the first 10 days, the peak bacterial concentration is roughly 1.75×10^8 cells/mL obtained at 0.05 mm/s or using 10.32 mm WAs; then the bacterial apoptosis gradually appears to be caused by leaching environment deteriorations. In the agglomerated heaps packed by 24.36 mm WAs, the bacterial concentration decreases from 1.16×10^8 cells/mL (at 14 days) to 1.06×10^8 cells/mL (at 21 days). It could be explained that after a longer period of bioleaching reaction, the crusts composited by amounts of passivation products widely attached on the mineral surface, which fill pores/voids and seriously restrict the continuation of bioleaching reactions. Further, the changes of peak bacterial concentration with the η in Figure 4c,d correspond well with copper extraction rate (Figure 3), which also follows the close relationship between copper extraction rate and bacterial concentration. In other words, a desirable bacterial concentration is obtained by using small-diameter WAs or appropriate high u, resulting in a higher copper leaching efficiency.

(3) **pH/Eh value**

The copper sulfide bio-/chemical leaching is a complex chemical reaction commonly occurring in acidic aqueous medium. The oxidation reduction potential (ORP) is closely related to soluble oxygen, and is also affected by the ratio of ferrous and ferric concentration [31]. Figure 5 shows that, regardless of R_g or u, the oxidation reduction potential (ORP, Eh) is higher when pH value is lower.

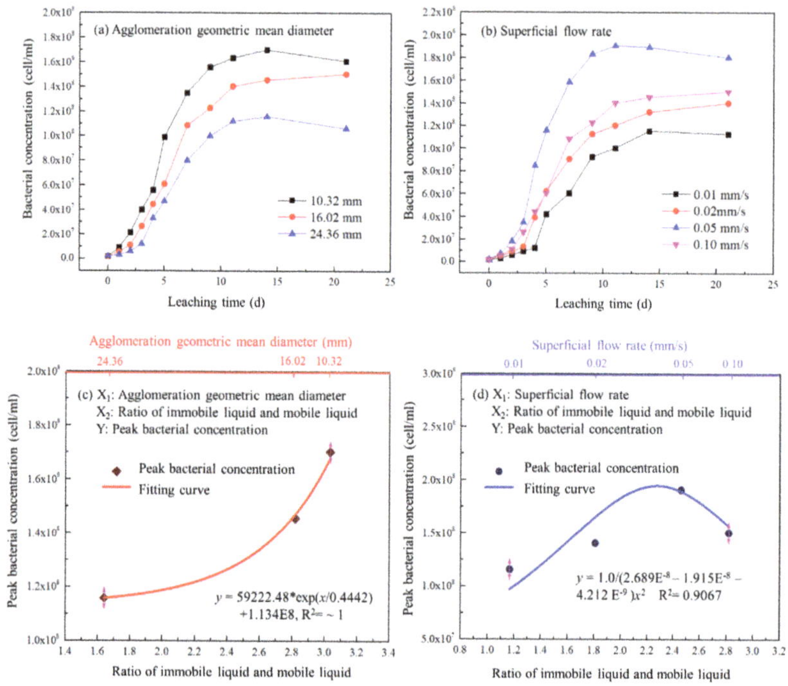

Figure 4. Bacterial concentration with leaching time under different packed and irrigation condition.

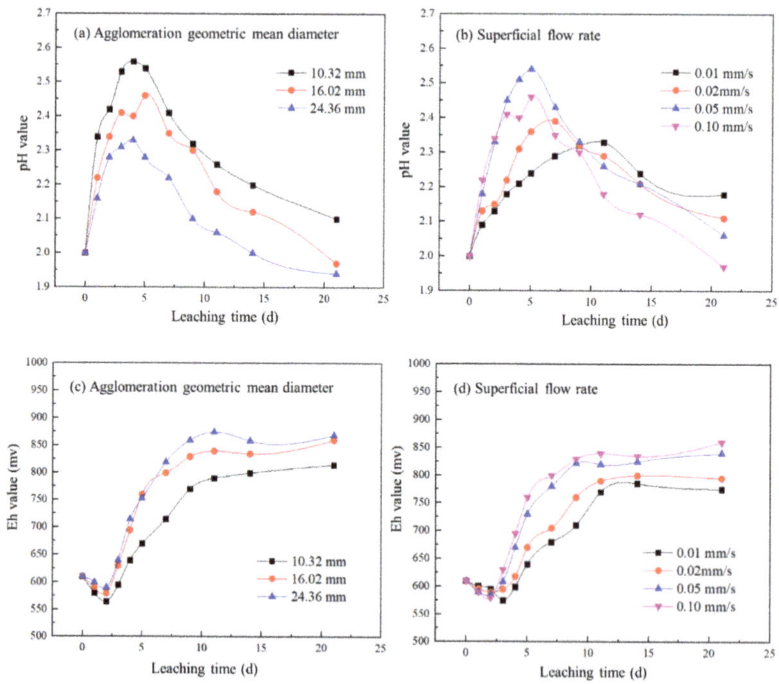

Figure 5. pH/Eh value with leaching time under the different packed and irrigation condition.

In the early leaching period (0–5 days), the pH value reaches nearly 2.60, and oxidation reduction potential (ORP) is generally lower than 650 mV (Figure 5), which corresponds with undesirable copper extraction rate (Figure 3). In the leaching process, H^+ is consumed by the copper oxides and alkaline gangues [32], the acidic leaching of copper oxides is shown in Equation (10). The Eh value increased to 750–850 mV after 10 days of leaching, and the copper minerals are quickly dissolved and leached, which is clearly echoed with a surge increase of copper extraction rate (Figure 3) and bacterial concentration (Figure 4).

$$Cu_2(OH)_2CO_3 + 4H^+ \rightarrow 2Cu^{2+} + 3H_2O + CO_2 \quad (10)$$

In copper sulfides leaching, the Fe^{3+} is reduced to produce passivation substances such as jarosite, polysulfide, etc. [33,34], and the S^{2-} is oxidized to produce S^0 and sulfate ions, which promotes the formation of H_2SO_4 (Equation (11)).

$$3Fe^{3+} + 2\,SO_4^{2-} + 6H_2O + M^+ \rightarrow MFe_3\,(SO_4)_2(OH)_6 + 6H^+ \quad (11)$$

These finding also inferred that the pH/Eh heavily affects the leaching, where the higher ORP promotes the mineral dissolutions. Combining with the existing research [35–37], this indicates that ORP affects the intermediate products formation, so it could limit the solute transfer and bioleaching efficiency. Referring to previous studies on liquid retention and hysteresis behavior, it clear that the liquid holdup of unsaturated ore packed beds is higher under a higher u or a smaller R_g condition; therefore, combining with results of Figure 5, the ORP is higher in agglomerated heaps where the immobile liquid has a higher proportion, especially in late leaching stage (Figure 5d).

3.2. Dynamic Liquid Retention Behavior under Different Leaching Condition

(1) **Effect of geometric mean diameter (R_g) on dynamic liquid retention**

To clearly reveal the dynamic liquid retention from the solute transfer, mobile and immobile liquid condition at 0 and 21 days are comparatively discussed in this paper. Figure 6a shows the effect of geometric mean diameter (R_g) of WAs on dynamic liquid retention. The ratio of immobile liquid and mobile liquid (η) continuously changes, and it represents the dynamics liquid retention during the agglomerated column leaching.

Figure 6a shows that if geometric mean diameter of packed WAs is fixed, the η decreases accompanied with leaching reaction. For instance, in the heaps packed by 10.32 mm agglomerations, the proportion of mobile liquid increases from 24.8% of leaching 0 d to 30.9% of 21 d leaching, with a net increase of 4.1%. It speculates that the bio/chemical leaching reaction leads to mineral interface damage, pore throat expansion, and even fluid flow path reconstruction, which promotes mobile fluid paths development, and then it manifests as the proportion increases of mobile fluid after leaching. Further, after the leaching reaction, the increase of mobile liquid proportion is more obvious in the agglomerated heap packed by the small-grained agglomerations. This is consistent with the experimental phenomenon that the pore structure in fine-grained agglomerated heap is well developed. Thus, the flow path is normally well expanded and not only along the former liquid breakthrough paths, which is easy to potentially interfer by leaching reactions and mineral dissolution. However, there is a fact not affected by leaching reaction, namely, the η will gradually decrease if the R_g of spherical agglomeration feeds increases. As seen in Figure 6a, the η of day 0 decreases from 3.672 to 1.629; similarly, the η of day 21 also decreases from 2.922 to 1.521. This means that, although mineral dissolution could accelerate the flow paths development and increase the mobile liquid flow as a result [38,39], the leaching reaction only weakly affects η and cannot fundamentally subvert the state of liquid retention in the agglomerated heaps. Thus, the liquid retention of agglomerated heaps is more sensitive to geometric mean diameter than leaching reaction. This infers that it is easier to control liquid holdup by regulating the packed feed diameter of WAs, rather than regulating the leaching reaction especially in industrial-scale leaching operation.

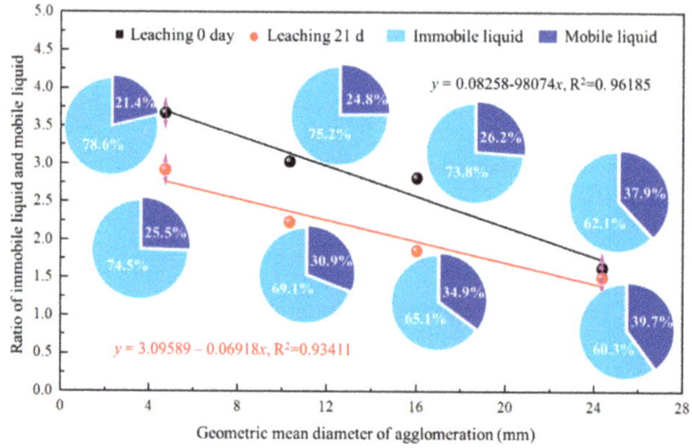

(**a**)

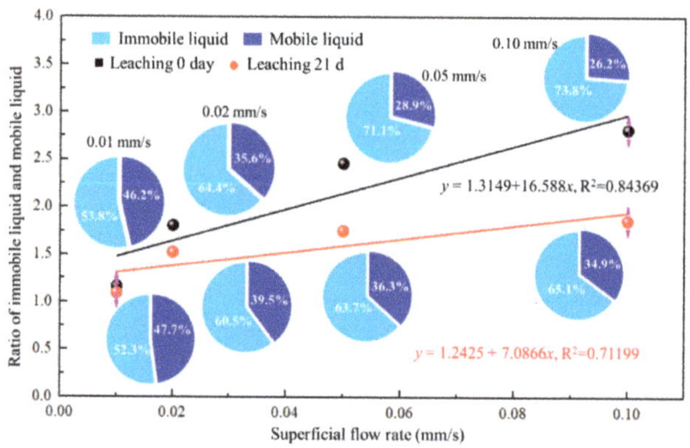

(**b**)

Figure 6. Ratio of immobile and mobile liquid under different geometric mean diameter and superficial flow rate. (**a**) Effect of different geometric mean diameter, (**b**) Effect of superficial flow rate.

(2) **Effect of superficial flow rate (u) on dynamic liquid retention**

Under the co-effects of capillary and gravity forces, the encapsulated liquid film on the agglomeration surface forms the potential fluid paths to promote liquid (or solute) transportation, while the incomplete liquid film causes intermittent liquid retention. The former liquid film is closely related to mobile liquid, and the latter liquid film is closely related to the immobile liquid in the agglomerated heaps. The corresponding relationship between the superficial flow rate (u) and ratio of immobile and mobile liquid (η) is shown in Figure 6b.

The leaching reaction promote the proportion of mobile liquid, eventually making the immobile liquid and mobile liquid to be more balanced. As Figure 6b shows, the η is positively related to u during the whole leaching process. In other words, the liquid exists as mobile liquid in agglomerated heaps instead residents in the stagnant regions, especially under a lower u. This does not mean that the mobile liquid is dominant, but it infers that the proportion of mobile liquid can be significantly increased when a lower u introduced. Taking the leaching on day 21 as an example, the mobile liquid occupied 34.9% at 0.10 mm/s; however, the mobile liquid only occupied 47.7% at 0.01 mm/s. This is because that the preferential flow paths quickly appear and the liquid retention easily reach the steady-state at a higher u (0.10 mm/s), leading to extensive unsaturated areas distributed in agglomerated heaps. Moreover, no matter how high u is, the η will certainly decrease accompanied with the mineral dissolution reactions. It has also been demonstrated that the leaching reaction could enlarge the flow paths and promote the proportion of mobile liquid as a result [40,41]. This decreasing of η is more pronounced at a higher u. Taking the 0.1 mm/s condition as an example, η significantly decreases from 2.816 to 1.866, and the net reduction is 0.95. However, the net reduction of η is only 0.067 at 0.01 mm/s, which is much lower than the value of the 0.10 mm/s condition.

3.3. Response of Solute Transport to Dynamic Liquid Retention of Agglomerated Heaps

(1) **Effect of superficial flow rate**

To better reveal the solute transport under the dynamic liquid spreading condition, the effect of irrigation condition, packed feed types, and agglomeration geometric mean diameter on solute transport behavior are considered in this paper, respectively. Figure 7a,b shows the effect of superficial flow rate (u) on the overflow conductivity and residence time distribution (RTD) with irrigation time.

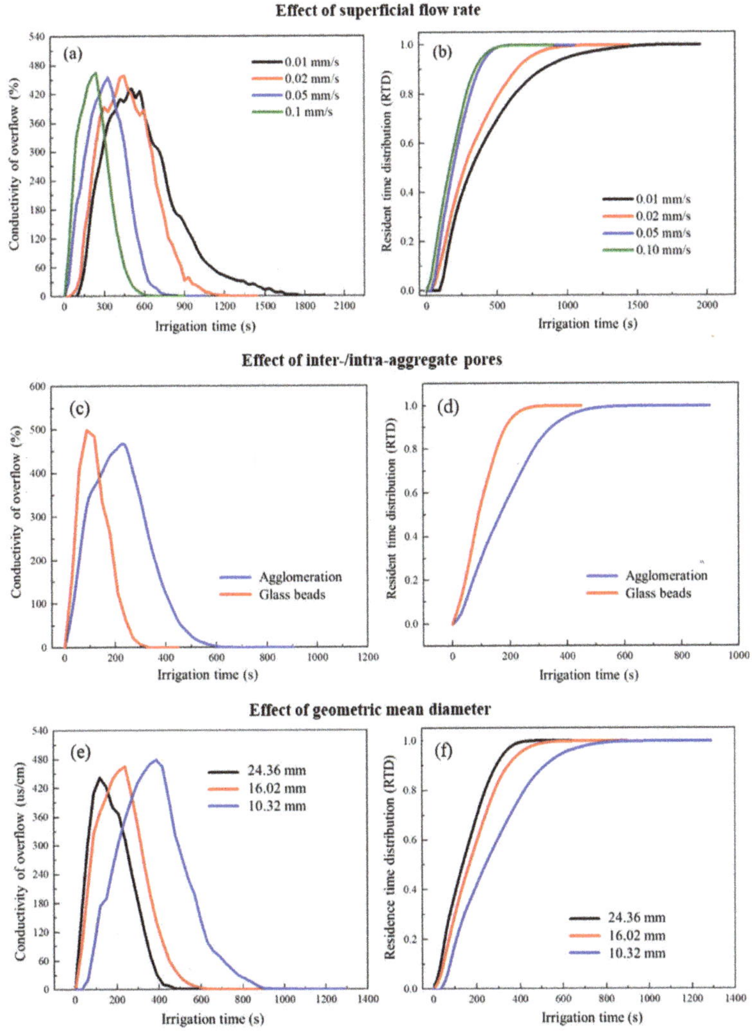

Figure 7. Effect of superficial flow rate, inter-/intra- particle porosity, geometric mean diameter on the overflow conductivity and RTD curve with irrigation time.

In Figure 7a, the appearance time of the conductivity pulse peak gradually lags as u decreases. Specifically, under 0.10 mm/s condition, the peak conductivity value of overflow is 466 us/cm at 240 s; when the u is reduced to 0.01 mm/s, the similar peak conductivity of overflow is obtained. In addition, based on the slope of the curve, the increasing and decreasing process of the overflow liquid conductivity at high u (0.10 mm/s) are almost symmetrical, but a "dragging phenomenon" is observed under low u (0.01 mm/s). This indicates that the solute is quickly washed away from the leaching reaction system at high u (0.10 mm/s), but slowly diffuses and stagnates in the pores/voids at low u (0.01 mm/s). The liquid dragging is mainly caused by coupled effects of adhesion forces and cohesion forces, which is recognized as the hysteresis behavior [42,43]. The degree of hysteresis is heavily influenced by the interaction surface features (contact angle, surface roughness, etc.) and wettability condition.

As seen in Equations (5) and (6), the residence time distribution (RTD) is shown in Figure 7b. The breakthrough time of solute pulse is positively proportional to the irrigation rate, which decreases from 120 s (0.01 mm/s) to 30 s (0.1 mm/s). Due to the consumed time of the liquid, breakthrough is closed related to the appearance time of preferential flow paths. This means that the shorter the breakthrough time is, the faster liquid retention steady-state reaches. Besides, based on the RTD curve, the solute residence time is around 660 s at 0.01 mm/s, which is much lower than 1770 s at 0.10 mm/s. It inferred that the ionic solute stagnates for a longer time at a low u (0.01 mm/s), which is caused by a more developed capillary diffusion under a slow flow rate. These findings also give a guidance for an industrial agglomerated heap, namely, advisably decreasing the irrigation rate (or u) could enhance capillary diffusion and promote mass transfer. In contrast, increasing the irrigation rate (or u) could quickly recycle the reserved valuable ionic solutes if the leaching reaction is carried out.

(2) **Effect of inter-/intra-aggregate pores**

Some previous studies have considered how the developments of inter- and intra-pores affect the retention potential of unsaturated packed beds. This paper focuses on the agglomerated heaps and promotes the understanding of inter-/intra- pores effects on solute transfer via pulse conductivity. Figure 7c,d indicates the effect of inter-/intra-aggregate pores on overflow conductivity with irrigation time via different packed feeds (glass beads, and WAs).

The results of Figure 7c show that consumed time of solute transports is 90 s in glass beads heap, which is much lower than 240 s in heaps of WAs. The intra-porosity of solid glass beads heap is 0%, which is far below that of agglomerated heaps. Based on Figure 7d, the RTD curve implies that the residence time of glass beads is 330 s, which is lower than the residence time (660 s) of agglomerated heaps. This fact confirms that the intra-pores promote liquid hysteresis behavior, strengthen the capillary diffusion, and cause more solutes to stagnate in the agglomerated heaps where the intra- and inter- pores are both well developed. Similarly, it is known the intra- porosity of crushed ore heaps is not as well developed as agglomerated heaps. Hence, this also explains why agglomerated heaps have better liquid holdup and leaching efficiency than crushed ore heaps.

(3) **Effect of geometric mean diameter**

Based on previous studies, the development of intra-/inter pores is very important for liquid hysteresis. Figure 7e,f shows the potential effect of geometric mean diameter on the overflow conductivity and solute residence time features.

Figure 7e indicated that the peak conductivity value and residence time of solute transport is heavily affected by the geometric mean diameter of packed feeds. The peak conductivity value is 480 us/cm (in 10.32 mm agglomerated heaps), which is much higher than 442 us/cm (in 24.36 mm agglomerated heaps). More potential flow paths exist in the agglomerated heaps with well-developed pores. Thus, the liquid dispersion and solute transport is more complicated and harder to predict. Based on the results shown in Figure 7f in this paper, the solute residence time in the heaps packed by 10.32 mm

WAs is highest (1050 s); in contrast, it decreases to 480 s in the heaps packed by 24.36 mm WAs. This obvious difference in residence time suggests that the well-developed pore structure (smaller diameter agglomerations) could have a higher liquid holdup value of (residual) steady-state and obvious hysteresis phenomenon. Thus, in the feed dumping procedure [44–46], appropriately reducing the feeds diameter of WAs effectively improves the heaps' porosity, ameliorating liquid retention, and solute transport.

4. Conclusions

This paper reveals the potential effects of leaching reaction on dynamic liquid retention and hysteresis behavior of agglomerated heaps. The main findings are as follow:

(1) The dynamic liquid holdup is carefully quantified via RILRCS system, the copper extraction rate is higher when the WAs diameter is smaller (10.32 mm). Increasing the surficial flow rate could decease the ratio of immobile and mobile liquid (η);
(2) The RTD results show that increasing intra- porosity of agglomerated heaps and superficial flow rate could extend solute residence time. However, the excessive superficial flow rate easily results in preferential flow formation and low copper leaching rate;
(3) Leaching reaction tends to promote the proportion of mobile liquid, and increase the peak liquid holdup value. The leaching reaction tends to develop the flow paths, which is mainly shown as the reduction of η.

From the results described in this paper, it can be seen that bioleaching reactions have a significant impact on liquid holding behavior. The concentration change of solute type caused by mineral reaction, reaction passivation, and interface liquid-solid contact angle will be the focus of follow-up research.

Author Contributions: Conceptualization, L.W. and X.Z.; methodology, LW., X.Z.; software, X.Z. and Z.Y.; validation, L.W. and S.Y.; data curation, L.W. and Z.Y.; writing—original draft preparation, L.W., X.Z. and Z.Y.; writing—review and editing, L.W. and S.Y.; visualization, L.W. and W.L.; supervision, S.Y., L.W. and W.L.; project administration, L.W. and S.Y.; funding acquisition, L.W. and S.Y. All authors have read and agreed to the published version of the manuscript.

Funding: This research was funded by the Program of the National Natural Science Foundation of China (52204124,52034001); China National Postdoctoral Program for Innovative Talents (BX20220036), China Postdoctoral Science Foundation (2022M710356), Fundamental Research Funds for the Central Universities (FRF-IDRY-21-010, QNXM20220001); open project of State Key Laboratory of Coal Mine resources and safety mining (SKLCRSM22KF006); open project of State Key Laboratory of Safety and Health for Metal Mines (2021-JSKSSYS-01), Open Project of Key Laboratory of Green Chemical Engineering Process of Ministry of Education (GCP202108).

Data Availability Statement: The data are not publicly available due to confidential reasons.

Acknowledgments: Thanks for the great efforts of editors and reviewers.

Conflicts of Interest: The authors declare no conflict of interest.

References

1. Yin, S.; Wang, L.; Kabwe, E.; Chen, X.; Yan, R.; An, K.; Zhang, L.; Wu, A. Copper Bioleaching in China: Review and Prospect. *Minerals* **2018**, *8*, 32. [CrossRef]
2. Rawlings, D.E. Heavy Metal Mining Using Microbes. *Annu. Rev. Microbiol.* **2002**, *56*, 65–91. [CrossRef]
3. Lizama, H.M. How Copper Dump Leaching Works. *Miner. Eng.* **2021**, *171*, 107075. [CrossRef]
4. van Staden, P.J.; Petersen, J. Towards Fundamentally Based Heap Leaching Scale-Up. *Miner. Eng.* **2021**, *168*, 106915. [CrossRef]
5. Yin, S.; Wang, L.; Chen, X.; Wu, A. Agglomeration and Leaching Behavior of Copper Oxides with Different Chemical Binders. *Int. J. Miner. Metall. Mater.* **2021**, *28*, 1127–1134. [CrossRef]
6. van Staden, P.J.; Petersen, J. The Effects of Simulated Stacking Phenomena on the Percolation Leaching of Crushed Ore, Part 2: Stratification. *Miner. Eng.* **2019**, *131*, 216–229. [CrossRef]
7. van Staden, P.J.; Petersen, J. The Effects of Simulated Stacking Phenomena on the Percolation Leaching of Crushed Ore, Part 1: Segregation. *Miner. Eng.* **2018**, *128*, 202–214. [CrossRef]

8. Lewandowski, K.A.; Eisele, T.C.; Komar Kawatra, S. Agglomeration for Copper Heap Leaching. In Proceedings of the XXV International Mineral Processing Congress 2010, Brisbane, Australia, 6–10 September 2010.
9. Wang, L.; Yin, S.; Wu, A.; Chen, W. Effect of Stratified Stacks on Extraction and Surface Morphology of Copper Sulfides. *Hydrometallurgy* **2020**, *191*, 105226. [CrossRef]
10. Wang, L.; Yin, S.; Wu, A.; Chen, W. Synergetic Bioleaching of Copper Sulfides Using Mixed Microorganisms and Its Community Structure Succession. *J. Clean. Prod.* **2020**, *245*, 118689. [CrossRef]
11. Maghsoudy, S.; Bakhtiari, O.; Maghsoudy, S. Tortuosity Prediction and Investigation of Fluid Flow Behavior Using Pore Flow Approach in Heap Leaching. *Hydrometallurgy* **2022**, *211*, 105868. [CrossRef]
12. Fernando, W.A.M.; Ilankoon, I.M.S.K.; Rabbani, A.; Chong, M.N. Applicability of Pore Networks to Evaluate the Inter-Particle Flow in Heap Leaching. *Hydrometallurgy* **2020**, *197*, 105451. [CrossRef]
13. Mitchell, R.J.; Mayer, A.S. A Numerical Model for Transient-Hysteretic Flow and Solute Transport in Unsaturated Porous Media. *J. Contam. Hydrol.* **1998**, *30*, 243–264. [CrossRef]
14. Midoux, N.; Charpentier, J.C. Apparent Diffusity and Tortuosity in a Liquid-Filled Porous Catalyst Used for Hydrodesulfuration of Petroleum Products. *Chem. Eng. Sci.* **1973**, *28*, 2108–2111. [CrossRef]
15. Shi, Z.; Zhang, Y.; Liu, M.; Hanaor, D.A.H.; Gan, Y. Dynamic Contact Angle Hysteresis in Liquid Bridges. *Colloids Surf. A Physicochem. Eng. Asp.* **2018**, *555*, 365–371. [CrossRef]
16. Fernando, W.A.M.; Ilankoon, I.M.S.K.; Rabbani, A.; Yellishetty, M. Inter-Particle Fluid Flow Visualisation of Larger Packed Beds Pertaining to Heap Leaching Using X-ray Computed Tomography Imaging. *Miner. Eng.* **2020**, *151*, 106334. [CrossRef]
17. Li, Z.; Liu, D.; Cai, Y.; Ranjith, P.G.; Yao, Y. Multi-Scale Quantitative Characterization of 3-D Pore-Fracture Networks in Bituminous and Anthracite Coals Using FIB-SEM Tomography and X-ray M-CT. *Fuel* **2017**, *209*, 43–53. [CrossRef]
18. Wang, L.; Yin, S.; Wu, A. Visualization of Flow Behavior in Ore Segregated Packed Beds with Fine Interlayers. *Int. J. Miner. Met. Mater.* **2020**, *27*, 900–909. [CrossRef]
19. Muzemder, A.S.H.; Singh, K. Intra-Pore Tortuosity and Diverging-Converging Pore Geometry Controls on Flow Enhancement Due to Liquid Boundary Slip. *J. Hydrol.* **2021**, *598*, 126475. [CrossRef]
20. Or, D.; Tuller, M. Liquid Retention and Interfacial Area in Variably Saturated Porous Media: Upscaling from Single-Pore to Sample-Scale Model. *Water Resour. Res.* **1999**, *35*, 3591–3605. [CrossRef]
21. Tuller, M.; Or, D. Unsaturated Hydraulic Conductivity of Structured Porous Media. *Vadose Zone J.* **2002**, *1*, 14–37. [CrossRef]
22. Or, D.; Tuller, M. Flow in Unsaturated Fractured Porous Media: Hydraulic Conductivity of Rough Surfaces. *Water Resour. Res.* **2000**, *36*, 1165–1177. [CrossRef]
23. Fagan-Endres, M.A.; Harrison, S.T.L.; Johns, M.L.; Sederman, A.J. Magnetic Resonance Imaging Characterisation of the Influence of Flowrate on Liquid Distribution in Drip Irrigated Heap Leaching. *Hydrometallurgy* **2015**, *158*, 157–164. [CrossRef]
24. Ilankoon, I.M.S.K.; Neethling, S.J. The Effect of Particle Porosity on Liquid Holdup in Heap Leaching. *Miner. Eng.* **2013**, *45*, 73–80. [CrossRef]
25. Ilankoon, I.M.S.K.; Neethling, S.J. Hysteresis in Unsaturated Flow in Packed Beds and Heaps. *Miner. Eng.* **2012**, *35*, 1–8. [CrossRef]
26. Wang, L.; Yin, S.; Deng, B. Understanding the Effect of Stepwise Irrigation on Liquid Holdup and Hysteresis Behavior of Unsaturated Ore Heap. *Minerals* **2021**, *11*, 1180. [CrossRef]
27. Wang, L.; Yin, S.; Wu, A. Ore Agglomeration Behavior and Its Key Controlling Factors in Heap Leaching of Low-Grade Copper Minerals. *J. Clean. Prod.* **2020**, *279*, 123705. [CrossRef]
28. Wang, L.; Yin, S.; Deng, B.; Wu, A. Copper Sulfides Leaching Assisted by Acidic Seawater-Based Media: Ionic Strength and Mechanism. *Miner. Eng.* **2022**, *175*, 107286. [CrossRef]
29. Ilankoon, S.K. Hydrodynamics of Unsaturated Particle Beds Pertaining to Heap Leaching. Ph.D. Thesis, Imperial College London, London, UK, 2012.
30. Toye, D.; Marchot, P.; Crine, M.; Pelsser, A.M.; L'Homme, G. Local Measurements of Void Fraction and Liquid Holdup in Packed Columns Using X-ray Computed Tomography. *Chem. Eng. Process. Process Intensif.* **1998**, *37*, 511–520. [CrossRef]
31. Hashemzadeh, M. Copper Leaching in Chloride Media with a View to Using Seawater for Heap Leaching of Secondary Sulfides. Ph.D. Thesis, The University of British Columbia, University Endowment Lands, BC, Canada, 2020.
32. Thomas, M. Understanding Gangue Acid Consumption in Copper Sulfide Heap Leaching: Predicting the Impact of Carbonates, Silicates and Secondary Precipitates. *Miner. Eng.* **2021**, *171*, 107090. [CrossRef]
33. Dreisinger, D. Copper Leaching from Primary Sulfides: Options for Biological and Chemical Extraction of Copper. *Hydrometallurgy* **2006**, *83*, 10–20. [CrossRef]
34. Tshilombo, A.F. Mechanism and Kinetics of Chalcopyrite Passivation and Depassivation during Ferric and Microbial Leaching. Ph.D. Thesis, The University of British Columbia, University Endowment Lands, BC, Canada, 2004.
35. Liu, R.; Hao, X.; Chen, Q.; Li, J. Research Advances of Tetrasphaera in Enhanced Biological Phosphorus Removal: A Review. *Water Res.* **2019**, *166*, 115003. [CrossRef]
36. Nicol, M.J.; Zhang, S. Anodic Oxidation of Iron(II) and Copper(I) on Various Sulfide Minerals in Chloride Solutions. *Hydrometallurgy* **2016**, *166*, 167–173. [CrossRef]
37. Nicol, M.J. The Electrochemistry of Chalcopyrite in Alkaline Solutions. *Hydrometallurgy* **2019**, *187*, 134–140. [CrossRef]
38. Wu, A.; Yin, S.; Qin, W.; Liu, J.; Qiu, G. The Effect of Preferential Flow on Extraction and Surface Morphology of Copper Sulphides during Heap Leaching. *Hydrometallurgy* **2009**, *95*, 76–81. [CrossRef]

39. Bouffard, S.C.; Dixon, D.G. Investigative Study into the Hydrodynamics of Heap Leaching Processes. *Metall. Mater. Trans. B* **2001**, *32*, 763–776. [CrossRef]
40. Chen, K.; Yin, W.; Ma, Y.; Yang, B.; Sun, H.; Rao, F.; Liu, J. Microstructure Analysis of Low-Grade Copper Ore Agglomerates Prepared by Geopolymerization. *Hydrometallurgy* **2021**, *200*, 105564. [CrossRef]
41. Yang, Y.; Yang, Y.; Gao, X.; Petersen, J.; Chen, M. Microstructure Evolution of Low-Grade Chalcopyrite Ores in Chloride Leaching—A Synchrotron-Based X-ray CT Approach Combined with a Data-Constrained Modelling (DCM). *Hydrometallurgy* **2019**, *188*, 1–13. [CrossRef]
42. Von Fraunhofer, J.A. Adhesion and Cohesion. *Int. J. Dent.* **2012**, *2012*, 951324. [CrossRef]
43. Extrand, C.W. Contact Angles and Their Hysteresis as a Measure of Liquid-Solid Adhesion. *Langmuir* **2004**, *20*, 4017–4021. [CrossRef]
44. Zhang, S.; Liu, W.; Granata, G. Effects of Grain Size Gradation on the Porosity of Packed Heap Leach Beds. *Hydrometallurgy* **2018**, *179*, 238–244. [CrossRef]
45. Ghorbani, Y.; Becker, M.; Mainza, A.; Franzidis, J.P.; Petersen, J. Large Particle Effects in Chemical/Biochemical Heap Leach Processes—A Review. *Miner. Eng.* **2011**, *24*, 1172–1184. [CrossRef]
46. Zhang, S.; Liu, W. Application of Aerial Image Analysis for Assessing Particle Size Segregation in Dump Leaching. *Hydrometallurgy* **2017**, *171*, 99–105. [CrossRef]

minerals

MDPI

Article

Improving the Efficiency of Downhole Uranium Production Using Oxygen as an Oxidizer

Bayan Rakishev [1], Zhiger Kenzhetaev [1], Muhametkaly Mataev [2] and Kuanysh Togizov [1,*]

[1] K.I. Satbayev Kazakh National Research University, 050013 Almaty, Kazakhstan
[2] Institute of High Technologies LLP, NAC Kazatomprom JSC, 050013 Almaty, Kazakhstan
* Correspondence: k.togizov@satbayev.university; Tel.: +7-7021710995

Abstract: The features occurring during borehole uranium mining in deposits with low filtration characteristics, as well as the conditions and reasons for the reduction of geotechnological parameters of uranium mining by the well are considered in this study. Core material samples were taken from the productive horizon of the Chu-Sarysui province deposit and granulometric compositions were established. The contents of uranium, aluminum, calcium, magnesium, iron and carbonate minerals in the samples were determined by atomic emission spectroscopy. The X-ray phase analysis method established the features and quantitative and qualitative characteristics of ore-containing minerals. A special technique has been developed for conducting experiments in laboratory conditions using core samples, where the intensity of uranium leaching in tubes is determined. The results of laboratory studies are analyzed and discussed and graphs are constructed, to show the dependencies of change in: the filtration coefficients of K_f; the uranium content in solution; the extraction coefficient; and the specific consumption of sulfuric acid on the values of L:S (the ratio of liquid to solid) in the experiments. The effectiveness of using a mild acidity regime, with the addition of oxygen as an oxidizer, is determined and shown. The values of the uranium content in the productive solution, with the addition of oxygen as an oxidizer, reached 220 mg/L, which exceeds the design parameters. The results of uranium extraction from ore show a positive trend, reaching 68%, with L:S from 1.7 to 3.0, low acidity values and the addition of oxygen as an oxidizer. The specific consumption of sulfuric acid reaches the minimum values when using leaching solutions with reduced acidity of 26 kg/kg. The obtained results, on the flow rate of the solution in the tube, the extraction of uranium from ore and the specific consumption of sulfuric acid, indicate a decrease in sedimentation in a porous medium and increased filtration characteristics, with reduced acidity values in the leaching solution.

Keywords: downhole uranium mining; X-ray phase studies; particle size characteristics; uranium leaching; tube

Citation: Rakishev, B.; Kenzhetaev, Z.; Mataev, M.; Togizov, K. Improving the Efficiency of Downhole Uranium Production Using Oxygen as an Oxidizer. *Minerals* **2022**, *12*, 1005. https://doi.org/10.3390/min12081005

Academic Editor: Kyoungkeun Yoo

Received: 27 June 2022
Accepted: 4 August 2022
Published: 10 August 2022

1. Introduction

Borehole mining of minerals, in particular uranium, involves the dissolution of a useful component by a moving solvent stream at the location of the ore body, followed by the removal and lifting of the formed compounds to the surface [1–3]. The positive aspects of the use of sulfuric acid solutions at enterprises in Kazakhstan are its low cost, widespread use in the national economy, and the possibility of complete dissolution of uranium mineralization [4,5]. However, there are negative aspects, such as the high reactivity of the interaction of sulfuric acid with carbonate-, and clay-minerals of ore-bearing rocks. When sulfuric acid interacts with carbonate minerals, gypsum is formed, and clay minerals swell and increase in size; these factors prevent the leaching process [6].

At mining enterprises, the number of production wells and technological blocks increases annually; this is caused by a gradual decrease in the productivity of the blocks being opened and a decrease in the utilization rate of wells, from 9.0 to 7.6. This is due to the accompanying difficulties of opening and mining peripheral parts of blocks with a ragged ore structure and high heterogeneity of the productive horizon, and low ore filtration coefficients, ranging from 2.0 to 4.0. At the same time, there are no economically justified, effective tools to improve the filtration characteristics of ores and prevent sedimentation for a long period, in difficult mining and geological conditions of well development.

The development of operational blocks in difficult mining and geological conditions is often accompanied by serious complications and an irreversible decrease in the permeability of the downhole zone of the productive reservoir, which dramatically increases the development time and leads to additional costs.

During uranium leaching, the clay content of rocks and ores, as well as the composition of clay cement, largely determines the primary permeability of rocks and reagent consumption. Due to clay minerals, sulfuric acid is neutralized, as a result of which the pH values of solutions do not reach the required parameters (<2.5). This causes an increase in the consumption of chemical reagents, energy, labor, and other operating costs [7,8]. Overcoming the solubility threshold and achieving the required uranium content in PS (Productive Solution) requires an increase in the time period of active leaching, which leads to a decrease in the technical and economic indicators of the process [9].

When designing the borehole development of uranium deposits and feasibility studies, studies of the mineral composition of ores and host rocks of the productive horizon were provided, as well as conducting experimental work on leaching uranium using sulfuric acid solutions with different acidities. Laboratory experiments on uranium leaching were carried out, to determine the effective parameters of the oxidizer application during the intensification of borehole extraction of uranium from ores [10,11]. The experiments included the establishment of mineral and granulometric characteristics of samples, and the leaching of uranium from the core in tubes using solutions with high and standard acidity. A solution with low acidity was prepared separately, but with oxygen feeding into the solution, for conversion from iron (II) to iron (III), for the purpose of effecting subsequent oxidation of uranium (IV) to uranium (VI).

2. Materials and Research Methods

2.1. X-ray Phase and Granulometric Studies

The research was carried out on the material of core samples from the uranium deposit of the Shu-Sarysu province. Previously, using an atomic emission spectroscope, with individual bound plasma, on the iCAP 7400 spectrometer, all private samples had been analyzed for the content of uranium, aluminum, carbonate, calcium, iron, and magnesium in the rock. The results of the analysis of private samples are shown in Table 1.

Table 1. Results of analysis of private samples [10].

Private Sample	U	CO_2	Al	Ca	Fe Total	Mg
1	0.0502	0.17	4.9640	1.2031	0.8031	0.3694
2	0.0568	0.18	4.5153	1.1953	0.8249	0.4780
3	0.0296	0.10	4.6824	0.8658	1.0250	0.3854
Average value, %	0.0455	0.15	4.7205	1.0881	0.8843	0.4109

The analysis results shown in Table 1 indicate that the CO_2 content averages 0.15% of the total sample weight, which indicates a low content of carbonate minerals in the ore and host rocks. However, the average Al content in process samples is 4.7%, which indicates the presence of a significant amount of feldspar and clay minerals. Table 2 shows data on the granulometric composition of the process sample.

Data from the Table 2 shows that the process sample consists mainly of a fine-grained fraction <0.35 mm—57.33%. This indicates an increased content of clay-silt particles, which prevents the filtration of solutions and the process of sulfuric acid leaching.

Table 2. Granulometric composition of the process sample.

Content, %, of Fractions with Fineness, mm								Total Indicator, %
Large				Medium			Small	
>2>	>1.6>	>1.4>	>1>	>0.8>	>0.5>	>0.35>	<0.35	
1.05	2.24	1.38	3.15	1.93	6.35	15.46	57.33	100

The mineralogical characteristics of the samples were studied using a DRON-3 X-ray diffractometer. The result of the sample diffractometry is shown in Figure 1, the results of the analysis of quantitative and qualitative characteristics are shown in Table 3.

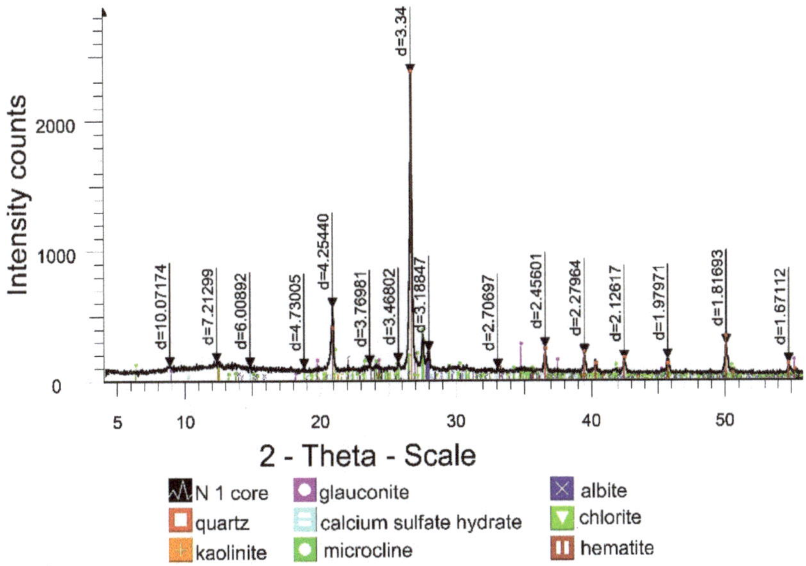

Figure 1. Diffractogram of the initial core material sample.

Table 3. Results of analysis of quantitative and qualitative characteristics of samples.

Mineral	Formula	Content, %
Quartz	SiO_2	58.6
Kaolinite	$Al_2 (Si_2O_5) (OH)_4$	10.3
Glauconite	$(K, Ca, Na)_{0.8}(Al, Fe, Mg)_2(Si, Al)4O_{10}(OH)_2\ nH_2O$	7.9
Calcium Sulfate	$Ca (SO_4) (H_2O)_{0.5}$	7.5
Microcline	$(K, Na) AlSi_3O_8$	6.4
Albite	$(Na, Ca) (Al, Si)_4O_8$	4.1
Chlorite	$(Mg, Fe)_5Al(Si_3Al)O_{10}(OH)_8$	2.8
Hematite	Fe_2O_3	2.4

According to the results of the analysis of the quantitative and qualitative characteristics of the samples, given in Table 3, it can be seen that the average quartz content is (58.6%). However, the discovered clay minerals: glauconite (7.9%), kaolinite (10.3%), microcline (6.4%) and albite (4.1%) cause difficulties in the leaching process, due to their swelling during reservoir water displacement and pH changes. The presence of gypsum (7.5%) also forms an obstacle, in the form of precipitation in the places of unloading, namely in the filter zones of wells, when the speed of movement of solutions changes. The predominance of fine fractions and fine-grained sands in the ores of the productive horizon causes widespread clogging of the pore space and difficulties in restoring the initial filtration of solutions. This explains the difficulties that arise in extracting uranium under similar conditions, with the current range of technological standards: LS (Leaching Solution) injection–PS pumping, and a decrease in the well utilization rate due to downtime for repair and restoration work (RRW) [12]. The X-ray diffractometer is not able to determine the content of uranium minerals, due to the small ratio (less than 1.0%).

2.2. Laboratory Studies on Leaching in Tubes from Core Samples

Tube leaching of uranium from samples in laboratory conditions enables the acquisition of more complete information about the process of dissolution of uranium minerals. The flow of the prepared solvent was carried out through a sample of ore material, through filled sealed tubes. The diagram of the laboratory installation is shown in Figure 2. To determine the mass of the sample in the corresponding tube, the technological sample was mixed and weighed on an electronic scale before loading. Filtration of solutions was carried out, with a constant injection of solvent and a pressure drop at the inlet and outlet. To determine the consumption of chemical reagents, a standard for the preparation of solutions with different acidities was developed. In order to study the flow rate and volume of dissolved uranium, relative to the initial sample, the filtered solution was collected in a container, for further analysis.

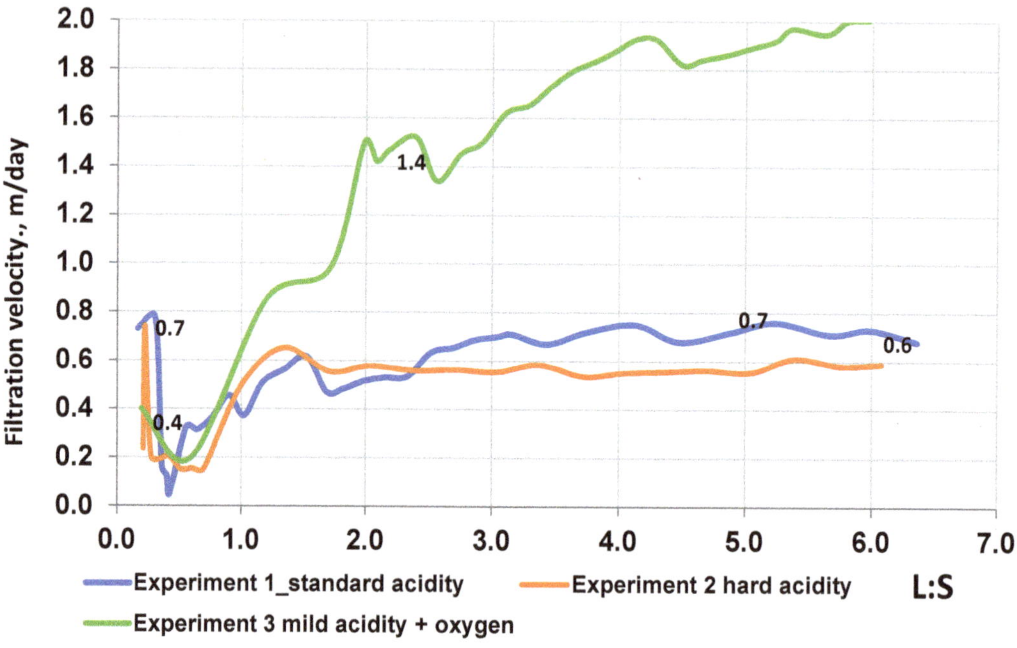

Figure 2. Dynamics of the change in the filtration rate of the solution depending on the L:S.

Before applying the solvent, reservoir water was injected into the tubes with a volume of one pore space. For Experiment 1 (tube 1), a standard was adopted with standard acidity in a leaching solution, with a change in the concentration of sulfuric acid in the LS, with a regime: 25–15–10–5–3 g/L when changing L:S from 0 to 2.5. For experiment 2 (tube 2), a standard with a strict acidity regime in the leaching solution was selected, providing for a change in the concentration of sulfuric acid with a regime: 25–20–15–10–5 g/L, with an increase in L:S from 0 to 2.5. In experiment 3 (tube 3), a standard with a mild acidity regime with the addition of oxygen to the solution under pressure from an oxygen cylinder, provided for the supply of sulfuric acid with a regime: 15–10–5–3 g/L, with a change in L:S from 0 to 2.5. Oxygen was injected into a tank containing leaching solution, as an oxidizer of iron (II), to effect the subsequent oxidation of uranium (IV). Table 4 shows the parameters of the acidity of the solutions in the experiments, according to the developed standards. The L:S indicator is the ratio of a unit volume of liquid leaching solution to a unit volume of solid (ore), it is one of the main indicators for calculation in geotechnological processes. The total time spent on conducting experiments depends on the rate of liquid filtration through the sample and the rate of dissolution of uranium minerals. The time spent on core samples from different horizons or deposits will be different. The rate of passage of solutions through the core material sample depends on the filtration characteristics of rocks, composition of the solution, and sedimentation capacity of the ores [13]. The leaching solution was prepared under laboratory conditions from reservoir water (the uranium deposit in question) and dosed with the addition of sulfuric acid, according to the standards.

Table 4. Parameters of acidity of solutions.

Ratio L:S	Acidity in LS, g/L, According to Experiments		
	1	2	3
0–0.2	25	25	15 + oxygen
0.2–0.8	15	20	10 + oxygen
0.8–1.5	10	15	10 + oxygen
1.5–2	5	10	5 + oxygen
2–2.5	3	5	3 + oxygen
2.5–3	0	0	0

According to the developed method of experimental studies, solutions for experiment 1, with standard acidity, provided for the preparation and supply of a solution with the values: 25–15–10–5–3 g/L in the L:S ranges: 0–0.2; 0.2–0.8; 0.8–1.5; 1.5–2; 2–2.5; when L:S > 2.5; with zero acidity. Experiment 2 provided for the preparation and supply of working solutions with a strict acidity regime, with a concentration of sulfuric acid: 25–20–15–10–5 g/L at the appropriate L:S ranges. In experiment 3, working solutions were prepared and fed under a mild acidity regime, with the values: 15–10–10–5–3 g/L when compressed oxygen was added in the L:S ranges: 0–0.2; 0.2–0.8; 0.8–1.5; 1.5–2; 2–2.5; when L:S > 2.5; without the addition of acid.

The parameters of the oxidizer's effect on the process of uranium dissolution from samples were determined using output solutions, which were recorded in a log, with volume determination in measuring vessels, for further calculations and analyses. Using the obtained data from the output solution, the volume of dissolved uranium was calculated, in addition to the values of L:S, pH, ORP, residual acidity, Fe^{3+} content, Fe^{2+} content and specific consumption of sulfuric acid per 1 kg of uranium. The content of uranium and iron in the solution was clarified and checked by titration, according to the method of chemical measurements (MVI) # 36-2019 # KZ06.01.00050-2019 from 11.07.2019.

3. Discussion or Results

The process of leaching uranium in natural permeability is a complex process involving several stages, one of which is heterogeneous, controlling at the liquid-solid phase boundary, containing a reagent capable of forming highly soluble compounds when interacting with uranium-containing minerals or rocks [14]. The calculated data obtained, based on the results of laboratory experiments, made it possible to plot: the change in the solution flow rate, as a function of L:S (Figure 2); the values of the uranium concentration in productive solutions, as a function of L:S (Figure 3); andthe extraction of uranium, as a function of

L:S (Figure 4); with the flow rate of the chemical reagent depending on L:S (Figure 5). The flow rate of the solution in the ore during the processes of interaction with uranium minerals, with further dissolution and transfer to the unloading and lifting zone, is one of the important parameters and is determined by the ore filtration coefficient [15,16]. The K_f ore filtration coefficient is determined in (m/day) and is calculated using the formula:

$$K_{\Phi} = \frac{\Delta VL}{\Delta t \Delta HS} \tag{1}$$

where: ΔV is the volume of filtered solvent; L is the length of the tube; Δt–is the sample measurement time; ΔH is the hydrostatic pressure drop; S is the cross-sectional area of the tube.

As can be seen from the graph, the changes in the filtration coefficient of K_f depend on the L:S. In experiment 1, when feeding working solutions with a standard mode of acidity, at first the K_f is 0.73 and rises to 0.78, but then it decreases to a minimum value of 0.11 at values of L:S, in the range from 0 to 0.41, with the corresponding acidity parameters of 25–15 g/L. This decrease indicates the formation of colmatation effects, during the interaction of sulfuric acid solutions with ore-containing rocks, that prevent the filtration of solutions. The subsequent gradual increase in the filtration coefficient on the tube from 0.45 to 0.62 at the L:S range, in the range of 0.4–1.4, is due to a decrease in acidity in the working solutions, from 15 to 10 g/L, and a partial reduction in the effect of colmatation. A further decrease in acidity in the solutions did not affect the filtration volume of the solution.

The filtration parameters in experiment 2, with a strict acidity regime of working solutions, approximately correspond to the values of the filtration coefficient in experiment 1, the only dissimilarity being due to a slight difference in acidity (5 g/L). However, the low acidity value in solutions in experiment 3 with the addition of oxygen as an oxidizer positively affected the filtration parameters of the experiment. In experiment 3, with an initial K_f of 0.42 and a further decrease to 0.25 with a change in L:S in the range of 0–0.5, acidity is observed in the working solution of 15 g/L, which indicates a slight formation of the colmatation. A subsequent increase of K_f from 0.25 to 0.9 at values of L:S from 0.5 to 1.5, with an appropriate acidity of 10 g/L, confirms a decrease in the effect of colmatation in the filtrate. The increased values of K_f in experiment 3, compared to experiments 1 and 2, indicate a proportional effect of acidity on the filtration rate and the formation of colmatation effects. The total time spent on conducting experiments, depending on the filtration rate of solutions in core samples, ranged from 320 to 428 h: a total of 320 h were spent on conducting experiment 1; 412 and 428 h were spent on experiments 2 and 3, respectively. A comparative analysis of the results of uranium leaching from core samples using various parameters of solutions allows us to determine the effectiveness of leaching.

These changes in the concentration of the useful component in the solution depending on the L:S, shows the maximum and minimum concentrations of uranium in the productive solution, with the corresponding volume of the output solution. The concentration of uranium in productive solutions of the C^{U}_{cp} and the ratio of L:S are determined by the formulas:

$$C^{U}_{cp} = \frac{\sum_{i=1}^{n} C^{U}_{i} \Delta V_i}{\sum_{i=1}^{n} \Delta V_i} \tag{2}$$

where n is the amount of sampling for a given value; ΔV is the amount of productive solution in the i-th sample; C^{U}_{i}—the concentration of uranium in the i-th specific sample. Counting is performed for all registered samples of productive solution n.

As can be seen from the graph (see Figure 3), in experiment 1, when leaching using solutions with a standard acidity regime, the maximum uranium content of 520 mg/L is achieved at L:S 0.561, after overcoming the threshold of 250 mg/L and reducing to 125 mg/L. This indicates the gradual overcoming of the solubility threshold and the achievement of active leaching with the required volume of solution and acid. However, there is a decrease in the concentration of uranium in the productive solution, from 520 mg/L to a low value of 30 mg/L, at a range of values of L:S from 0.7 to 1.5. This may be due to a decrease in the acidity of working solutions in 10 g/L. In experiment 2, in the process of leaching under a strict acidity regime, the uranium content in the productive solutions reaches a peak of 670 mg/L within the range of L:S 0.7, with a further sharp decrease in the uranium content in the productive solutions from 670 to 130 mg/L, with an increase in L:S to 1.0, which indicates a decrease in acidity from 25 to 15 g/L. In experiment 3, when leaching uranium with a mild acidity regime of solutions and the addition of oxygen, the uranium content gradually increases to 220 mg/L with a corresponding increase in L:S from 0 to 2.0. This also confirms the gradual overcoming of the solubility threshold by weak sulfuric acid solutions, followed by oxidation and conversion of tetravalent uranium into a soluble form, while creating favorable conditions for stable active leaching and oxidation. The uranium content in the solutions in all experiments showed similar maximum peaks, approximately corresponding to the initial uranium content, depending on the acidity of the leaching solutions. This indicates the neutralization of sulfuric acid during interaction with ore-containing rocks and the influence of an oxidizer during the transfer of uranium mineralization into solution (experiment 3).

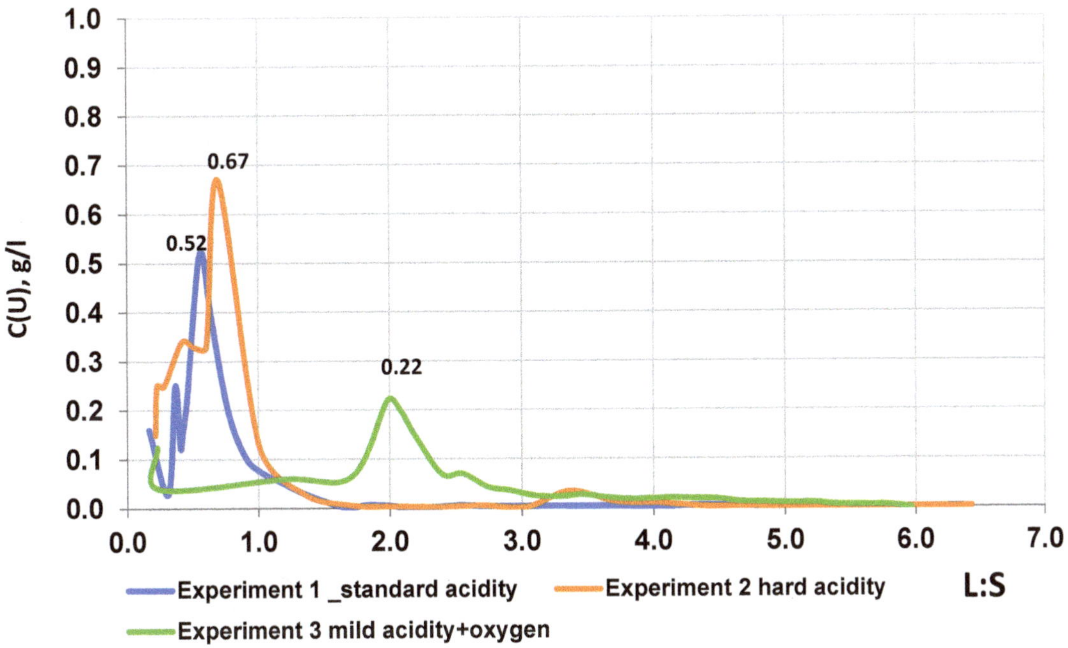

Figure 3. Uranium concentration values in productive solutions as a function of L:S.

The effectiveness of the parameter of the proposed solution, with reduced acidity and with the addition of an oxidizer, was determined on the basis of a comparative analysis with the parameters of uranium extraction under standard and rigid modes of acidity of solutions. For this purpose, experimental data were collected and processed, calculations were made and schedules for uranium extraction were constructed. Uranium extraction depends on the uranium content in the productive solution and the volume of the most productive uranium; these data are the most informative and fundamental in assessing the solubility and stability of uranium minerals. Uranium extraction is calculated as the ratio of the mass of uranium in the output solutions to its initial calculated mass in the sample:

$$\varepsilon = \frac{\sum_{i=1}^{n} C_i^U \Delta V_i}{M_p C_{initial\ core\ sample}^U} \tag{3}$$

where $C^{U}_{initial\ core\ sample}$—the concentration of uranium in the initial sample; C^{U}_{i}—is the concentration of uranium in the productive solution of the *i*-th sample; M_p is the mass of ore in the initial sample; ΔV_i is the volume of the productive solution in the *i*-th sample.

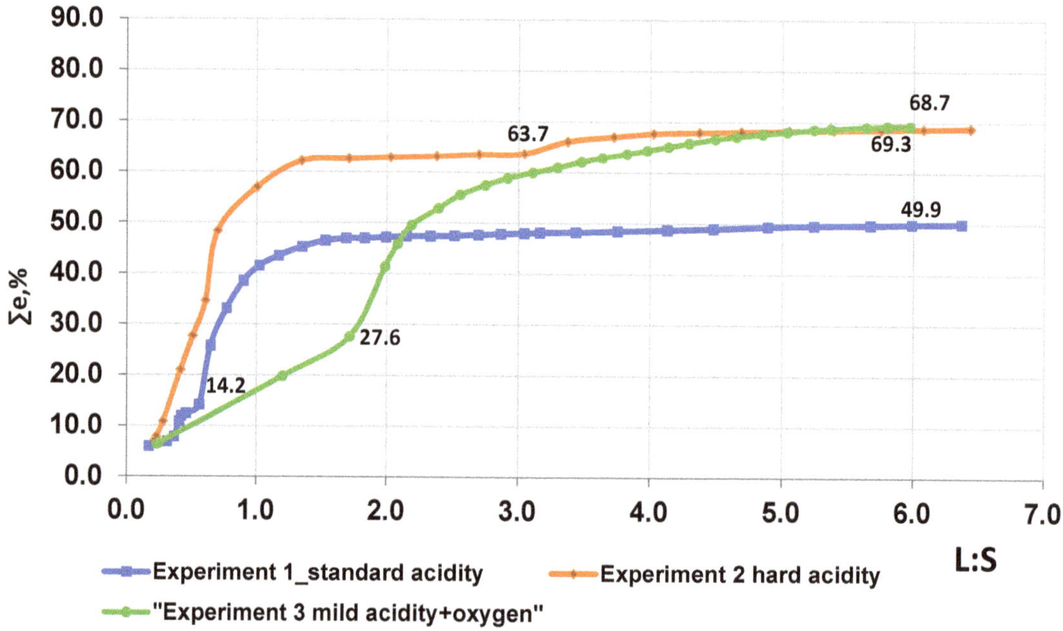

Figure 4. Uranium extraction depending on L:S.

The graph in Figure 4 shows that the maximum extraction values in Experiment 1 reached 49% of the mass of uranium contained in the sample; this indicates insufficient solvent capacity of working solutions with standard acidity and low filtration values. A sharp increase in extraction occurs in the range of L:S from 0.5 to 0.9 to 38.6%, followed by a slowdown in extraction at L:S > 1; the acidity of working solutions in this range was maximum: 25–15 g/L. The extraction data correlate with the graph of the uranium content in the solution, where the peak of the maximum uranium content in the PS falls in the range of L:S from 0.5 to 0.8. A further slowdown in the extraction of the useful component is due to a decrease in the uranium content in the solution, due to a decrease in the acidity of working solutions from 15 to 10 g/L. In experiment 2, the increase in uranium extraction occurs at L:S parameters from 0 to 1.0 with the achievement of 56% extraction, followed by a slowdown of 63%, which indicates intensive extraction of uranium in solutions with a strict acidity regime. In the range of L:S from 0 to 1.0, the acidity of the solutions supplied to tube 2 was maximum and amounted to: 25–20–15 g/L. Intensive extraction is due to the high uranium content in the solution: 320–670 mg/L and the corresponding filtration coefficient of the solution is 0.34–0.44 in the range of L:S from 0 to 0.9. The extraction of the useful component in experiment 3, when feeding working solutions with a mild acidity regime with the addition of oxygen, reached a maximum value of 68% with a change in L:S to 3.0. However, the active extraction of uranium in Experiment 3 also occurred in the range of L:S from 1.7 to 3.0, from 27.6 to 60.3%, respectively. The intensive extraction of uranium in experiment 3 in the range L:S 1.7–3.0 is due to the achievement of the solubility threshold and the creation of conditions for active leaching, with the oxidation of iron (II) and the interaction of iron (III) for the oxidation of uranium (IV) [17,18]:

$$4Fe^{2+} + O_2 + 4H^{+} \rightarrow 4Fe^{3+} + H_2O \quad (4)$$

$$UO_2 + 2Fe^{3+} + 3SO_4{}^{2-} \rightarrow UO_2(SO_4) + 2Fe^{2+} + 2SO_4{}^{2-} \tag{5}$$

To determine the economic efficiency of using an oxidizer in the intensification of uranium leaching, and a comparative analysis of solvent consumption, graphs were constructed based on the experiments performed. The consumption of sulfuric acid as a solvent per kilogram of extracted uranium P_K is calculated as the ratio of the total mass of the reagent consumed during the experiment to the calculated mass of extracted uranium:

$$P_K = \frac{\sum_{i=1}^{n}(C_o^K - C_i^K)\Delta V_i}{\sum_{i=1}^{n} C_i^U \Delta V_i} \tag{6}$$

where C_o^K is the initial concentration of the reagent in the leaching solution (determined when using a mother liquor with the initial acidity); C_i^K—the actual concentration of the reagent in the i-th sample of the leaching solution (actual); C_i^U the actual concentration of uranium in the productive solution of the i-th sample; M_p the total mass of ore used in the experiment; ΔV_i the total volume of leaching solution in the i-th sample.

Reagent costs per unit of processed ore mass (calculated to determine the acid capacity of ore) are determined by the formula:

$$P_K = \frac{\sum_{i=1}^{n}(C_o^K - C_i^K)\Delta V_i}{M_P} \tag{7}$$

where C_o^K is the initial concentration of the solvent reagent in the leaching solution (determined when using sorption curves with residual acidity); C_i^K—is the concentration of the solvent reagent in the i-th sample of the leaching solution (actual); M_p is the mass of ore used in the experiment; ΔV_i is the volume of the leaching solution in the i-th sample.

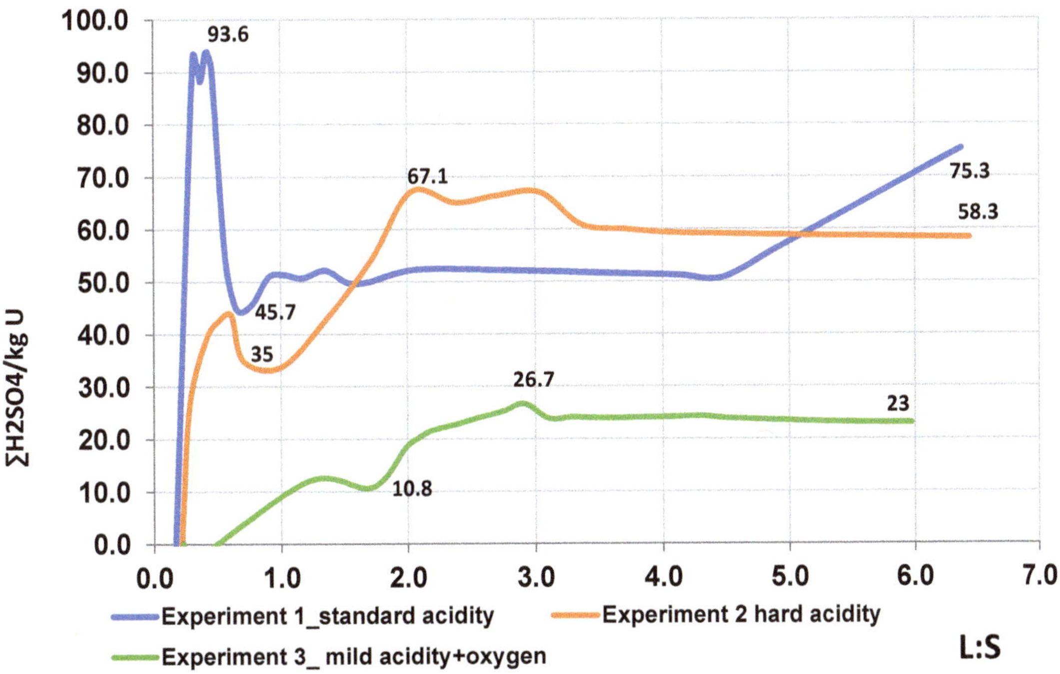

Figure 5. Consumption of sulfuric acid depending on L:S of the process.

The graph in Figure 5 shows that the consumption of sulfuric acid in Experiment 1 reaches maximum values of 90–93 kg/kg U in the L:S range: 0.2–0.7, after which it sharply decreases to values of 48–50 kg/kg U. A sharp increase in the consumption of the solvent reagent is associated with the acidity in the leaching solution and low parameters of uranium extraction in the range of L:S from 0 to 0.8. After increasing the extraction of uranium from 25.8 to 47% in the L:S range from 0.647 to 1.689, there is a noticeable decrease in the consumption of the solvent reagent. The maximum values of the reagent consumption in experiment 2, when feeding working solutions with a hard acidity regime, reach 67 kg/kg U at L:S 2.0. A gradual increase in the specific consumption in experiment 2, with a hard acidity regime of working solutions, is due to a slowdown in the extraction of uranium in the corresponding range L:S. The maximum values of the specific consumption of sulfuric acid in experiment 3 are not able to reach the corresponding indicators in experiment 2 and amount to 26.7 kg/kg U with a change in L:S from 2.0 to 3.0, after which they decrease to the minimum values of 23 kg/kg U. Reduction of reagent consumption in the L:S range >3.0 is due to intensive uranium extraction in this interval. In a comparative analysis of the solvent reagent consumption, the lowest values are observed in Experiment 3, with a mild acidity regime of working solutions and with the addition of oxygen as an oxidizer. The decrease in reagent consumption is caused by intensive uranium extraction, due to the oxidizing effect of iron (III) on uranium (IV) in its conversion to uranium (VI), during its further dissolution. In the hard mode of acidity (experiment 2), the specific acid consumption is higher than in the soft mode of acidity, with the addition of an oxidizer, but lower than in the standard mode of acidity (experiment 1), due to the intensive extraction of uranium from the tube.

4. Conclusions

X-ray phase studies of ore-bearing rocks indicate their diverse mineral composition, and the presence of calcium sulfate will contribute to chemical colmatation in wells. The predominance of a fine fraction of fine-grained sands (57% of the total mass) prevents the filtration of solutions in the productive horizon. The data indicate a complex structure of the productive horizon with low filtration parameters. As practice shows, in such conditions, the leaching processes are hindered by the low uranium content in the PS and low parameters of the turnover of solutions, which requires continuous intensification of the process by hydrodynamic methods.

The data indicate the direct influence of the concentration of sulfuric acid on the dissolution of uranium minerals and the content of uranium in productive solutions. However, the influence of filtration characteristics on the overall indicators of the process (extraction, specific acid consumption) is no less important when solving issues of production intensification. The addition of oxygen as an oxidizer to sulfuric acid solutions, with reduced acidity, positively affects the intensification of uranium leaching. Reducing the acidity of working solutions significantly increases the filtration characteristics of samples, due to the oxidizing ability of iron (III) to uranium (IV); the uranium content in the solutions reach the design values of 300 mg/L. A decrease in the consumption of the solvent reagent (37 kg ^{H2SO4}/kg ^{U}) indicates the economic feasibility of using an oxidizer at the first stage of well production, in areas with a predominance of ores containing uranium (IV). In addition, the low acidity of the leaching solution will not create sedimentation, will lead to intensive development of blocks and reduce operating costs, electricity costs and labor costs.

These studies allowed us to establish that the uranium content in the productive solution directly depends on the concentration of acid in the leaching solution. However, an increase in the concentration of the solvent reagent does not always lead to an increase in its consumption. The analysis of the values of uranium extraction against time allows us to determine the optimal concentration and consumption of reagents during borehole production. The optimal concentration is one in which an excess leads to a sharp consumption of acid and a slight increase in the intensity of uranium extraction. The value of the optimal acid concentration depends on the properties of the specific ore minerals in the test samples.

The use of air at the enterprises' uranium mining, in areas with low filtration characteristics of ores, has become possible thanks to special devices, the design of which allows for dosing of air into the LS pipeline. This technology does not need capital investments and allows for reductions in the consumption of chemical reagents, as well as increases in the extraction of uranium, by improving the filtration characteristics of ores and host rocks. Scientists and specialists at Satbayev University have already developed this area and have achieved positive results [19].

Author Contributions: Conceptualization, B.R. and Z.K.; methodology, Z.K. and M.M.; software, M.M. and K.T.; validation, Z.K., M.M. and K.T.; formal analysis, Z.K. and M.M.; investigation, Z.K. and K.T.; resources, Z.K. and K.T.; data curation, B.R.; writing—original draft preparation, Z.K. and K.T.; writing—review and editing, Z.K. and K.T.; visualization, Z.K. and K.T.; supervision, B.R.; project administration, B.R.; funding acquisition, B.R. All authors have read and agreed to the published version of the manuscript.

Funding: This research is funded by the Science Committee of the Ministry of Education and Science of the Republic of Kazakhstan (Grant No. AP08856422—"Development of an Innovative Technology for the Intensification of Downhole Uranium Mining Using Hydrodynamic Decalmatation Device in Combination with a Complex of Chemicals for Multifunctional Purposes").

Data Availability Statement: Not applicable.

Acknowledgments: The authors express their gratitude to the Limited Liability Partnership "Institute of High Technologies" at JSC "NAC "Kazatomprom" for providing the computer program and for the possibility of using the laboratory facilities.

Conflicts of Interest: The authors declare no conflict of interest. The funders had no role in the design of the study; in the collection, analyses, or interpretation of data; in the writing of the manuscript, or in the decision to publish the results.

References

1. Khawassek, Y.M.; Taha, M.H.; Eliwa, A.A. Kinetics of leaching process using sulfuric acid for Sella uranium ore material, South Eastern desert, Egypt. *Int. J. Nucl. Energy Sci. Eng.* **2016**, *6*, 62–73. [CrossRef]
2. Zakia, S.A.; Rashad, M.M.; Mohamed, S.A.; Mira, H.E.; Abd el Wahab, G.M. Kinetics of uranium leaching process using sulfuric acid for Wadi Nasib ore, South western Sinai, Egypt. *Aswan Univ. J. Environ. Stud.* **2020**, *1*, 171–182. [CrossRef]
3. Atia, B.M.; Gado, M.A.; Cheira, M.F. Kinetics of uranium and iron dissolution by sulfuric acid from Abu Zeneima ferruginous siltstone, Southwestern Sinai, Egypt. *Euro-Mediterr. J. Environ. Integr.* **2018**, *3*, 39. [CrossRef]
4. Chen, J.; Zhao, Y.; Song, Q.; Zhou, Z.; Yang, S. Exploration and mining evaluation system and price prediction of uranium resources. *Min. Miner. Depos.* **2018**, *12*, 85–94. [CrossRef]
5. Rakishev, B.R.; Mataev, M.M.; Kenzhetaev, Z.S. Analysis of mineralogical composition of sediments in in-situ leach mining of uranium. *MIAB* **2019**, *7*, 123–131. [CrossRef]
6. Kenzhetaev, Z.; Nurbekova, M.; Togizov, K.; Abdraimova, M.; Toktaruly, B. Methods for intensification of borehole uranium mining at the fields with low filtration characteristics of ores. *Min. Miner. Depos.* **2021**, *15*, 95–101. [CrossRef]
7. Rakishev, B.R.; Bondarenko, V.I.; Matev, M.M.; Kenzhetaev, Z.S. Influence of chemical reagent complex on intensification of uranium well extraction. *Nauk. Visnyk Natsionalnoho Hirnychoho Univ.* **2019**, *6*, 25–30. [CrossRef]
8. Panfilov, M.; Uralbekov, B.; Burkitbayev, M. Reactive transport in the underground leaching of uranium: Asymptotic analytical solution for multi-reaction model. *Hydrometallurgy* **2016**, *160*, 60–72. [CrossRef]
9. Beleckij, V.I.; Bogatkov, L.K.; Volkov, N.I. *Spravochnik po Geotehnologii Urana*; Energoatomizdat: Moskva, Russia, 1997.
10. Rakishev, B.; Mataev, M.M.; Kenzhetayev, Z.; Altaybayev, B.; Shampikova, A. Research into leaching of uranium from core samples in tubes using surfactants. *Min. Miner. Depos.* **2020**, *14*, 97–104. [CrossRef]
11. Kenzhetaev, Z.; Kuandykov, T.; Togizov, K.; Abdraimova, M.; Nurbekova, M. Selection of rational parameters for opening and drilling of technological wells underground uranium leaching. *News NAS RK Ser. Geol. Tech. Sci.* **2022**, *3*, 115–127. [CrossRef]
12. Kuandykov, T.; Nauryzbayeva, D.; Yelemessov, K.; Karmanov, T.; Kakimov, U.; Kolga, A. Development and justification of a hydro-impulse method for increasing ore permeability in conditions of uranium borehole production. *News NAS RK Ser. Geol. Tech. Sci.* **2020**, *6*, 126–133. [CrossRef]
13. Rakishev, B.; Yazikov, E.G.; Mataev, M.M.; Kenzhetayev, Z. Studies of uranium leaching from core sample in tubes using an oxidizer. *Gorn. Zhurnal* **2021**, *9*, 84–89. [CrossRef]

14. Mamytbekov, G.K.; Kozhakhmetov, S.K. Borehole Filter. Innovation Patent of the Republic of Kazakhstan No. 924, 5 April 2012; bulletin No. 3, 15 March 2013.
15. Yusupov, K.A.; Aliev, S.B.; Dzhakupov, D.A.; Elzhanov, E.A. Application of ammonium bifluoride for chemical treatment of wells in underground uranium leaching. *Gorn. Zhurnal* **2017**, *4*, 57–60. [CrossRef]
16. Golik, V.I.; Dmitrak, Y.V.; Brigida, V.S. Impact of duration of mechanochemical activation on enhancement of zinc leaching from polymetallic ore tailings. *Nauk. Visnyk Natsionalnoho Hirnychoho Univ.* **2020**, *5*, 47–54. [CrossRef]
17. Rakishev, B.R.; atayev, ..; Kenzhetayev, Z.S.; Shampikova, A.H.; Toktaruly, B. Innovative methods for intensifying borehole production of uranium in ores with low filtration characteristics. *News NAS RK Ser. Geol. Tech. Sci.* **2020**, *6*, 213–219. [CrossRef]
18. Mamytbekov, G.K.; Kozhakhmetov, S.K.; Aitenov, S.M.; Raimkhanov, A.E.; Shavanda, V.V. Mobile Modular Installation of Cavitation Action for Development and Cleaning of Wells. Patent for Invention of the Republic of Kazakhstan No. 26619, Application No. 2012/0299, 11 March 2012. E21B 37/00 (2006.01); E21B 37/06 (2006.01) publ. 25 December 2012; byul. No. 10 of 15 October 2014.
19. Aben, E.; Toktaruly, B.; Khairullayev, N.; Yeluzakh, M. Analyzing changes in a leach solution oxygenation in the process of uranium ore borehole mining. *Min. Miner. Depos.* **2021**, *15*, 39–44. [CrossRef]

minerals

Review

Review on the Art of Roof Contacting in Cemented Waste Backfill Technology in a Metal Mine

Fengbin Chen [1,2], Jiguang Liu [1,2], Xiaowei Zhang [3], Jinxing Wang [1,2,*], Huazhe Jiao [1,2,*] and Jianxin Yu [1,2]

[1] School of Civil Engineering, Henan Polytechnic University, Jiaozuo 454003, China; fbchen@hpu.edu.cn (F.C.); liujiguang1998@126.com (J.L.); jianxinyu@hpu.edu.cn (J.Y.)
[2] Collaborative Innovation Center of Coal Work Safety and Clean High Efficiency Utilization, Jiaozuo 454000, China
[3] Jiaozuo Qianye New Material Co., Ltd., Jiaozuo 454150, China; zhangxiaowei@163.com
* Correspondence: wjx@hpu.edu.cn (J.W.); jiaohuazhe@hpu.edu.cn (H.J.)

Abstract: The backfilling mining method can effectively solve the environmental and safety problems caused by mining. It is the key technology to realize green mining. Scientific development has accelerated the pace of research on the rational utilization of mine solid waste and improved the research level of backfilling technology. The development history of the backfilling mining method is introduced in the present paper, and it is determined that roof-contacting backfilling is the key technology of mine-solid-waste backfilling mining. This paper introduces three calculation methods of similar roof-contacted backfilling rates. In this paper, the relationship between the characteristics of backfilling slurry made from solid waste from mines and the roof-contacted backfilling rate is systematically analyzed, such as the flow law in stope (gravity gradient), bleeding shrinkage, and natural sedimentation of backfilling slurry. It is pointed out that the characteristics of the stope, such as washing-pipe water, water for the leading way, filling pipeline, and shape of the backfilling stope, are closely related to the roof-contacted backfilling rate. The influential relationship between objective factors, such as human factors, limited auxiliary leveling measures, and backfilling "one-time operation" in the backfilling process, and high-efficiency top filling are considered, and a schematic diagram of the influencing top-filling rate and structure is drawn. At the same time, this paper summarizes the improvement measures of roof connection from three aspects. These include the use of expansive non-shrinkable materials, forced roof-contacted technology, and strengthening management level. It is pointed out that the roof-contacted filling technology is still facing severe challenges, and the research on the backfilling connection technology needs to be strengthened.

Keywords: solid-waste filling; roof-contacted rate; influencing factors of roof connection; regulation and improvement of roof-connection measures

Citation: Chen, F.; Liu, J.; Zhang, X.; Wang, J.; Jiao, H.; Yu, J. Review on the Art of Roof Contacting in Cemented Waste Backfill Technology in a Metal Mine. *Minerals* **2022**, *12*, 721. https://doi.org/10.3390/min12060721

Academic Editor: Mostafa Benzaazoua

Received: 24 April 2022
Accepted: 1 June 2022
Published: 4 June 2022

Publisher's Note: MDPI stays neutral with regard to jurisdictional claims in published maps and institutional affiliations.

1. Introduction

Mineral resources are the precious materials given to mankind by nature and the basis for the survival and development of human society. They are closely and positively proportional to the enhancement of national economic strength, civilization and progress, social stability, and the improvement of national living standards [1–3]. Whether from the Stone Age or the Information Age, a kind of mineral raw material with good performance and strong functionality is produced in each historical development stage. Therefore, without the development and utilization of mineral resources, human society will not be able to progress [4–7]. However, the safety and environmental problems accompanying the mining process restrict the sustainable development of mineral resources [8–10].

China's economic take-off is inseparable from the great help of mining. Due to the combination of extensive type and excess capacity production, the area of land destroyed by mining in China has reached 2 million hm^3. The amount of tailings formed by mining and beneficiation has reached 14.6 billion tons, occupying a land area of 8700 km^2, equivalent to the area of four Shenzhen cities [11]. About 5–7 billion tons of tailings are generated globally every year [12–14]. The generation of such a large quantity of mill tailings adversely affects the environment, including the air, water, and soil [15–18], as shown in Figure 1. At the same time, in the process of underground mining, a large number of mined-out areas are created, which not only threatens the safety of underground operations, but also includes the hidden danger of inducing mine earthquakes and surface collapse [19–21]. To solve a series of problems caused by mining, such as surface stacking, solid waste stacking, and goaf treatment, it is necessary to break through the bottleneck constraints of resources and the environment, adhere to energy conservation and emission reduction, and develop a mining circular economy [22,23]. With the idea of "one filling to treat three wastes, one waste to treat two hazards", the filling technology creatively uses mine solid waste efficiently, eliminates the tailings pond, and governs the goaf, forming a mining method with a high recovery rate and low dilution rate [24–26]. It is an important technical means to move towards the win–win situation of building "environmentally friendly" and "safe and efficient" mines [18,27–30]. Nowadays, there are many studies on the strength of backfilling massif, but there are relatively few studies on how to improve the roof-contacted backfilling rate to address the overall performance of the backfilling massif, which is an urgent problem that needs to be solved.

Figure 1. Surface disposal of tailings: (**a**) Wet disposal and (**b**) dry disposal.

2. Development of Backfilling Technology and Roof-Contacting Backfilling Technology

The backfilling mining method has a development history of more than half a century. At first, it aimed at the simple treatment of mine solid waste, such as waste rock. Nowadays, backfilling mining has gradually developed into a comprehensive technology to improve the environment, control the ground pressure, and reduce the poor-loss index to form a complete mining process [31,32]. According to different backfilling materials, backfilling mining has experienced dry backfilling, hydraulic sand backfilling, and cemented backfilling [33–38]. From dry backfilling to cemented backfilling, backfilling technology has developed rapidly and gradually developed to include paste filling, as shown in Figure 2.

Due to the different mining methods, in some upward-filling mining methods, the roof-contacted backfilling rate is not strictly required, such as the upward-approach filling mining method. However, in the continuous mining and continuous backfilling method and open stoping following the backfilling mining method, the roof-contacted backfilling rate is of great concern, which is directly related to the ability of the upper surrounding rock to be a form of support, and thus ensure the safety of the stope [39–41]. Over the years, the relevant experts believe that, for the conventional high-concentration cemented filling, the influences of roof-contacting backfilling are mainly gravity gradient, bleeding shrinkage, and natural sedimentation of backfilling slurry. [42–44]. In order to achieve an efficient one-time roof connection, a series of processes, such as setting multiple backfilling discharge openings and the shape of stope backfilling adjustment, are proposed, but the practical application effect is poor.

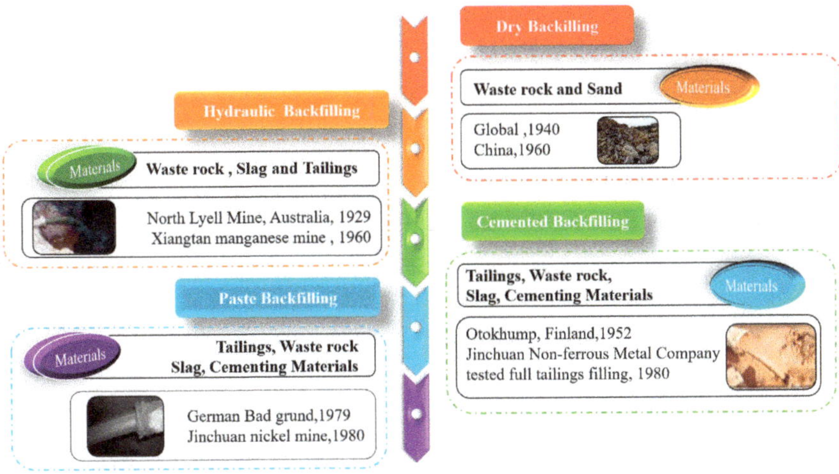

Figure 2. Development history of backfilling mining method.

Therefore, in order to improve the roof-contacted backfilling rate, the safety of the mining operation, and reduce the environmental burden, this paper analyzes the factors affecting the roof-contacted backfilling rate under the conventional high-concentration cemented filling, and summarizes the corresponding improvement measures.

3. Factors Affecting Roof-Contacted Backfilling Rate and Improvement Measures

3.1. Similar Roof-Contacted Rate

The roof contact of the backfilling massif is an important part of backfilling work. The effect of the roof connection of the backfilling massif can be expressed by the backfilling roof-contacted ratio. This definition can be intuitively understood as the ratio of the area of the filling body contacting the roof to the area of the whole roof. However, in practical applications, this method is difficult. In order to simplify the calculation, three kinds of similar methods are proposed in the literature to calculate the roof-contacted backfilling rate, as shown in Table 1 [45].

Table 1. Similar roof-contacted rate.

Method	Calculation Formula	Explanation
Similar volume ratio	$\epsilon = \frac{v}{V} \times 100\%$	ε—Similar backfilling roof-contacted rate v—Volume of filling body V—Volume of mined ore
Average height ratio	$\epsilon = \frac{h}{H} \times 100\%$	H—Average height of measured area before filling h—Average height of filling body in the measured area
Cross-sectional area ratio	$\epsilon = \frac{s}{S} \times 100\%$	s—The area of the top of the filling body connected to the top S—Area of goaf roof

3.2. The Influencing Factors of Roof-Contacted Backfilling Rate

3.2.1. Slurry Characteristics

(1) Gravity gradient

In the past, filling and mining experts have performed a large number of similar simulations and field industrial tests for the roof connection of the backfilling massif, conducted a lot of research on backfilling slurry accumulation contour and slope angle, and summarized and proposed a significant amount of application experience [46–48]. If self-flowing transportation is adopted, the backfilling slurry shrinkage can be divided into two processes, namely, the unrestricted sediment diffusion process and the restricted upward-stacking process.

The unrestricted sediment diffusion process refers to the filling slurry that does not make contact with the boundary of the short side of the corresponding route after the filling slurry is filled into the goaf, and it can be regarded as the free flow of slurry on an infinite plane. It can be observed from Figure 3 that the curve law of free flow on the infinite plane has the characteristics of normal distribution. The sediment diffusion movement of aggregates with different particle sizes in the filling slurry can be regarded as sedimentation diffusion events, and the successive sedimentation along the goaf presents probabilities.

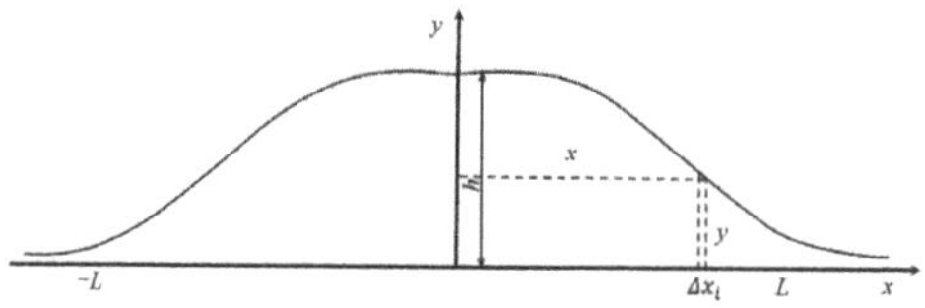

Figure 3. Sediment diffusion slope curve with unrestricted slurry.

The individual aggregate molecules of the backfilling slurry are not enough to affect the whole backfilling massif. k (k = 1, 2, ... , n) is used to represent the aggregate with different sizes constituting the filling slurry. According to the Lyapunov central limit theorem, sedimentary events within Δx_i are always independent of each other. No matter what distribution the random variable k has, when n in $\sum_{k=1}^{n} x_k$ approaches infinity, it approximately conforms to the normal distribution $N(0,1)$. From this, it is calculated that the slope curve equation of slurry infinite diffusion sediment is

$$y = he^{-\frac{x^2}{2\sigma^2}} \tag{1}$$

where h is the maximum height of unrestricted diffusion sediment. σ^2 represents the mean squared deviation.

The mean squared deviation reflects the steepness and slowness of the slurry sedimentation slope, which is jointly determined by the slurry concentration, particle-size distribution, and the content of cementitious materials. The specific value of each mine should be estimated by the histogram of the density function through experiments. The filling slurry diffuses from the initial falling point to the surroundings to form a sedimentary body with an approximate radius of L. The slope of the sedimentary body is approximately normally distributed. There is a certain functional relationship between the unrestricted diffusion radius L of the backfilling slurry in the stope and the sedimentary height h; that is, h = (L).

The restricted upward-stacking process is a process in which the filling slurry enters the goaf, diffuses to the relatively short-side boundary, and accumulates upward until it is relatively close to the roof of the backfilling stope. After the backfilling is completed, by observing the final shape of the backfilling massif, it can be found that the effect of the backfilling slurry roof-connection in an area centered on the filling pipe's orifice is remarkable, while a certain sedimentary slope is formed at the edge of the area far from the pipe orifice, which is related to the backfilling slurry concentration, and mobility and operation times.

On the basis of the three-dimensional scanning of the basic-settlement-layer contour, Wang Xinmin et al. [47] used the model to estimate the roof-contacted area and roof-contacted backfilling rate, which provided a basis for the next safety production of the mine. Lu Hongjian et al. [49] studied the flow-trajectory model of backfilling parts in the filling stope of Shirengou Iron Mine, detected the surface contour of the backfilling massif by using three-dimensional laser scanning technology, and analyzed the characteristics of the backfilling-slurry-flow sediment slope curve, as shown in Figure 4. Tang Li et al.

studied [50] the problem of filling and roof connection in the Jinchuan No. 2 mining area. By taking the rod ground sand-based cemented filling slurry as an example, using the theoretical model, parameters, such as the optimal size of the stope, the optimal position of the filling pipeline, and the optimal number of pipeline movements, were studied to ensure the roof-connection effect of the filling body.

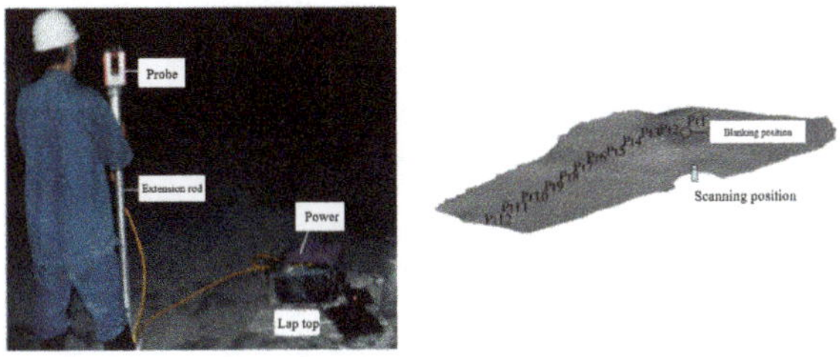

Figure 4. Flow path and slope monitoring of filling slurry in stope.

"The infinite sedimentary diffusion body model" assumes that the backfilling massif is the mean slurry without segregation, so it has good applicability when the stope size is small. However, in relatively large stopes, coarse aggregate of backfilling slurry sedimentate faster than fine aggregate, as shown in Figure 5 [51], so there are more coarse aggregates at the filling pipeline, and fine particles move and sedimentate to the far end with the movement of the slurry. The newly injected slurry flows and sedimentates along the slope of the deposited slurry. During the backfilling process, the gravity water accumulates at the far end of the underground stope and is discharged.

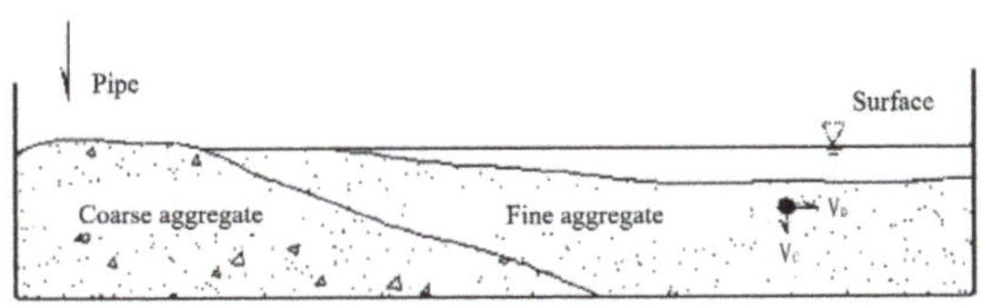

Figure 5. Sedimentation model of backfilling slurry with different aggregate sizes.

Hua-fu Qiu [52] modified the original model to some extent, and considered that the backfilling slurry expanded to both sides after entering the underground stope, and the mass fraction gradually decreased. The backfilling slurry still conforms to the normal distribution in the processes of flow and sedimentation. The modified slope curve model relationship is as follows:

$$y = a + he^{\frac{(x+b)^2}{2\delta^2}} \tag{2}$$

where y—slope height; h—sedimentation height of filling slurry; δ^2—mean square deviation; and a, b—undetermined constant.

The undetermined constants a and b are related to the location of the backfilling area. When the backfilling slurry flows and sedimentates on the wireless plane, both a and b are 0. At this time, the slope curve model degenerates into the original infinite plane accumulation model.

(2) Dewatering and sedimentation

Goaf filling and dewatering are an essential part of backfilling work transported by pipeline hydraulic gravity. A filling body is a kind of loose body, so its water content is complex, which can be divided into adsorbed water, capillary water, and gravity water [53]. Gravity water is the main object of dewatering in the backfilling process. This part of water exists in the large pores in the solid aggregate of backfilling. It has the general characteristics of water and can flow freely between the backfilling aggregate and flow downward under the action of gravity [54].

Dewatering technology can be divided into two categories: external action and non-external action. The external effects mainly include the electro-osmosis method, negative pressure method, and pressure ventilation. The non-external effects mainly include the chain dewatering method, setting the dewatering closed wall and dewatering well, increasing the installation spacing and pipe diameter of a dewatering pipe, increasing the backfilling water overflow pipeline, and improving aggregate gradation [55].

Zhang Aiqing et al. [55] improved the common dewatering pipe without considering the external effect, increasing the number of dewatering pipes, and shortening the spacing of dewatering pipes, designed a new root-like dewatering pipe based on bionics, and conducted dewatering tests of the new root-like dewatering pipe and the common dewatering pipe, respectively. It was concluded that the new root-like dewatering pipe can significantly improve the dewatering rate compared tot the common dewatering pipe, as shown in Figure 6. Wang Bingwen et al. [56] explored the relevant laws of electro-osmotic dehydration and consolidation of filling slurry, and conducted the test with a self-made electro-osmotic dehydration test and natural dehydration test device, as shown in Figure 7. The results show that the electroosmosis method can not only accelerate the drainage speed, but also improve the strength of the test block for the full tailings non-cemented filling slurry without cementitious material.

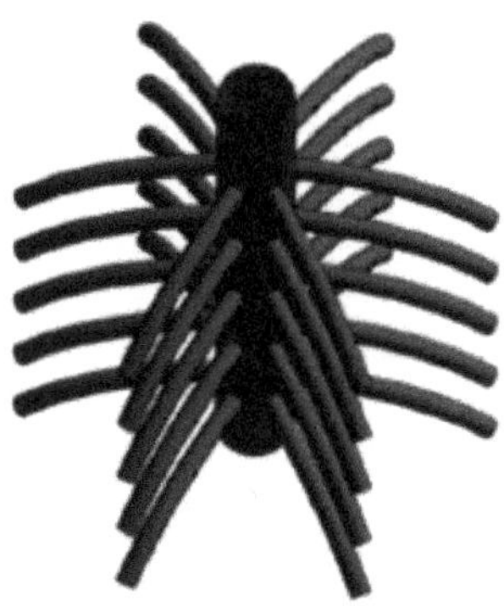

Figure 6. New type of dehydration pipe.

Figure 7. The test device.

(3) Slurry shrinkage

The shrinkage of backfilling slurry is an inevitable phenomenon in the backfilling process, which is mainly in the form of pipeline hydraulic gravity and heterogeneous flow transportation. In order to ensure hydraulic gravity transportation, the water content of the slurry is much greater than that required by the hydration process of the cementitious material, and a large quantity of water needs to be removed in the filling process [57–59]. The shrinkage of backfilling slurry is mainly composed of bleeding shrinkage and seepage shrinkage.

Bleeding shrinkage is due to the fact that the backfilling slurry is in the state of supersaturated water. When the cemented filling slurry enters the goaf, the coarse and fine aggregates of the backfilling slurry sink one after another, forcing the rich water to separate on the surface of the backfilling massif. When the water is removed, a space is formed between the backfilling massif surface and the roof of the goaf, thus affecting the roof-contacted effect in filling process. Seepage shrinkage refers to the transition from a supersaturated state to a saturated or wet state after the water in the backfilling slurry is discharged by runoff. Following this, the gravity water in the gap between the solid aggregates is discharged by seepage, the aggregates are rearranged, the porosity between the solid aggregates is reduced, and the backfilling massif is subject to secondary sedimentation. In order to solve the bleeding problem of low-concentration self-flowing tailings filling slurry, Liu Juanhong et al. [60] conducted experiments with solid concentrations of 57%, 60%, 63%, and 66%, respectively. The results are shown in Figure 8., which verify that the bleeding and shrinkage of filling slurry causes the filling body to be unable to connect and compact.

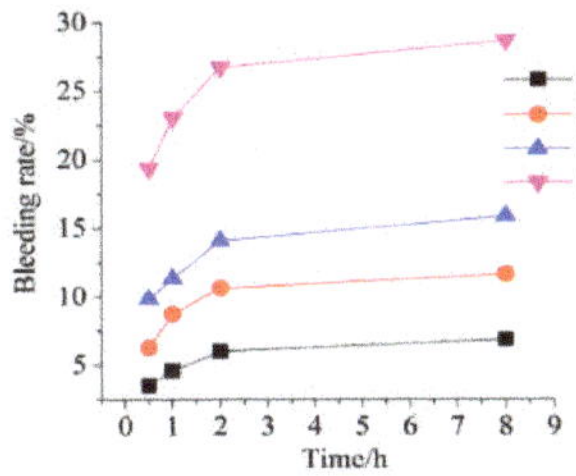

Figure 8. Effect of the different solid concentrations on slurry bleeding rate.

3.2.2. Stope Characteristics

(1) Washing-pipe water and water for the leading way

In the hydraulic filling mode of gravity transportation, in order to prevent the residue of filling slurry left in the pipeline from sticking and ensure the smooth outflow of slurry, 5~10 min of water for the leading way and washing-pipe water are discharged before and after each filling. A large quantity of water cannot rapidly dewater from the stope in the stope. After the goaf is filled and dehydrated, a space is formed, which is difficult to ensure the roof connection of the backfilling massif.

(2) Backfilling pipe

During the backfilling process of the goaf, the backfilling pipeline must be hung in the safety zone of the highest point of the goaf. Due to the limitation of stope conditions, it is difficult for the backfilling pipeline to be hung in the highest position. At this time, relevant measures, such as cutting, need to be taken to ensure that the backfilling pipe is safely set at the highest point of the stope. During the design, the position of the filling hole must be designed according to the mobility of the backfilling slurry. Certain measures, such as multiple backfilling, zone filling, reasonable setting of exhaust pipe and mobile backfilling pipe can be adopted to ensure the dense connection of the top. A mobile backfilling pipe is used in the Sanshandao gold mine to ensure that the filling massif is roof-contacted and dense. Chen Qiusong et al. [61] used paste slurry as the research object, based on the actual stope size, and applied the similarity theory to design the simulated stope size, as shown in Figure 9. Based on the test results, they proposed reasonable suggestions for the position of the discharge port during staged backfilling to improve the backfilling roof-contacted effect.

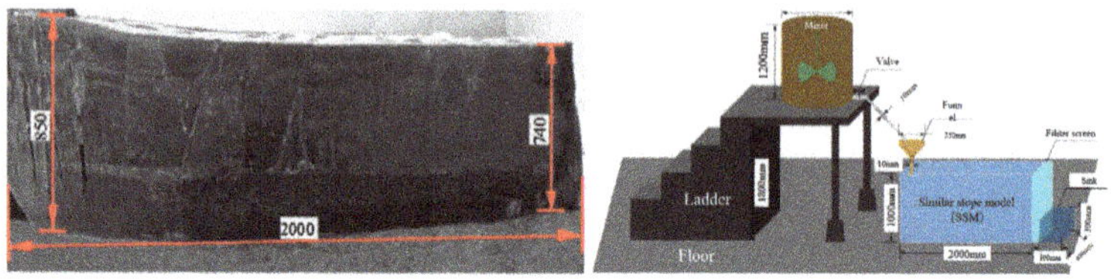

Figure 9. Simulated stope backfilling massif.

(3) Shape of filling field

There are often no strict requirements for the geometry of the roof of the goaf in the design. Due to blasting and other reasons, local over-and under-excavations exist, resulting in the irregular shape of the roof. In addition, the filling method is inappropriate, forming a relatively natural slope angle at the blanking point, and after the blanking opening is blocked at the top of the filling body, some goafs cannot achieve the predetermined goal.

To date, for high-grade ore mining, stope structure tends to be narrow and long in order to improve mining recovery. Narrow long stope refers to the stope with the ratio of stope length to stope width (ratio of length to width) greater than 5~10 by the layered filling method or subsequent filling method. The southeast orebody (760ML-3MU-3 #) in the Chambishi Copper Mine, Zambia, China is segmented and subsequently filled with a length of 123 m, a width of 9 m, a height of 7–9 m, and a aspect ratio of about 14 [62]. It is a great challenge for filling slurry to roof connection once and effectively in such a long mining field with a large aspect ratio.

3.2.3. Objective Factors

(1) Human factors

In traditional engineering design and construction, they are often conducted through experience. Due to different engineering conditions, there will be a lot of blind performances in the construction process, resulting in simple filling facilities and unreasonable hanging of backfilling pipelines, resulting in a poor backfilling effect and multiple backfilling operations, making it difficult to ensure the stability of the backfilling massif, prolonging the production cycle and increasing the backfilling cost. In addition, the experience level of operators, the understanding of operation time, and the adaptability to backfilling technology also affect the roof-contacted backfilling rate.

(2) Limited auxiliary leveling measures

The backfilling process of goaf is similar to the common concrete-pouring operation, which also requires the pouring slurry to fill the predetermined space. However, in the process of backfilling, the underground stope is closed and the auxiliary leveling measures are limited. Under the existing technical conditions, the roof connection backfilling technology is similar to being performed in a "black box", which has a certain impact on the effect of the roof-contacted backfilling. It is difficult to perform manual auxiliary leveling, such as concrete pouring, and it can only rely on self-leveling. The operators cannot perceive the actual situation of the backfilling slurry, so they can only wait until the solidifying reaches the specified age, check the stope shape, adjust the position of the filling pipeline, and perform subsequent filling operations [63,64].

(3) "One-time" operation

The filling facilities are arranged in advance, and the backfilling pipeline is hung at the designated position of the roof of the underground stope according to the design requirements. After constructing the backfilling retaining wall, the operators exit the area to be filled and conduct the backfilling operation. Therefore, once the relevant backfilling equipment is determined, it cannot be moved and regulated at will during the backfilling operation. Therefore, the backfilling operation is called a "one-time" operation, and the layout parameters of backfilling equipment directly affect the roof-contacted effect of the filling body.

In order to improve the roof-contacted backfilling rate, this section sorts out and analyzes the influencing factors of the characteristics of filling slurry, stope characteristics, and related objective conditions. The structural diagram of the influencing factors of the roof-contacted backfilling rate is shown in Figure 10.

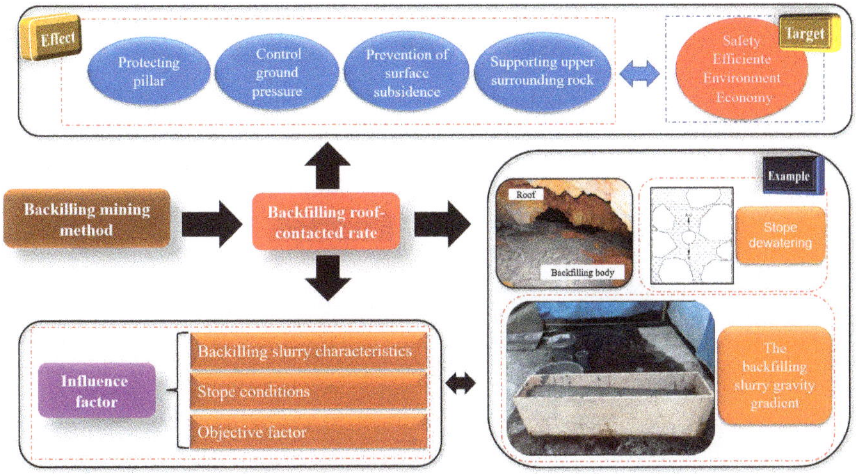

Figure 10. Structural diagram of influencing factors of roof-contacted rate.

3.3. Improvement Measures

3.3.1. Expansion and Non-Shrinkage Material

Compared with the traditional passive roof connection, the backfilling technology of expansion and non-shrinkage material is an effective method. When the backfilling slurry enters the goaf, the volume of the backfilling massif expands in a short time by using its expansion performance, so as to achieve the effect of backfilling massif roof connection. Expansion filling materials can be divided into two types according to different expansion sources. One is the gas-phase expansion caused by an external expansion agent (foaming agent) in a physical way and the generation of bubbles in the slurry through a chemical reaction; the other is solid expansion. Chemical foaming refers to the chemical reaction between a foaming agent and corresponding substances in the backfilling slurry material to release gas. With the condensation of the backfilling slurry material, the gas forms bubbles in the backfilling slurry material, which promotes the expansion of the backfilling slurry volume

(1) Expansive filling material

Inspired by expansive concrete, filling workers at home and abroad introduced the expansion technology of cement-based materials into mine backfilling [65–67]. While selecting appropriate filling materials and the ratio, they modified the filling slurry to cause it to have a certain expansion performance [68,69]. It is a rational idea to adopt the expansion and non-shrinkage material filling technology in the backfilling process. Lan Wentao et al. [70] used HPG (hemihydrate phosphogypsum) as the main raw material, and used its gelling activity to prepare a new type of multiphase, condensable, water-swelling material. The material is mainly composed of four materials: HPG (hemihydrate phosphogypsum), SAP (amorphous coagulant), GPA (gas phase introduction agent), and HA (hydrophobic agent). The formed filling body is solid, liquid, and gas. The outstanding three-phase features are solidification under the condition of a non-solid volume ratio of 87.6%; high early strength; and, with a certain expansion performance, it can realize the "active roof connection" of the backfilling massif.

Bentonite is mainly composed of montmorillonite clay minerals and belongs to a natural pozzolanic material. It can be divided into three types: nano bentonite, calcium bentonite, and organic bentonite [71]. Among them, sodium bentonite has the characteristics of high dispersion, high water absorption, and multiple large expansions (20~30 times). Using this characteristic, adding it to the backfilling material can expand the backfilling massif. In 1995, Professor Siriwardane et al. [72]. discussed the problem of adding fly ash filling material into bentonite to avoid overburden, collapse, and land subsidence. Through indoor experiments, numerical simulation, and large-scale field practice, the results show that, after adding bentonite, the fly ash filling material has good adhesion and rheological properties, so that the slurry can be filled into the goaf smoothly, and the backfilling slurry has certain expansion properties. Satter Barat et al. [73] mixed bentonite with tailings to study its strength performance. The test proved that bentonite could be used as backfilling material. Bentonite not only has a good expansion performance, but also has significant adsorption on heavy metal ions, which has great environmental benefits.

(2) Foaming expansion filling material

Based on the research results in the field of foamed mortar, it has been introduced into the filling field and achieved a good backfilling and roof-connection effect [74,75]. After the foaming agent is mixed with other filling materials, the filling slurry produces a strong alkali-solution environment, and the foaming agent produces tiny bubbles in the strong alkali environment, as shown in Figure 11. According to the way that filling materials produce bubbles, they can be divided into chemical foaming and physical foaming. Physical foaming is made by mechanical agitation or foaming agent, which has a certain tension of foam, and then the foam is blown into the slurry. With the condensation of the slurry, an expansion material with uniform porosity is formed. Chemical foaming refers to the chemical reaction between the foaming agent and the corresponding substances in the filling slurry material to release gas. With the condensation of the slurry material, the gas solidifies in the slurry material to form bubbles, which expand the volume of the backfilling slurry.

Figure 11. The pores in backfilling body.

Taking hemihydrate phosphogypsum as a raw material, Rong K et al. [76] studied the performance-change law of hemihydrate phosphogypsum expansive material under the combined action of multiple factors, and conducted random tests. The results show that increasing the parameters of gas-phase air entraining agent can increase the expansion rate of backfilling slurry, which is more conducive to the roof-contacted backfilling massif, but reduces the strength and durability of the filling body. Adnan Colak et al. [77] used sodium lauryl sulfate and nonylphenol ethoxylated foaming agent to produce bubbles in gypsum. In order to promote foam and bubble formation, retarder citric acid and tackifier carboxymethyl cellulose were used. The results show that foams or bubbles are only intermediate products of expansive mortar materials, and the ultimate goal is that the volume of mortar after filling will result in volume expansion due to the bubbles formed inside the mortar. When a variety of expansion agents, such as foaming agents, are added to fill the filling slurry, the phenomenon of first shrinkage and then expansion can be achieved in the process of solidification and condensation. Compared with the final volume without an expansion agent, the filling slurry expands, as shown in Figure 12.

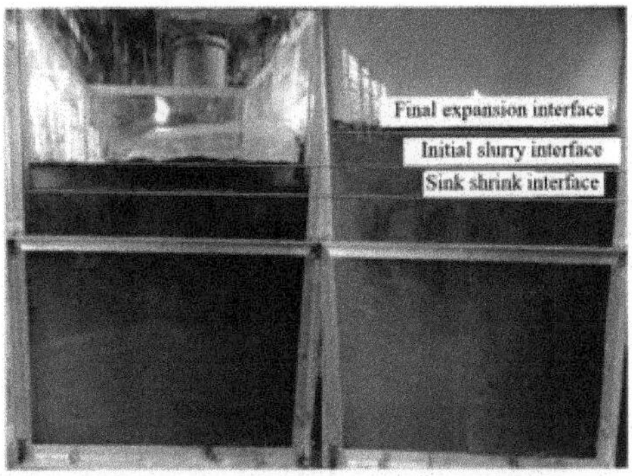

Figure 12. Effect of backfilling body with expansion agent.

In mine backfilling, due to the addition of expansive agent, the internal structure of the backfilling massif is relatively weak, and the increase in volume is due to the decrease in volume density, namely, the increase in porosity. In the backfilling massif expansion technology, the expansion of the backfilling massif is often accompanied by the deterioration of the strength of the backfilling body, and the high-dose expansive agent causes fatal damage to the stability of the backfilling massif. In the stope with a large aspect ratio, the slurry flow forms a certain sedimentary slope, and the later expansion leads to partial roof connection. Therefore, only relying on expansive filling cannot completely deal with the problem of filling body roofing.

3.3.2. Forced Roof-Contacted Measures

When the above methods are not enough to achieve the predetermined backfilling target, it is a favorable way to apply the improvement measures in Table 2, but each of these improvement measures is restricted by goaf condition and can only solve specific situations.

3.3.3. Strengthen-Management Level

Due to the limitation of goaf conditions, the uneven concave or convex shape of the stope roof (local over-excavation and under-excavation), the inconsistent filling sequence, the mixing of pipe washing water, and the randomness of manual operations affect the roof-contacted rate of the filling body. The ratio of filling slurry can be optimized and its roof connection performance can be increased by optimizing the filling and discharge process, means of standardization of operation parameters, and other processes. Pay more attention to filling quality management and hire professionals engaged in filling management. After finding the problems, take the initiative to perform some rectification and disposal measures. Moreover, achieve on-site supervision and tracking, and perform reasonable disposal and accurate analysis, and the evaluation of on-site problems, as shown in Table 3.

Table 2. Partial improvement measures.

Forced Roof-Contacted Measures	Characteristic
Manual roof connection	High labor intensity, low work efficiency, and poor working conditions.
Mechanical roof connection	It is used in metal mines, such as the segmented filling method and route filling method. It is mainly used in the field of cemented slurry of high-concentration coarse aggregate and non-cemented filling of waste rock.
Forced-caving roof connection	It is a common method for slightly and gently inclined ore bodies.
Natural-caving roof connection	The physical requirements for the high performance of ore and rock are guaranteed by a high management level. Applied to its own low-strength characteristics and ore bodies with joint fissure distribution.
Slurry-pressure roof connection	It is mainly used for the up- and down-filling mining methods. It is divided into residual pressure roof-connection of the filling system and pressure pump injection roof-connection.
Slurry self-flowing roof connection	Use the slurry to level under its own gravity or use the height difference to extrude the slurry to connect the roof.

Table 3. Ways to improve management.

Operating Time	Measures
Before backfilling operation	Creating good conditions for stope-filling top pick. Select the appropriate slurry concentration and packing materials. The use of intumescent material additives.
During backfilling operation	Eliminating the influence of water. Noting the empty top pressure. Reducing worker error and strictly quality-controlling projects.
After backfilling operation	Leakage of slurry is prohibited. Prevent the influence of water on slurry. Improving roof monitoring.

In order to improve the backfilling roof-contacted rate, this section summarizes the improvement measures in three aspects: using expansive materials, forced roof-connection technology, and strengthening management. The structure chart of influencing factors of roof-connection improvement measures is shown in Figure 13.

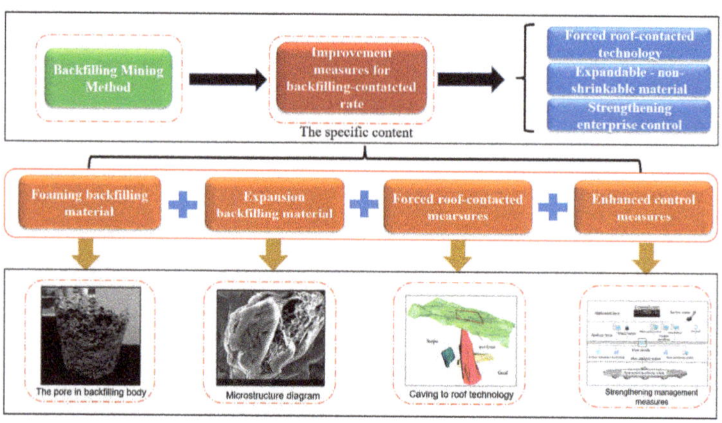

Figure 13. Influence structure of roof-contacted improvement measures.

4. Conclusions

(1) Backfilling technology is an important technical method for creating a win–win situation of "environmentally friendly" and "safe and efficient" mines. Roof-contacting backfilling is the key factor of the backfill mining method, which is directly related to the support capacity of upper surrounding rock and guarantees the safety of stope. With the increase in mining depth and the deterioration of mining conditions, more attention must be paid to the roof-contacted backfilling rate in the future.

(2) The roof connection of the backfilling massif is an important part of backfilling work. In this paper, the method of calculating similar roof-contacted backfilling rates was introduced. For conventional high-concentration cementitious backfilling, the main influencing factors, improvement measures, and auxiliary measures of the roof-contacted backfilling rate were summarized in detail.

(3) It is still a challenge for the backfilling massif to connect to the roof efficiently in a mining field with a large aspect ratio. During the flow process of filling slurry in the underground stope, the yield surface position dynamically changes, and there is no directly test method to detect the yield surface position. In future research, in the process of backfilling slurry flow in underground stopes, relevant research on the position of the slurry yield surface should be strengthened to make up for the deficiency of theoretical models in parameter corrections.

(4) The roof connection of the backfilling massif is a systematic project. In the design of the roof-connection scheme, stope design, mining process requirements, slurry performance, and roof auxiliary technology should be compared and selected.

Author Contributions: Conceptualization, F.C. and H.J.; methodology, J.W.; formal analysis, X.Z.; investigation, H.J. and X.Z.; writing—original draft preparation, J.L.; writing—review and editing, H.J., F.C. and J.Y. All authors have read and agreed to the published version of the manuscript.

Funding: This research was funded by the National Natural Science Foundation of China (No. 51834001).

Data Availability Statement: Data is contained within the article.

Conflicts of Interest: The authors declare no conflict of interest.

References

1. Worlanyo, A.S.; Li, J.F. Evaluating the environmental and economic impact of mining for post-mined land restoration and land-use: A review. *J. Environ. Manag.* **2020**, *279*, 111623. [CrossRef] [PubMed]
2. Xi, Z.; Tang, S.; Wang, J. Pore structure and fractal characteristics of Niutitang shale from China. *Minerals* **2018**, *8*, 163. [CrossRef]
3. Xiao, W.; Chen, W.Q.; Deng, X.Y. Coupling and coordination of coal mining intensity and social-ecological resilience in China. *Ecol. Indic.* **2021**, *131*, 108167. [CrossRef]
4. Wang, L.M.; Yin, S.H.; Wu, A.X. Ore agglomeration behavior and its key controlling factors in heap leaching of low-grade copper minerals. *J. Clean. Prod.* **2021**, *279*, 123705. [CrossRef]
5. Jia, H.X.; Li, T.J.; Wang, A.J.; Liu, G.W.; Guo, X.Q. Decoupling analysis of economic growth and mineral resources consumption in China from 1992 to 2017: A comparison between tonnage and exergy perspective. *Resour. Policy.* **2021**, *27*, 102448. [CrossRef]
6. Wolfbauer, C.A. Mineral resources for agricultural uses. *Agric. Energy.* **1997**, *50027*, 301–314. [CrossRef]
7. Wang, L.M.; Yin, S.H.; Deng, B.N.; Wu, A.X. Copper sulfides leaching assisted by acidic seawater-based media: Ionic strength and mechanism. *Miner. Eng.* **2022**, *175*, 107286. [CrossRef]
8. Li, J.; Sun, W.; Li, Q.; Chen, S.; Yuan, M.; Xia, H. Influence of Layered Angle on Dynamic Characteristics of Backfill under Impact Loading. *Minerals* **2022**, *12*, 511. [CrossRef]
9. Nassani, A.A.; Aldakhil, A.M.; Zaman, K. Ecological footprints jeopardy for mineral resource extraction: Efficient use of energy, financial development and insurance services to conserve natural resources. *Resour. Policy.* **2021**, *74*, 102271. [CrossRef]
10. Wang, L.; Yin, S.; Deng, B. Understanding the Effect of Stepwise Irrigation on Liquid Holdup and Hysteresis Behavior of Unsaturated Ore Heap. *Minerals.* **2021**, *11*, 1180. [CrossRef]
11. Cheng, H.Y.; Wu, A.X.; Wu, C.S.; Zhu, J.Q.; Li, H.; Liu, J.; Niu, Y.H. Research status and development trend of solid waste filling in metal mines. *China J. Eng.* **2022**, *44*, 11–25. [CrossRef]
12. Edraki, M.; Baumgartl, T.; Manlapig, E.; Bradshaw, D.; Franks, D.M.; Moran, C.J. Designing mine tailings for better environmental, social and economic outcomes: A review of alternative approaches. *J. Clean. Prod.* **2014**, *84*, 411–420. [CrossRef]
13. Qi, C.; Fourie, A. Cemented paste backfill for mineral tailings management: Review and future perspectives. *Miner. Eng.* **2019**, *144*, 106025. [CrossRef]

14. Pal, K.; Vardhan, H.; Aruna, M. Investigation of contaminant transport in groundwater from the tailings pond of uranium mine: A case study. *Int. J. Min. Miner. Eng.* **2010**, *2*, 290–309. [CrossRef]
15. Belem, T.; Benzaazoua, M. Design and application of underground mine paste backfill technology. *Geotech. Geol. Eng.* **2008**, *26*, 147–174. [CrossRef]
16. Sivakugan, N.; Rankine, K.J.; Rankine, K.S. Study of drainage through hydraulic fill stopes using method of fragments. *Geotech. Geol. Eng.* **2006**, *24*, 79–89. [CrossRef]
17. Ercikdi, B.; Cihangir, F.; Kesimal, A.; Deveci, H.; Alp, I. Utilization of industrial waste products as pozzolanic material in cemented paste backfill of high sulphide mill tailings. *J. Hazard. Mater.* **2009**, *168*, 848–856. [CrossRef]
18. Yin, S.H.; Shao, Y.J.; Wu, A.X.; Wang, H.J.; Liu, X.H.; Wang, Y. A systematic review of paste technology in metal mines for cleaner production in China. *J. Clean. Prod.* **2020**, *247*, 119590. [CrossRef]
19. Jiao, H.; Zhang, W.; Yang, Y.; Yang, L.; Hu, K.; Yu, J. Pore structure evolution and seepage characteristics in unclassified tailing thickening process. *Minerals* **2022**, *12*, 164. [CrossRef]
20. Lv, S.R.; Lv, S.J. Research on governance of potential safety hazard in Da'an Mine goaf. *Procedia Eng.* **2011**, *26*, 351–356. [CrossRef]
21. Jiang, L.C.; Yang, C.; Jiao, H.Z. Ultimately exposed roof area prediction of bauxite deposit goaf based on macro joint damage. *Int. J. Min. Sci. Technol.* **2020**, *30*, 699–704. [CrossRef]
22. Dold, B. Submarine. Tailings Disposal (STD)—A Review. *Minerals* **2014**, *4*, 642–666. [CrossRef]
23. Chen, X.; Zhang, J.; Jiao, H.; Hu, K.; Wan, L.; Ruan, Z.; Yang, L. Mechanism of rake frame shear drainage during gravity dewatering of ultrafine unclassified tailings for paste preparation. *Minerals* **2022**, *12*, 240. [CrossRef]
24. Jiao, H.Z.; Chen, W.L.; Wu, A.X.; Yu, Y.; Ruan, Z.E.; Honaker, R.; Chen, X.M.; Yu, J.X. Flocculated unclassified tailings settling efficiency improvement by particle collision optimization in the feedwell. *Int. J. Miner. Metall. Mater.* **2021**, *25*, 124023. [CrossRef]
25. Jiao, H.Z.; Wu, Y.C.; Wang, H.; Chen, X.M.; Li, Z.; Wang, Y.F.; Zhang, B.Y.; Liu, J.H. Micro-scale mechanism of sealed water seepage and thickening from tailings bed in rake shearing thickener. *Miner. Eng.* **2021**, *173*, 107043. [CrossRef]
26. Yang, Y.; Zhao, T.; Jiao, H.; Wang, Y.; Li, H. Potential Effect of Porosity Evolution of Cemented Paste Backfill on Selective Solidification of Heavy Metal Ions. *Int. J. Environ. Res. Public Health.* **2020**, *17*, 814. [CrossRef]
27. Wu, A.X.; Ruan, Z.E.; Bürger, B.; Yin, S.H.; Wang, J.D.; Wang, Y. Optimization of flocculation and settling parameters of tailings slurry by response surface methodology. *Miner. Eng.* **2020**, *156*, 106488. [CrossRef]
28. Qi, R.; Li, S.; Qu, L.; Sun, L.; Gong, C.Z. Critical factors to green mining construction in China: A two-step fuzzy dematel analysis of state-owned coal mining enterprises. *J. Clean. Prod.* **2020**, *273*, 122852. [CrossRef]
29. Miao, X.X.; Zhang, J.X.; Feng, M.M. Waste-filling in fully-mechanized coal mining and its application. *J. China Univ. Min. Technol.* **2008**, *18*, 479–482. [CrossRef]
30. Rybak, J.; Kongar-Syuryun, C.; Tyulyaeva, Y.; Khayrutdinov, A.M. Creation of backfill materials based on industrial waste. *Minerals* **2021**, *11*, 739. [CrossRef]
31. Chen, Q.S.; Tao, Y.B.; Zhang, Q.L.; Qi, C.C. The rheological, mechanical and heavy metal leaching properties of cemented paste backfill under the influence of anionic polyacrylamide. *Chemosphere.* **2022**, *286*, 131630. [CrossRef] [PubMed]
32. Behera, S.K.; Mishra, D.P.; Singh, P.; Mishra, K.; Mandal, S.K.; Ghosh, C.N.; Ritesh, K.; Mandal, P.K. Utilization of mill tailings, fly ash and slag as mine paste backfill material: Review and future perspective. *Constr. Build. Mater.* **2021**, *309*, 125120. [CrossRef]
33. Chen, L.J.; Jiao, D.Y. A design procedure for cemented fill for open stoping operations. *Min. Sci. Technol.* **1991**, *12*, 333–343. [CrossRef]
34. Zhang, Q.L.; Li, Y.T.; Chen, Q.S.; Liu, Y.K.; Feng, Y.; Wang, D.L. Effects of temperatures and pH values on rheological properties of cemented paste backfill. *J. Cent. South Univ.* **2021**, *28*, 1707–1723. [CrossRef]
35. Yilmaz, E.; Belem, T.; Benzaazoua, M. Specimen size effect on strength behavior of cemented paste backfills subjected to different placement conditions. *M. Bull. Geol. Surv. India Ser. B* **2015**, *185*, 52–62. [CrossRef]
36. Deng, D.Q.; Liu, Z.L.; Yao, K.I.; Song, D.Z. A practice of ultra-fine tailings disposal as filling material in a gold mine. *J. Environ. Manag.* **2017**, *196*, 100–109. [CrossRef]
37. Ding, Y.W.; Du, C.F.; Lin, Y.F.; Chang, B.M. Impact factors of hydration heat of cemented tailings backfill based on multi-index optimization. *Case Stud. Therm. Eng.* **2020**, *18*, 100610. [CrossRef]
38. Wang, L.M.; Yin, S.H.; Wu, A.X. Visualisation of flow behavior in segregated packed beds with fine interlayers. *Int. J. Miner., Metall. Mater.* **2020**, *27*, 900–909. [CrossRef]
39. Yang, L.; Qiu, J.P.; Sun, X.G.; Xing, J. Research and application on strength model of cemented backfill pillar for stage subsequent filling mining method. *J. Cent. South Univ. Technol. Nat. Sci.* **2018**, *49*, 2316–2322. [CrossRef]
40. Xu, B.; Lu, B.; Wang, Y.X.; Wang, Z.X. Analysis of deformation characteristics of immediate roof based on filling rate and time effect. *Arab. J. Geosci.* **2022**, *15*, 303. [CrossRef]
41. Jiang, L.C.; Jiao, H.Z.; Wang, Y.D.; Wang, G.G. Comprehensive safety factor of roof in goaf underdeep high stress. *J. Cent. South Univ.* **2021**, *28*, 595–603. [CrossRef]
42. Cheng, H.Y.; Wu, A.X.; Zhou, S.P.; Wang, Y.M. Experiment and numerical simulation of forced air dewatering in non cemented filling stope. *China J. Nonferrous Met.* **2017**, *27*, 811–817. [CrossRef]
43. Guo, F.H.; Guo, Y.; Zhang, Y.X.; Liu, H.; Li, J.; Li, P.; Wu, J.J. Dewatering mechanism of gasification fine slag by coupled mechanical force fields and its potential guidance for efficient dewatering process. *Fuel Process. Technol.* **2020**, *205*, 106459. [CrossRef]
44. Yuhi, M.; Mei, C.C. Slow spreading of fluid mud over a conical surface. *J. Fluid Mech.* **2004**, *22*, 337–358. [CrossRef]

45. Kouamel, K.J.A.; Feng, Y.; Jiang, F.X.; Zhu, S.T. A study of technical measures for increasing the roof-contacted ratio in stope and cavity filling. *J. Mater. Sci. Res.* **2016**, *5*, 54. [CrossRef]
46. Liu, Z.X.; Dang, W.G.; He, X.Q. Undersea safety mining of the large gold deposit in Xinli District of Sanshandao gold mine. *Int. J. Miner. Metall. Mater.* **2012**, *19*, 574–583. [CrossRef]
47. Wang, X.M.; Zhu, Y.Y.; Jiang, Z.L.; Liu, Q.; Wan, X.H. Stability of filling materials with different roof-contacted filling ratios in upward filling stoping method. *Sci. Technol. Rev.* **2014**, *20*, 37–43. [CrossRef]
48. Ford, J.R.; Price, S.J.; Cooper, A.H.; Waters, C.N. *An Assessment of Lithostratigraphy for Anthropogenic Deposits*; Special Publications; Geological Society London: London, UK, 2014; Volume 391, pp. 55–89. [CrossRef]
49. Lu, H.J.; Liang, P.; Nan, S.Q.; Song, A.D.; Hu, Y.J. Research on the flow path of stope filling slurry and analysisof filling body characteristics. *Metal Min.* **2016**, *10*, 31–34. [CrossRef]
50. Tang, L.; Xiao, W.G. Study on the flow law of filling slurry in stope. *R&D Min.* **2005**, *25*, 7–9. [CrossRef]
51. Long, T.; Wang, J.X. Experimental study on dewatering of backfill stope in a Lead-zinc mine. *Min. Technol.* **2009**, *6*, 5–8. [CrossRef]
52. Qiu, H.F.; Liu, L.; Sun, W.B.; Zhang, X.Y. Experimental study on strength distribution of backfill in goaf. *J. Cent. South Univ.* **2018**, *49*, 2584–2592. [CrossRef]
53. Zhang, L.; Lv, L.X.; Wu, C.X. Study on dewatering of full tailings backfill in a copper mine. *Nonferrous Met. Eng.* **2014**, *66*, 107–110. [CrossRef]
54. Zhang, A.Q.; Wang, Y.M.; Sun, H.X.; Shao, C.Y.; Zhang, J.T. Research on strengthening of dehydration test and law for non-cemented backfill. *Ind. Miner. Process.* **2018**, *47*, 50–54. [CrossRef]
55. Zhang, A.Q.; Wu, A.X.; Wang, Y.M.; Yin, S.H.; Li, J.Y. Structural parameters and model construction of a new root-like dehydration tube. *J. Fuzhou Univ. Nat. Sci. Ed.* **2021**, *49*, 407–412. [CrossRef]
56. Wang, B.W.; Li, Q.M.; Gao, L.J.; Xiong, T.Y.; Li, Y.N.; Wang, Z.Y. Chin. Experimental study on electroosmosis dehydration and consolidation of all tailings filling slurry. *J. Rock Mech. Eng.* **2019**, *1*, 3163–3170. [CrossRef]
57. Wang, L.M.; Yin, S.H.; Wu, A.X.; Chen, W. Effect of stratified stacks on extraction and surface morphology of copper sulfides. *Hydrometall.* **2020**, *191*, 105226. [CrossRef]
58. Zhang, A.Q.; Li, J.Y.; Wang, Y.M. Study on hydrodynamics model of total tailings filling slurry dehydration. *Adv. Civ. Eng.* **2021**, *2021*, 9225718. [CrossRef]
59. Jiao, H.Z.; Wang, S.F.; Yang, Y.X.; Chen, X.M. Water recovery improvement by shearing of gravity-thickened tailings for cemented paste backfill. *J. Cleaner Prod.* **2020**, *245*, 118882. [CrossRef]
60. Liu, J.H.; Wu, R.D.; Wu, A.X.; Wang, S.Y. Bleeding characteristics and improving mechanism of self-flowing tailings filling slurry with low concentration. *Minerals* **2017**, *7*, 131. [CrossRef]
61. Chen, Q.S.; Zhang, Q.L.; Fourie, A.; Chen, X.; Qi, C.C. Experimental investigation on the strength characteristics of cement paste backfill in a similar stope model and its mechanism. *Constr. Build. Mater.* **2017**, *154*, 34–43. [CrossRef]
62. Zhang, B.; Wang, Y.; Wu, A.X.; Hu, G.B.; Wang, H.J.; Wang, Y.M.; Zheng, X.M. Large flow paste self-flow filling technology and its application in southeast ore body of Chambishi copper mine. *Min. Techol.* **2021**, *21*, 160–163. [CrossRef]
63. Shao, Y.J. Study on flow characteristics of paste slurry in stope and filling and roof connection technology. *USTB* **2021**, *6*, 14. [CrossRef]
64. Jiao, H.Z.; Han, Z.Y.; Chen, X.M.; Yang, Y.X.; Wang, Y.F. Fleexural toughness evolucion of basalt fiber reinforced shotcrete based on NMR technology. *J. Coal Sci. Eng.* **2019**, *44*, 2990–2998. [CrossRef]
65. Yin, S.H.; Wu, A.X.; Hua, K.J.; Wang, Y.; Zhang, Y.K. The effect of solid components on the rheological and mechanical properties of cemented paste backfill. *Miner. Eng.* **2012**, *35*, 61–65. [CrossRef]
66. Hefni, M.; Hassani, F. Experimental development of a novel mine backfill material: Foam mine fill. *Minerals* **2020**, *10*, 564. [CrossRef]
67. Amran, Y.H.M.; Farzadnia, N.; Ali, A.A.A. Properties and applications of foamed concrete: A review. *Constr. Build. Mater.* **2015**, *101*, 990–1005. [CrossRef]
68. Wyrzykowski, M.; Lura, P. Moisture dependence of thermal expansion in cement-based materials at early ages. *Cem. Concr. Res.* **2013**, *53*, 25–35. [CrossRef]
69. Hodhod, O.A.; Salama, G. Simulation of expansion in cement based materials subjected to external sulfate attack. *Ain. Shams. Eng. J.* **2014**, *5*, 7–15. [CrossRef]
70. Lan, W.T.; Wu, A.X.; Wang, Y.M. Experimental study on factors affecting the filling performance of composite condensate expansion materials. *Adv. Eng. Sci.* **2019**, *51*, 192–198. [CrossRef]
71. Mirza, J.; Riaz, M.; Naseer, F.; Rehman, A.N.; Khan, Q.A. Pakistani bentonite in mortars and concrete as low cost construction material. *Appl. Clay Sci.* **2009**, *45*, 220–226. [CrossRef]
72. Siriwardane, H.J.; Kannan, R.S.S.; Ziemkiewicz, P.F. Use of waste materials for control of acid mine drainage and subsidence. *Environ. Sci. Eng.* **2003**, *129*, 910–915. [CrossRef]
73. Satter, B.; Piltan, T.S.; Nozar, S.; Asadi, S. Stabilization of iron ore tailings with cement and bentonite: A case study on Golgohar mine. *B Eng. Geol. Environ.* **2020**, *79*, 4151–4166. [CrossRef]
74. Hefni, M.; Hassani, F.; Nokken, M.; Kermani, M.; Vatne, D. Investigation into the Development of Foam Mine Fill. In Proceedings of the 11th International Symposium on Mining with Backfill, Perth, Australia, 20–22 May 2014; pp. 49–59.

75. Ruan, Y.M.; Luo, X.; Lin, S.; Pei, X.K. Experimental study on simulation filling of new underwater cementitious filling material (NWC-FM). *Urban Rail Transit.* **2022**, *2022*, 2618478. [CrossRef]
76. Rong, K.; Lan, W.T.; Li, H. Industrial experiment of goaf filling using the filling materials based on hemihydrate phosphogypsum. *Minerals* **2020**, *10*, 324. [CrossRef]
77. Çolak, A. Density and strength characteristics of foamed gypsum. *Cem. Concr. Compos.* **2000**, *22*, 193–200. [CrossRef]

minerals

MDPI

Review

Accessing Metals from Low-Grade Ores and the Environmental Impact Considerations: A Review of the Perspectives of Conventional versus Bioleaching Strategies

Rosina Nkuna [1], Grace N. Ijoma [1], Tonderayi S. Matambo [1] and Ngonidzashe Chimwani [2,*]

[1] Institute for the Development of Energy for African Sustainability, University of South Africa, Christiaan De Wet, Pioneer P.O. Box X6, Florida 1710, South Africa; rosinamakofane@gmail.com (R.N.); nkechiijoma@gmail.com (G.N.I.); matamts@unisa.ac.za (T.S.M.)

[2] Department of Mining Engineering, University of South Africa (UNISA), Florida Campus, Private Bag X6, Johannesburg 1710, South Africa

* Correspondence: ngodzazw@gmail.com; Tel.: +27-731-838-174

Abstract: Mining has advanced primarily through the use of two strategies: pyrometallurgy and hydrometallurgy. Both have been used successfully to extract valuable metals from ore deposits. These strategies, without a doubt, harm the environment. Furthermore, due to decades of excessive mining, there has been a global decline in high-grade ores. This has resulted in a decrease in valuable metal supply, which has prompted a reconsideration of these traditional strategies, as the industry faces the current challenge of accessing the highly sought-after valuable metals from low-grade ores. This review outlines these challenges in detail, provides insights into metal recovery issues, and describes technological advances being made to address the issues associated with dealing with low-grade metals. It also discusses the pragmatic paradigm shift that necessitates the use of biotechnological solutions provided by bioleaching, particularly its environmental friendliness. However, it goes on to criticize the shortcomings of bioleaching while highlighting the potential solutions provided by a bespoke approach that integrates research applications from omics technologies and their applications in the adaptation of bioleaching microorganisms and their interaction with the harsh environments associated with metal ore degradation.

Keywords: low-grade ore; metal recovery; metallurgical methods; bioleaching; environmental pollution

Citation: Nkuna, R.; Ijoma, G.N.; Matambo, T.S.; Chimwani, N. Accessing Metals from Low-Grade Ores and the Environmental Impact Considerations: A Review of the Perspectives of Conventional versus Bioleaching Strategies. *Minerals* **2022**, *12*, 506. https://doi.org/10.3390/min12050506

Academic Editors: Shenghua Yin and Leiming Wang

Received: 23 March 2022
Accepted: 14 April 2022
Published: 20 April 2022

1. Introduction

Initially, only precious metals were prioritized, but with the development of various industries and their demand for unique metals for production activities, the focus shifted to include the assorted metals demanded [1]. For many years, we have witnessed the development and improvement of technologies to accommodate the mining and processing of various types of metals [2,3]. Furthermore, these technological advancements have been adapted to facilitate the extraction of minerals of interest from low-grade ores, which is an expected consequence of the alarming depletion rates of high-grade ore reserves. This will increase the exploitation of previously underutilized low-grade ores [4,5]. Surprisingly, in the nineteenth and early twentieth centuries, low-grade ores were considered worthless, and mining activities were concentrated on ore bodies containing at least 5% or more of the metal of interest [6]. Such decisions were based not only on the venture's efficiency and profitability but also on the technological capability of the time [7]. However, the current state of affairs, as well as the need to meet rising demand, have altered industrial outlooks, with process considerations now including ore bodies containing less than 1% of the metal of interest.

Low-grade ores are obtained from a variety of sources, including newly discovered reserves or mine waste containing previously overlooked metals, most likely as a result of mining activities that focused on a different target metal during the initial mining exploration. The target metal was frequently at a higher concentration, making extraction easier [8,9]. Regardless of the source, conventional mining of low-grade ores has been shown to cause more environmental issues than it is worth by generating more waste, which eventually leads to soil and water pollution [10,11], and, eventually, climate change. These environmental problems are the effects of mining activities on the biophysical environment, which consequently lead to environmental degradation. Furthermore, the presence of mine tailings is regarded as a health risk as well as an environmental challenge, and the reduction in toxic element concentrations through re-mining processes is regarded as an environmental beneficiation initiative. This initiative is a necessary strategy for mitigating the long-term environmental damage caused by transformation reactions and the production of acid mine drainage (AMD). In the presence of water-activating microbial solubilization activities, these unavoidable biochemical reactions occur [12]. However, re-extraction processes, particularly those using conventional methods, may prove unprofitable due to low yields and may release toxic gases into the environment [13].

Conversely, when addressing the challenges of mining low-grade ores, we must consider both the economic and opportunity costs. The latter is the justification that focuses on environmental sustainability in all mining strategies, which is an important consideration for future generations. As a consequence, in order to address the challenges of mining low-grade ores, current technological developments must incorporate an environmentally friendly approach as well as seek alternatives that ensure profitability. This review will look at the progress that has been made in this regard.

In addition to investigating these advancements, there has been a noticeable shift in focus in recent years toward biotechnological methods to address environmental pollution issues associated with traditional smelting methods and the use of toxic chemicals. However, in industrial-scale mining, these biotechnological alternatives have yet to attain popularity or commercial acceptance. The stalemate in the situation encountered thus far in improving the bioleaching capabilities of the microorganisms used for this operation is likely one of the major reasons that have kept the biotechnological approach a niche application since its discovery [6,14]. The allure of biomining, however, remains its undeniable environmental friendliness when compared to process operations such as smelting and electrowinning. Because of this appeal, as well as the current inclination toward environmental considerations, bioleaching will continue to be the focus of research efforts aimed at unlocking its potential for metal extraction from ores. Furthermore, bioleaching offers more appealing advantages, such as energy efficiency and broad-spectrum applicability to a diverse range of ore types and grades, as well as the processing of ores or mixtures containing very low concentrations of target metals [15,16]. However, the reliance on microorganisms as the primary role players in bioleaching degradation processes carries the burden of inefficiency caused by a plethora of multifactorial effects, with the ultimate outcome being attributable to poor process kinetics [13]. Metal toxicity, which is a direct result of their activities in the environment, is typically a major challenge for even these robust microorganisms. The increasing concentration of bioleached metals impedes further activities and hurts the entire biochemical process [17,18]. Despite this significant disadvantage and the slow kinetics, the method is regarded as a promising process for recovering valuable metals from mineral sulfides. As a result, it is still regarded as a credible alternative method to traditional leaching methods, particularly for low-grade ores [19–25]. As a result, attention is still firmly focused on developing strategies to address these slow kinetics issues. Several studies have revisited this problem in recent years using newly available techniques, some of which will be discussed in this review.

2. Current Sources of Metals

The global demand for a wide range of metals used in various industries has risen exponentially. As a result, the mining industry is addressing this challenge by devising strategies to boost output. According to Johnson [1], metal demand will more than double in the next two decades, if not sooner. This is a likely driver of the current exploitation of low-grade ores. Consequently, focused research has been conducted to develop methods for assessing the potential environmental impact of mining process variables, such as ore grade and deposit type. Such assessments can provide information on the amount of ore that can be extracted, as well as the amount of tailings that can be produced for one unit of metal, based on the deposit's grade [26]. However, the viewpoint that low-grade ore mining and processing has the potential for additional waste generation and profitability limitations is juxtaposed.

Aside from the information derived from the nature and quality of low-grade ore bodies, their constitutions present additional challenges to mining considerations due to their mineralogical complexity, which is frequently manifested in polymetallic states [6,11]. A polymetallic ore can contain several metals in its composition, such as Cu, Zn, Pb, Ag, and Au, and as a result, extracting a single metal from it may not be economically viable and may cause several environmental issues [10,11]. This implies that when dealing with low-grade ore mining, some of the challenges can be overcome by developing and implementing efficient technology to recover a group of metals rather than just one [10]. Despite these obstacles, profitable low-grade ore mines, such as the Aitik mine in Gällivare, Sweden, are currently in operation. Although there are conflicting reports on the actual copper content of the ore, with Mozaffari [27] reporting 0.4% copper content and Johnson [6] reporting 0.24% copper content, both reports show that recovery is feasible and successful for copper ore with a copper content less than 1%. Another example is Talvivaara, a low-grade ore mine in Finland with a Ni-Zn-Cu-Co complex ore deposit that began operations in 2009. The mine's ore deposition averages 0.22 wt.% Ni, 0.50 wt.% Zn, 0.13 wt.% Cu, and 0.02 wt.% Co [28,29]. This mine's profitability was linked, among other things, to the easy access to the ore located on the land surface, which made it less expensive to mine [6].

Mining companies have also evolved a focus on waste from mineralogical and metallurgical processes as potential metal sources [30]. This waste, which was previously generated during the processing of high-grade ores, was thought to have no economic value and was discarded. However, mining companies no longer take this stance in recent times. The characterization of these mine wastes, which have revealed significant quantities of potentially valuable metals, is now the focus of research [31]. Furthermore, advances in flotation technology, an efficient technique for concentrating target minerals, have improved the re-extraction of precious metals from mine tailings, implying that their use as a raw material in mining is economically viable. The low cost of mineral beneficiation associated with mine tailings is an advantage, as the ore is already milled and available on the surface, eliminating the capital costs of excavation required to access the metals of interest [32,33]. This competitive advantage was demonstrated in the mining of cobalt in the Democratic Republic of Congo's province of Katanga, where cobalt is mined from mine tailings containing 0.2%–1% cobalt, which is characteristically in the range of low-grade ores [33,34]. According to Van Zyl et al. [35], there is currently approximately 75 major mine tailing re-mining projects underway around the world. The ERGO project, located in Johannesburg, South Africa, is one successful venture that re-mines gold from mine tailings [35]. Other examples include the Kaltails project, which was established to reprocess tailings from mine dumps in Kalgoorlie, Western Australia, in order to obtain gold [36,37]. The continued survival of these re-mining projects can only mean that the savings from avoiding excavation and milling processes allow the mining projects to be viable and profitable.

3. Technological Improvement in Metallurgical Ores Processing Methods

The pertinent enquiry of what technology to use during extraction of metals from low-grade ores must, without prevarication, factor in the environmental impact and product recovery efficiency. Environmental impact can also be exacerbated by the type of mining process. In open-pit mining, at least 1–3 times quantities of overburden materials are produced than the ore extracted; however, in underground mining, the produced barren rock is often less than 10% of the volumes of ore produced [26,38]. This is due to the high cost of underground mining, which makes it necessary to only focus on the orebody. The overall mining process and environmental impact are shown in Figure 1. In summary, the process of ore mining and metal processing involves ore excavation, pulverization and concentration, conversion of concentrates into oxides, reduction, and refining. Although process optimization initially targeted high-grade ores, there is still room for improvement in relation to processing low-grade ore/tailings. This is because of the fast-paced depletion of high-grade ores on the one hand, and the difficulties faced by the mineral processing industries to efficiently process low-grade ores and metal-contaminated sites on the other hand. The problem has been worsened by the failure of the conventional processing methods to address environmental pollution [39], which includes contaminated sediments and soils, aquatic environment, soil nutrients, and vegetative cover, leading to severe public health concerns.

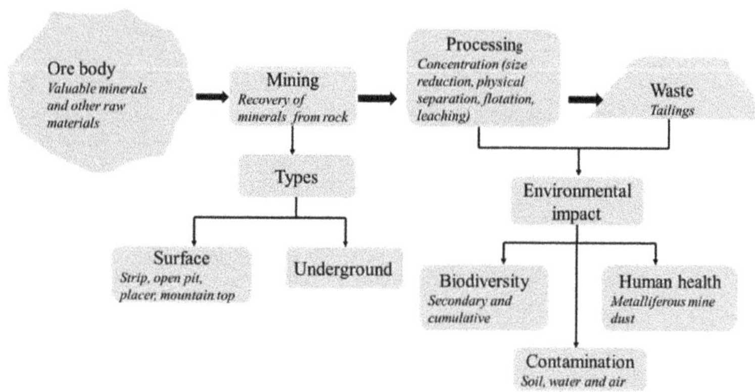

Figure 1. The process of mining from ore exploration, mining, waste generated, and environmental impact.

In this regard, significant progress has been made in terms of research. The abundance of literature demonstrates that researchers have made commendable efforts to improve the processing of various types of low-grade ores. Spooren et al. [40] observed in their review that the efficient treatment of diverse and complex low-grade ores necessitates the use of a new metallurgical systems toolbox comprising existing and innovative mineral processing, metal extraction, metal recovery, and matrix valorization processes. Tabelin et al. [41] focused on ways to address challenges in ore processing, metal extraction, and E-waste management using emerging technologies and sustainable socio-environmental strategies in another review.

Mochizuki and Tsubouchi [42] upgraded several types of low-grade iron ores from various countries to the desired high-grade ores by removing gangue components (Si, Al, and P) via hydrothermal treatment with various solvents. Mokmeli [43] investigated the economics of low-grade chalcopyrite deposits hydrometallurgical treatment as a function of ore grade and extraction recovery. Following the drawbacks of the Bayer process for leaching mineral acids, which include high energy consumption, waste generation with gaseous emission, and concentrated acid/base emission [44–46], the use of microorganisms has received considerable attention. In this regard, Shah et al. [47] investigated the recovery of aluminum from low-grade bauxite via the bio-hydrometallurgical route, discovering that high yields of organic acid and other metabolites produced by a marine-based *Aspergillus niger* enabled the sustainable and economical production of aluminum from bioleaching low-grade bauxite ore. Brinci and Gok [48] also upgraded the low-grade boehmitic bauxite by removing silica via floatation, resulting in a high-grade bauxite concentrate with an A/S ratio of 8.54. A Bauxite A/S ratio of less than 8 results in the sintering process, which is well known for being energy intensive and emitting gases [49–51]. Screening/washing, gravity separation, and flotation are the other beneficiation methods used to reduce the reactive silica content in bauxite [52]. The siliceous gangue minerals, kaolinite and illite, are known to be closely associated with bauxite ore, and this complex association has a negative impact on mineral particle liberation, which is linked to a low A/S ratio [48].

Heap leaching is another traditional method for extracting precious metals with the help of microorganisms. It is typically used in acid leaching of copper oxide ores, oxidative acid leaching of secondary copper-sulphide ores, and cyanide leaching of gold ores [53]. However, as observed by Ghorbani et al. [54], its implementation is hampered by the difficulty of maintaining high heap permeability, particularly when leaching fine-grained ores. As a result, researchers are focusing more on bioleaching, which is thought to be more sustainable for extracting metals such as copper, gold, and zinc. Biological leaching is less expensive than the energy-intensive pyrometallurgical process or other leaching methods that use chemicals under harsh conditions. Using gold as a case study, Calvo et al. [55] investigated how the decline in gold ore grade affects energy expenditure and the associated environmental impact using current technology. The authors conclude from the case study that the lower the ore grade, the greater the energy requirements for ore processing. They also point out that technological advancements only partially reduce energy costs and that high energy demands necessitate the use of fossil fuels, which raises GHG emissions.

The feasibility of beneficiating low-grade skarn phosphate from the Mactung Tungsten deposit was investigated by Li et al. [56]. The anticipated benefit is two-fold: extracting phosphorous for the production of phosphoric acid and fertilizer, as well as cleaning the environment. Phosphate retention in mine tailings is harmful to the aquatic ecosystem because it causes eutrophication of lakes and algae blooms. Thus, both direct and reverse flotation were effective in producing acceptable concentrates from low-grade phosphate rocks containing carbonate and silicate gangue minerals. Additionally, with regard to phosphates, Arroug et al. [57] investigated acid leaching methods for the beneficiation of rejected low-grade phosphate washing sludges. Tailings from phosphate washing sludges have low P_2O_5, high carbonates, and silica content. The authors discovered that the type and concentration of organic acid, as well as the ore's milling conditions, had a significant impact on the ore's upgrade. Above 7%, concentrations of acid were found to attack phosphate minerals, as well as dissolve carbonates.

Sudagar et al. [58] investigated the effect of low-grade metakaolin on compressive strength and heavy metal absorption, and the authors concluded that low-grade kaolins can be useful as construction materials and adsorbents. Magnetic separation techniques were used to beneficiate low-grade rare earth ore [59], and it was discovered that effective upgrading of the rare earth element ore required a combination of dry and wet high-intensity magnetic separation techniques followed by froth flotation. Because rare earth elements are mostly found in deposits that are not concentrated enough to be economically exploited, upgrading is required [60]. The importance of rare earth elements cannot be overstated because they are used to make permanent magnets used in wind turbines and traction motors for electric vehicles. As a result, froth flotation is an important step in concentrating metals and a mixture of chemicals that act as collectors or depressants and are used to improve selectivity during target metal separation from gangue material, as described by Tian et al. [61]. Concentrates from flotation can be subjected to pyrometallurgy, hydrometallurgy, and bioleaching processes for further processing of metals of interest [62], as detailed in Figure 1.

However, it is important to emphasize that the current pyrometallurgical and hydrometallurgical processes cannot efficiently recover metals from low-grade ores because they were designed for high-grade ores [63]. This is due to the difficulty of striking a balance between extracting valuable metals from very low concentrations in the matrix material while lowering the concentration of potentially hazardous metals to stringent concentration limits and achieving extraction with minimal impact on the physicochemical properties of the matrix materials because additional processing, such as valorization, would be required [40].

3.1. Pyrometallurgical Process

Pyrometallurgy is a traditional method for extracting and purifying metals through high-temperature calcination, roasting, reduction, and refining. The entire process consumes a lot of energy and adds a lot to the carbon footprint. Although pyrometallurgical energy requirements vary depending on ore grade and mineralogy, the current exploitation of low-grade ores may have a significant impact on mining process energy requirements [64–66].

It is undeniable that the mining industry is under economic pressure to reduce energy consumption and associated greenhouse gas emissions [5]. The major barrier is in the pyrometallurgy process, where the use of higher-temperature furnaces and blast furnaces is an essentially integrated aspect of metal recovery [65–67]. Previously, Agarwal et al. [68] estimated that an electric furnace requires 1200 kWh per ton of dry deep-sea manganese nodule ore. Concerns about pyrometallurgical process air pollution have prompted governments to impose stringent restrictions through emission regulations and the implementation of acceptable limits to ensure significant levels of environmental protection. Mining companies are required to comply throughout their business cycle, from the beginning to the end, where they are expected to implement land reclamation and clean-up activities. It is also necessary to obtain legal documents pertaining to pollution mitigation and prevention in accordance with each country's mining legislation. The majority of these laws uphold environmental considerations, and they even require the use of best practices and best available technology (BAT) to ensure the protection and restoration of the sites of mining operations [69]. So far, technological advancement in pyrometallurgical processes has sought to reduce energy consumption and carbon emissions while maintaining process efficiency [65]. South African company Evraz Highveld Steel and Vanadium Limited is an example of a company with a technological modification designed with such a consideration [70]. Their initial process employed submerged-arc furnaces (SAF), a technology that has since revealed a number of technical challenges. Submerged-arc furnaces were converted to open slag baths (OSB) for maximum iron and vanadium production in 2004. The OSB furnaces, on the other hand, are limited by their high energy consumption due to their open-arc configuration. This configuration and its implications for energy demand entail a relatively high electricity cost, which fails to balance profitability associated with vanadium (product) recovery. This has prompted a paradigm shift toward improved equipment and operational processes. Several important energy-saving factors have been identified. One such contentious factor identified was the effect of coal source, which was mitigated by limiting suppliers to prevent plant instability caused by the varied composition of the coal, which was linked to poor metallurgical performance, poor metallurgical control of the OSB, and inconsistencies in kilns for the solid-state reduction process. Finally, when compared to SAF, the system was able to achieve improved product recovery while using less energy per ton of metal produced. Furthermore, the use of OSB allowed for lower coal consumption and the use of less expensive coal, such as seam 2 coals with lower reactivity than seam 5 coal with higher reactivity used in SAF.

Blast furnaces are critical pieces of equipment for iron production and are commonly used in mining countries. These massive furnaces have recently been upgraded to accommodate volumes in excess of 5000 m^3. In Korea, for example, a 6000 m^3 blast furnace was built. Upscaling furnaces has been shown to cause non-uniformity, which increases energy consumption and, ultimately, disrupts secure, stable operation. The development of oxygen blast furnaces (OBF), which are thought to be superior to conventional furnaces, can be viewed as a solution to energy-saving efforts. The latter has the potential for greater energy flexibility as well as other environmental advantages. The main distinction between these furnaces is that conventional blast furnaces use hot air with less than 5% oxygen and a high nitrogen (N_2) concentration, whereas the OBF uses 99.0% pure oxygen [71]. Nonetheless, when compared to conventional blast furnaces, the use of oxygen in blast furnaces was found to double productivity due to the absence of N_2 [72]. Surprisingly, Takahashi et al. [72] make the rather broad but dire assumption that energy-saving efforts related to conventional blast furnaces have likely been exhausted and that the possibility of developing additional energy-saving methods is unlikely. Further technological advancements in the OBF have resulted in the addition of a top gas remover process (TGR-OBF) with the ability to significantly reduce carbon dioxide emissions [65]. Currently, the OBF is the most productive technology, with the potential for significant reductions in energy consumption and environmental impact [8]. Although technological advances are important, particularly when they can significantly reduce negative environmental impact and energy demand, the US DOE (2007) warns that these effects are limited. Recently, Chetty et al. [64], proposed the incorporation of mineralogical monitoring as an important step toward understanding furnace energy needs assessments, based on the understanding that the composition of the ore plays an important role in furnace efficiency. Their research utilized quantitative mineralogy to differentiate ore types with similar grades and investigated their energy consumption in furnaces. Their findings suggested that this would lead to better decision making regarding the handling of ore variability, allowing furnaces to run more efficiently and thus reducing energy consumption.

The advancement of new technology has resulted in improved methods that have added environmental benefits, rendering some older technologies obsolete and unpopular. A good example is manganese extraction, which is conducted in the presence of a reductant due to its stability in acid or alkaline oxidizing conditions. Normally, coal or carbon is used as the reductant to convert manganese dioxide to manganese oxide; however, this reaction is energy intensive, with several authors, including, Abbruzzese et al. [73], Jiang et al. [74], and Welham [75], indicating that the temperature requirement is between 800 and 1100 °C. Surprisingly, the use of other reductants, such as cornstalk, bagasse, ammonium sulfate, pure sulfur, and so on, reduced the reaction temperatures to a range of 450–570 °C [76]. Despite the lower energy requirements, the process remains unpopular due to the high amount of fine dust smoke produced by the plants, as well as the high investment and operating costs [77,78]. Furthermore, applying this process to the extraction of low-grade ores will present additional technological, economic, and environmental challenges [79]. Because low-grade ores are polymetallic, the various metals in the ore must be recovered in order for the process to be profitable. This becomes an issue in the case of low copper-bearing minerals with high arsenic and other impurities. If copper concentrates contain more than 0.2 percent arsenic, they are considered "dirty" and are rejected by smelters [41,80].

3.2. *Hydrometallurgical Process*

Concerns about air pollution caused by pyrometallurgical processes have prompted governments to impose stringent restrictions through emission regulations and the implementation of acceptable limits to ensure significant levels of environmental protection. Compliance is required for mining companies throughout their business cycle, from the start of operations to the end, where they are expected to implement land reclamation and clean-up activities. It is also necessary to obtain legal documents pertaining to pollution mitigation and prevention in accordance with mining legislation specific to each country. Most of these laws have an environmental theme and even require the use of best practices and best available technology (BAT) [81].

In terms of energy consumption, hydrometallurgical techniques appear to be the best alternative to smelting and refining routes, and they have great potential for treating various concentrates while also demonstrating increased metal recovery and reduced air pollution [82]. Hydrometallurgy is a process that uses chemical reactions to extract metals of interest from ores, concentrates, and recycled materials. Sulfuric acid (H_2SO_4), hydrochloric acid (HCl), ferric sulfate, nitric acid (HNO_3), ferric chloride, sodium chloride, and other inorganic chemicals are used in the leaching of metals from various raw materials [78,83–86].

Metal recovery in hydrometallurgy is frequently achieved through the use of H_2SO_4, but this has a negative impact on the environment occurring from accidental spillages. The likelihood of ground water contamination occurring from H_2SO_4, which are transported through pipes that burst and leaks that cause seepages into surrounding land, is very high [86]. However, technological advancements have resulted in their replacement with chemicals less harmful to the environment. Furthermore, hydrometallurgical processes have evolved in recent years to meet the challenge of mining low metal concentrations in ores. Historically, the majority of zinc metal was extracted from zinc sulfide ores. However, due to the current depletion of zinc sulfide ores, the exploitation of zinc oxide ores has increased [85,87]. Following gangue material separation using conventional flotation methods, the concentrated ore was leached with a solvent to prepare the leached solution for subsequent electrowinning [77,85]. However, low-grade zinc oxide ores, particularly those containing high concentrations of iron, silicon, chloride, and calcium, are processed in a different manner. This process uses a lot of acid, and the purification process is complicated, since it is difficult to separate zinc oxide from the slurry due to the formation of silica gel, which reduces zinc recoveries [77]. Although pyrometallurgical processes can achieve high zinc recoveries from oxidized zinc ores, the associated high energy and production costs have prevented widespread industrial acceptance [77]. To overcome the hydrometallurgical process's poor zinc leaching, Santos et al. [88] and Ju et al. [89] propose that impurities be kept insoluble, which can be accomplished by using alkaline agents, such as sodium hydroxide and ammonia solutions, in a selective alkaline process. To prevent the solubility of impurities, such as Fe_2O_3, SiO_2, MgO, and CaO, this process requires a high pH of about 6–7. Treating ore containing high alkaline gangue, such as zinc oxide ore, with alkaline solution reduces the consumption of leaching agents when compared to sulfuric acid leaching, and it also reduces the complexity of the purification process [90]. Despite research into the use of specific alkaline processes [88,89,91–93], challenges leading to poor zinc leaching were still encountered, and Rao et al. [77] reported further enhancement of zinc extraction from low zinc oxide ore, through the addition of organic ligand–nitrilotriacetic acid, improves the stability of zinc complexes.

Another case in point is the extraction of alumina (Al_2O_3) from bauxite, a major raw material [94]. Gibbsite, boehmite, and diaspora are all types of alumina minerals, with diaspora being the most common in China. The Bayer process has always been the principal method for producing Al_2O_3 from bauxite raw material [95], although the bauxite ores must have a sufficient alumina (A) to silicon (S) ratio to use this process. However, due to the abundance of bauxite ores with high silica and low alumina (low-grade ore), this becomes a challenge [96]. Several studies have been carried out in order to improve the Al_2O_3 grade of such ores and meet the Bayer process requirements [49,95,97]. There has been significant progress, with the inclusion of an additional desalination step used to raise the Al_2O_3 grade to above 60%. Desalination, also known as ore grade improvement, is accomplished through washing or flotation. The latter is thought to be preferable and recommended. However, because carbonate minerals have similar flotation properties to alumina minerals, a single flotation process is not recommended for bauxite containing carbonate minerals. Guan et al. [95] recently reported an additional step after flotation in which hydrochloric acid leaching is used to dissolve the carbonate minerals, increasing the A/S ratio even further. Sukla et al. [97] proposed bio-beneficiation (the use of microorganisms) as a potential solution to the problems associated with conventional methods due to the limitations that may still be encountered during conventional bauxite purification.

Hydrometallurgical leaching is reliant on the use of chemical reagents, and for a time, sulfuric acid was the primary acid used in the industrial extraction of metals such as nickel from saprolitic ores. The availability and low cost of sulfuric acid were two factors that contributed to its widespread use [98]. Because of their recyclable nature, other acids, such as hydrochloric acid and nitric acid, have recently been considered as alternatives to sulfuric acid for nickel leaching [99]. Furthermore, comparative studies were conducted to compare other acids to sulfuric acid in the leaching of metals from low-grade ores. Astuti et al. [99] investigated the efficacy of citric, sulfuric, nitric, hydrochloric, lactic, and oxalic acids in the leaching of low-grade saprolitic ores from two different mining areas in Indonesia. Surprisingly, citric acid and sulfuric acid were more effective in nickel than other acid solutions, providing motivation for the potential use of citric acid in nickel leaching. Table 1 lists various leaching acids and their functional abilities under various conditions. This table emphasizes the high recovery of target metals in short periods, ranging from a few minutes to a few hours.

Table 1. Application of hydrometallurgy for leaching valuable metals from low-grade ore and mine tailing.

Ore Type	Leaching Reagent	Leaching Efficiency	Common Parameter	Leaching Period	Reference
		Low-Grade Ore			
Low-grade vanadium-bearing titanomagnetite	Hydrochloric acid (HCL)	99.4% vanadium and 4.2% iron	Selective extraction using mixed solvents	10 min	[100]
Low-grade zinc silicate ore	H_2SO_4	94% of Zn	The effect of particle size and concentration of H_2SO_4	180 min	[85]
Low-grade zinc oxide ore	NH_4Cl–NH_3 solution	90.3% Zn	Effect of particle size and influence of NTA concentration	60 min	[77]
Low-grade chalcopyrite concentrate	Sulfuric acid electrolyte	96% Cu	Effect of the particle size, nitrite concentration, and acid concentration	2 h	[101]
Low-grade manganese dioxide ores	H_2SO_4	92.8% Mn and 24.6% Fe	The effect of sulfuric acid concentration and particle size	60 min	[102]
Low-grade Ni–Mo ore	Sodium hypochlorite sodium hydroxide and sodium carbonate	96.3% Mo	Sodium hypochlorite and sodium carbonate concentrations	5 h	[103]
		Mine tailings			
Tailings of oxidized ores	Sodium hydrosulfide (NaSH) and potassium amylxanthate (KAX)	45% Cu and 83% Co	Size distribution of particles	14 min	[30]
Copper sulfide tailings	98% H_2SO_4	98.45% Cu, 21.41% Zn, 56.13% Mn, and 17.25% Fe	Sulfuric acid concentration and leaching time	2 h	[104]
Pyrite flotation tailings	H_2SO_4 and ferric iron	79.6% Cu and 43.7% Zn	Reagent concentrations	2 h	[105]
Molybdenite flotation tailings associated with galena	Sodium hydroxide	98% Pb and 98% Mo	Concentration of sodium hydroxide	1 h	[91]

3.3. Environmental Impact of Waste Generated from Conventional Processing Methods

Ample evidence of mining's impact on surrounding ecosystems demonstrates the process's many devastating consequences, including changes in the interspecific interactions of resident organisms [7,106]. Even with advancements in conventional methods for mitigating these environmental impacts, waste generated during the extraction and processing of mineral ores is unavoidable. As a result, careful planning for these wastes and their management is essential. Most governments address this by imposing regulatory measures and policies governing mine waste treatment and storage. However, oversight and adherence continue to be challenges.

Mine tailings and overburden are the two major types of solid waste associated with all mining processes (waste). Their environmental effects vary, but mine tailings are thought to be more problematic [26,38]. These mine tailings have been linked to several documented cases of water pollution, with most environmental regulatory bodies agreeing that water contamination from mining should be considered one of the world's top three ecological security threats [6,7,107–109].

The presence of amalgamated sulfide minerals, such as pyrite and pyrrhotite, in most precious metals contributes to the toxicity of mine tailings. These sulfide-containing minerals are responsible for acid production, which promotes metal solubilization in tailings [12]. Although the sulfide minerals coexist with the carbonate minerals that are responsible for neutralizing the produced acid [110], these carbonate minerals are usually insufficient to completely neutralize the concentration of acids present. The most likely reason is that they are not present in sufficient quantities to adequately counteract acidic effects. Ultimately, the failure to completely neutralize all H^+ within mine tailings that come into contact with water results in the production of AMD. The gold and coal deposits in South Africa are an example of this, where the neutralization effect is overwhelmed by the production of this acid, as demonstrated by the visible common problem of AMD in South Africa [111]. Similarly, even when complete neutralization is achieved in the presence of sufficient carbonates, the product derived from it is referred to as contaminated neutral drainage (CND) [112]. CNDs are still regarded as a problem for the environment because toxic elements, such as Se, As, Ni, Zn, and Co, are frequently present in high concentrations in these bodies of water, posing a health risk to microorganisms, plants, and animals [12,113]. This is exacerbated by environmental phenomena, such as bioaccumulation and biomagnification [114,115].

It is critical to recognize the contribution of comminution to chemical reaction rates through the crushing and milling of rocks and ore. As a result of the increased surface area, favorable reaction kinetics in the presence of oxygen and water are facilitated. The tailings' sulfide components undergo an oxidative reaction that uses oxygen to generate sulfuric acid, which promotes metal leaching or solubilization [38,116]. This phenomenon is thought to be a natural part of the weathering process. This weathering process is accelerated, however, due to the increase in surface area (fine particles). Because of this rapid reaction, AMD is considered one of the most serious environmental issues in the mining industry [108,117,118]. The continuous production of acid from mine waste causes a drop in pH, which creates favorable conditions for autochthonous microorganisms found in these wastes. These microorganisms contribute significantly to metal solubilization processes by accelerating acid generation and, in most cases, enhancing metal mobilization to other parts of the environment [108,119,120].

Water polluted by AMD is not safe for drinking. Furthermore, the purification process is considered costly and is influenced to some extent by the type and concentration of toxic metals that must be removed from the polluted water. The rising cost of the process frequently forces the abandonment of this polluted water source, which has a heavy toll on residents in settlements near mining operations, which is exacerbated when other sources of potential potable water are not nearby [121]. Contaminated water may also pose a risk when used for irrigation because some plants have the ability to bio-accumulate metals that will eventually affect humans or animals who consume them [122].

Mine tailing stockpiling remains the preferred method of storage and/or disposal. Often, designated mine dumps are established in the vicinity of major mining cities (Figure 2). Unfortunately, this method of storage (disposal) introduces pollution to areas that are not necessarily close to mines, with winds dispersing dust particles containing toxic metals, such as Hg, Cr, Pb, Zn, Cu, Ni, Cd, As, Co, and Sn. The accumulation of mine tailings around cities and suburbs is extremely hazardous to human health, with some of them containing significant amounts of radionuclides [123]. Most pertinent to note is the cyanide pollution that is frequently associated with gold extraction processes [107].

Figure 2. Mine tailing dumps. Pictures taken in Gauteng province, South Africa.

Mine tailings have been discovered to be inadvertently storing low-grade ores. Although the grade of the mineral they contain may be too low for economical extraction at the moment, there is a possibility that these wastes will hold economic value in their mineral content in the future [124]. This understanding is most likely why some mining companies practice continued storage, even after tailings have been re-extracted and reprocessed. According to Blight [124], some South African gold mine tailings have been re-mined and reprocessed at least three times in the last century. As a result, they are stored in a location where, if the need to re-extract arises, an ease of access is guaranteed, as opposed to when the tailings are rendered inaccessible by being buried beneath old mines. However, reprocessing tailings generates waste, which poses environmental problems. Consequently, not all reprocessed tailings are permanently stored for future reprocessing. Some are investigating its use in the production of bricks as a building material in the construction industry [125]. According to Beulah et al. [126], although the bricks are more expensive than traditional bricks, this cost-effective method of producing brick is environmentally friendly. Another example is the use of tailing as a substitute for fine aggregates in the manufacture of cement [127].

For those tailings that are continuously stored, certain metals demand may still be met, to some extent by re-extracting and reprocessing tailings. The key to this re-use strategy, however, is the adoption of storage methods that reduce the environmental impact of these old and new mine tailings. Combining technological advances and waste management strategies should be integrated to ensure that both aspects work in synergy [128]. Most government policies address this issue by requiring waste management licenses for the storage of mine tailings and dump sites as part of the overall regulation of mining activities.

4. Environmental Considerations and Alternative Metal Ore Extraction Methods

So much has been done to reduce the environmental impact of traditional methods, and progress has been made over the years [70,77,85,87,99]. Despite these efforts, pollution and its negative effects on surrounding ecosystems are an unavoidable by-product of mining activities. As a result, research into mitigation and amelioration strategies is ongoing. The biomining strategy, which takes advantage of microorganisms' bioleaching activities, is one such intervention. Several authors agree that the potential benefits of bioleaching in the exploitation of low-grade ores will be significant, particularly in addressing challenges identified by the use of conventional methods [15,63,129–131]. To support this claim, studies comparing the metal extraction efficiency of chemical and bioleaching methods have been conducted. Nguyen and Lee [132] investigated the efficiency of chemicals (sulfuric acid (H_2SO_4) and hydrogen peroxide (H_2O_2) and bioleaching (mixed cultures) processes in the removal of arsenic (As) and other heavy metals from mine tailings. In comparison to chemical leaching, bioleaching achieved higher removal efficiency for all metals (except zinc) after 300 h of incubation.

Previously, Bayat and Sari [133] compared the efficacy of the bioleaching process using *At. ferrooxidans* for the recovery of metals from sludge to the use of H_2SO_4 and ferric chloride while holding all other conditional parameters constant. Bioleaching had the highest metal removal efficiency, followed by sulfuric acid and ferric chloride, according to their findings. However, the bioleaching process, required a longer operational period. This was similar to the findings of Szolucha and Chmielewski [134] when using bioleaching to treat low-grade uranium ore, with uranium recovered at 75%–15% w/w in 55 days compared to 64%–13% w/w in 31 days when H_2SO_4 and H_2O_2 were used. A review of several studies revealed similar trends of longer operation time (Table 1). However, in terms of residence times, chemical leaching has been observed to range from a few minutes to hours, as opposed to the days required for the bioleaching process. Critically, the bioleaching process exposes mineral sulfides to acids, and the presence of these acids promotes further metal leaching.

Microbial communities rely on biological activity, which is frequently regulated by various proteins, including enzymes, which are influenced by a variety of parameters, including pH. Moreover, any significant change in environmental conditions often necessitates a period of acclimation and adaptation, which slows several biochemical functions and, as a result, has a significant impact on process outcomes. Despite the longer operating time required, bioleaching is still regarded as a viable alternative to traditional chemical reactions [132,133]. The allure of bioleaching among researchers stems from the credible consideration of its status as a characteristic of "green technology," involving microorganisms, mostly autotrophs, with the ability to fix CO_2, much like plants, whereas processes such as smelting are notoriously associated with CO_2 emissions (Johnson 2014). Bioleaching also has a lower carbon footprint than traditional methods due to its ability to occur at atmospheric pressure and lower temperatures (20–80 °C), as well as lower associated energy costs [135]. Despite the fact that the bioleaching process produces an acid by-product (which is usually recovered to prevent environmental pollution), it is still considered relatively eco-friendly due to the moderate concentration of acids produced by microorganisms. The low-cost advantage of bioleaching can be realized in both low-grade ores and mine tailing leaching, with the latter achieving a higher yield. However, it is important to note that the scaled-up bioleaching process undergoes similar pre-requisite energy-consuming blasting and grinding treatments, as shown in Tables 1 and 2. This step is critical for the efficiency of the process.

Dump leaching, heap leaching, in situ leaching, and tank leaching are commercial bioleaching processes for extracting metals, such as copper, cobalt, and uranium [136]. Nonetheless, bioleaching's demonstrated capabilities have not been sufficient to gain widespread acceptance or consideration as a replacement for conventional methods. This is because of the slow dissolution kinetics. Furthermore, the possibility of compromised metal leaching yield due to high metal concentration and its associated toxicity, inhibiting microbial activity or survival, has limited the bioleaching process's large-scale application by Pathak et al. [15] and Clark et al. [137] who reviewed heap leaching strategies and described the limitations of slow kinetics in the bioleaching process. It is estimated that 300 days are required to obtain results with laboratory-scale bioleaching experiments and that when scaled up, it will usually take at least 900 days to obtain reasonable metal extraction. Furthermore, the immediate inclusion of freshly dug-up mineral heaps, which is standard industrial practice, can be hampered by the long lag phase of bacterial growth, with incubation and biological treatment lasting up to three years [138]. This lengthy incubation period has discouraged the use of bioleaching and reduced its feasibility for commercial scale metal sulfide mining processes [13]. To alleviate the lengthy lag phase associated with the heap leaching strategy, perhaps a preparatory treatment should include the introduction and mixing of small inoculating heaps from old sites to new heaps, with an incubation period provided before their addition to older heaps. This has the potential to improve bioleaching efficiency by shortening the lag phase and acclimation periods required by microorganisms. In particular, the constraints and limitations associated with bioleaching necessitate multi-directional solution-driven research aimed at addressing these challenges in order to improve this bioprocess.

Table 2. Application of acidophilic microorganisms in bioleaching of low-grade ores.

Ore Type	Microorganisms	Amount Leached	Common Parameters	Leaching Period	Reference
		Low-Grade Ore			
Sulfidic tailing of Golgohar Iron Mine	Mixed culture of *Leptospirillum ferriphilum*, *Acidithiobacillus caldus*, *Sulfobacillus* sp., and *Ferroplasma* sp.	53% Cu, 84% of Ni, 56% of cobalt, and 35% Fe	Bacterial inoculation, media pH, ore particle size, stirring rate, and pulp density	21 days	[4]
Low-grade uranium ore	*Aspergillus niger*	71.4% U	Media pH and pulp density	7 days	[139]
Low-grade manganese ore	Pure and mixed cultures of *Acinetobacter* sp. MSB 5 and *Lysinibacillus* sp.	70% Mn from *Acinitobacter* sp., 67% Mn from *Lysinibacillus* sp., and 74% Mn from mixed culture	Bacterial inoculation, media pH, ore particle size, stirring rate, and pulp density	21 days	[140]
Low-grade laterite ores	*Aspergillus* and *Penicillium strains*	36% Ni, 54% Co, and 0.76% Fe	Media pH, ore particle size, stirring rate, and pulp density	60 days	[141]
Oxide low-grade ores (mining residue)	*Aspergillus niger*	68% Cu, 46% Zn, and 34% Ni	Media pH, ore particle size	14 days	[142]
Low-grade granitic chalcopyrite	*Sulfolobus*	85% Cu	Media pH, ore particle size, stirring rate, and pulp density	30 days	[143]
		Mine tailings			
Lead-zinc mine tailings	*Acidithiobacillus ferrooxidans*	0.82% Pb, 97.38% Zn, and 71.37% Fe	Media pH, ore particle size, stirring rate, and pulp density	50 days	[144]
Concentrated arsenic mine tailings	*Acidithiobacillus ferrooxidans*	>85% of As	pH and pulp density	20 days	[145]
Mine tailings	Indigenous sulfur-oxidizing bacteria	96.13% Cu, 97.59% Zn, and 40.43% Pb	Effects of initial pH	13 days	[146]
Mine tailings	Single and mixed cultures of *Acidithiobacillus thiooxidans* and *Acidithiobacillus ferrooxidans*	34.4% 31.5%, and 47.1% of Cu for *A. thiooxidans*, *A. ferrooxidans*, and mixed cultures, respectively	Bacterial inoculation, media pH, ore particle size, and stirring rate	6 h	[132]

Particle size below 74 μm, stirring rate above 120 rpm, pH below 2.5 and 6.5 for bacteria and fungi, respectively.

5. Technological Improvements in Bioleaching Processes

There are significant differences in developments related to traditional mining methods versus the use of bioleaching for the same endeavor. As previously discussed, the primary focus of attention for development in conventional mining is on modifications to related equipment to improve extraction efficiency, sometimes with environmental consideration. Perhaps this is not the case for bioleaching strategies, because the capability of microorganisms is limited not only by the type of ore or potential environmental damage but also by the microbial enzymatic reactions (specificity) involved and the ability of microorganisms to withstand high concentrations of metals deposited in the resident environment, as well as the pH [147,148]. Consequently, technological advances in bioleaching are the culmination of research and investigation aimed at improving the diverse capabilities of various bioleaching microorganisms. Progress in this area of research, however, is frequently hampered by the investigative tools and techniques available at the time of research. As a result, the study of bioleaching processes has evolved in tandem with advances in molecular biology and biochemistry techniques. Furthermore, as is often the case with scientific investigations, each new discovery presents new challenges in our understanding of aspects of bioleaching mechanisms. Other factors critical to bioprocess effectiveness complicate matters in some cases. This section explains some of the aspects of microbial metal dissolution mechanisms that have opened up new research avenues.

5.1. *Research Progress and Biomining Process Improvement*

The evolution of bioleaching can be traced back to early studies that assumed natural leaching processes were devoid of biological contributions and were instead influenced by a chemical weathering process [149]. However, convincing evidence of microorganisms' active role in the previously assumed natural chemical weathering process, demonstrated in the mid-twentieth century, triggered a cascade of subsequent research [150]. This includes everything from growing and identifying microorganisms to applying them in bioleaching [151–153] and, later, understanding the molecular factors that allow them to survive and grow in such harsh environments [154]. Nevertheless, molecular research and evolving technology have revealed that their roles are far from being fully understood.

5.1.1. Accumulation of Metals in Solution, Toxicity and Its Effects on Microbial Activity

The dependence of microbial activity on various environmental factors (such as nutrients and trace elements) and its impact on process efficiency is limiting. Adeleke [155] explains that this problem could be solved by incorporating bioaugmentation strategies that provide nutrition and energy requirements in processes that use microbial leaching to extract metals of interest. This is significant because, while metal ore or mine tailings contain a variety of compounds that contribute to nutritional growth, they are frequently deficient in some vital elements required for proliferation and biomass growth. Increased biomass is critical for peak enzyme production, which is required to catalyze chemical reactions. Meeting the nutrient growth requirement is thus essential for bioprocess efficiency. Furthermore, synergistic co-factor effects in bioleaching environments have been identified as important. Co-factors, for example, include some nutrients that, while only required in trace amounts, play a critical role in improving enzyme activities. Nickel, for example, has been shown to play an important co-factor role in a number of enzymatic reactions [156]. Thus, it is critical to maintain a balance of these trace element requirements for specific bioleaching microorganisms, particularly in conditions where tailing analysis has revealed a trace element deficiency. The progressive leaching process and the resulting low pH typically increase metal toxicity while also inhibiting enzyme (protein) activity. The nickel requirement is the challenge posed by the metal's increase. The ability of such an increase to replace metals in metalloproteins, thereby affecting their ability to bind to active sites of enzymes such as dioxygenase or sulfite oxidase, causes indirect oxidation stress in cells [18]. When Chen et al. [17] investigated the toxicity of nickel to *At. thiooxidans*, they discovered that a high nickel concentration of about 600 mM inhibited bacterial growth and its sulfur reduction ability. Although metal toxicity does not completely inhibit acidophilic bacteria growth, it can significantly reduce growth at high concentrations, as demonstrated by Ramos-Zúñiga et al. [157]. The effect of cadmium at concentrations ranging from 0 to 200 mM was also tested on *A. ferrooxidans* strains at 75, 100, and 200 mM, and a significant growth reduction was observed when compared to cells grown in the absence of cadmium. It corroborates previous research and observations by Sampson and Phillips [158], who found that increasing concentrations of nickel, cobalt, and copper inhibited the oxidation ability of mesophilic cultures of *Thiobacillus ferrooxidans* and *Leptospirillum ferrooxidans*, with copper having the greatest inhibition. Chen et al., [159] investigated other metals and their toxicity concentrations to bioleaching microorganisms, finding that 300 and 400 mg/L Cu (II) and Cd (II) inhibited the growth of Thiobacillus thiooxidans, respectively. Pradhan et al. [160] discovered that 5 g/L and 0.03 g/L of V^{4+} and Mo^{6+}inhibited the growth of *At. ferrooxidans*, respectively. As a result, the efficacy in the bioleaching process necessitates a high tolerance to metal toxicity [161].

Another study that contributed to our understanding of the bioleaching process was the role of evolutionary adaptations and their significance in microbial survival in the harsh environment of mine tailings and metal ores. It became clear that a thorough understanding of these evolutionary adaptations was required to connect the dots regarding metal toxicity and tolerance. Some studies have shown that developed mechanisms linked to the evolution of these bioleaching microbial communities can be used to combat metal toxicity. Early research isolated nickel-resistant microorganisms from harsh environments and assessed their toxicity adaptation mechanisms [162]. Extracellular detoxification, intracellular sequestration, cation transport system modification, and active transport by efflux pumps are among the mechanisms identified [17].

To summarize, increased biomass is required for the bioleaching process to be effective. When toxic element concentrations and pH levels reach extremes, bacterial growth is inhibited, and the process is compromised. Although microorganisms have evolved to meet the challenges of metal toxicity in their environments, they have not been able to overcome the extreme increases in metal concentrations that are frequently achieved during bioleaching activities. Our understanding of the tolerance levels for the various environments and metals to which these organisms are exposed will allow us to develop metal toxicity management strategies in biomining operations.

5.1.2. Improving and Adapting Bioleaching Organisms for Biomining Operations

Although bioprospecting in mines allows for the collection of microorganisms with natural bioleaching capabilities, these desired qualities are usually enhanced in laboratory environments to improve biomining process efficiency [147]. This is accomplished in microorganisms by selectively pre-adapting them through repeated culturing in media containing gradually increasing concentrations of the target metal. This procedure aims to improve the tolerance and metal dissolution ability of microorganisms prior to their use in bioleaching systems, as well as their efficiency under scaled-up process conditions [148,163].

Kim et al. [164] demonstrated the importance of pre-adaptation in their study by comparing the abilities of an adapted and un-adapted bacteria strain in leaching spent catalysts, such as Ni^{2+}, V^{4+}, and others. Adaptation was achieved in their study by serial sub-culturing with increasing concentrations of metals until maximum tolerance capacities were achieved. The comparison of the bioleaching process using these two groups of organisms revealed that adapted bacteria leached over 90% of Ni^{2+} and V^{4+} in 40 h, whereas un-adapted bacteria leached around 70% of Ni^{2+} and V^{4+} during the same period. In another study comparing the bioleaching abilities of adapted and un-adapted *A. ferrooxidans*, Xia et al. [165] discovered that adapted bacteria leached 47.5% of Cu^{2+} while un-adapted bacteria leached 39.5%. Adaptation was achieved in this study by serially increasing chalcopyrite pulp densities from 1% to 5%, and bioleaching experiments were carried out under 5% chalcopyrite conditions. Ultimately, this adaptation strategy is based on reducing the time required for the required lag phase in bacterial growth, and there is evidence that a pseudo-continuous-type approach significantly improves biomass generation and substrate utilization efficiency in most fermentation and biodegradation processes [165]. The main disadvantage of this strategy, which incorporates serial sub-culturing to improve tolerance and adaptation, is that it is time consuming and may introduce mutations that change the physiological advantages derived from previous exposure to harsh environments [147,166].

Most studies on metal toxicity, tolerance, and adaptation found that a full complement of these desirable traits is not typically found in a single native microorganism but rather in a consortium, which is typical of the operating conditions found in natural environments. Table 2 shows that the efficacy of bioleaching from a consortium is greater than that of a single species. Early studies and applications tended to concentrate on achieving bioleaching with a single population of organisms [139,141]. However, recent research suggests that this approach is unrealistic, as there is now abundant evidence of the importance of interspecific interactions and how they are critical to metabolite production and degradation processes in the natural environment [4,167,168]. These communities of microorganisms make significant contributions to the degradation of complex compounds through the use of their individual enzymes. As a result, a practical approach to bioleaching strategies must apply biomimicry of what is common in the natural environment to the degradation of these ores by allowing biodegradation to occur in a similar manner. Few studies have used techniques such as polymerase chain reaction (PCR) and quantification/real-time polymerase chain reaction (qPCR/RT-PCR) to investigate community dynamics in relation to metal toxicity/tolerance and adaptation in bioleaching processes [169–171]. The application of these techniques has enabled the real-time monitoring and quantification of the dynamics of important genes or microbial communities as metal concentrations rise. This is critical for understanding the active roles that individual organisms play within population dynamics during bioleaching activities. A few recent studies have shown that microbial genes coding for ferrous iron and sulfur oxidation play an important role in bioleaching processes [17,172,173]. This knowledge is important in the selection and application of microbial consortia, particularly in enrichment processes, to ensure the ideal balance is achieved for the organisms that are critical to bioleaching efficiency within a determined population.

5.1.3. Expanding our Understanding of Bioleaching Using (Meta) Genomics

Genomics is a branch of molecular biology that focuses on understanding the structure, function, and evolution of genomes, allowing for unprecedented insights into microorganism genetic and metabolic potential [116]. It is especially useful when studying organisms isolated from niche environments that have evolved survival traits. Genomics studies enable us to identify similarities and differences between the organism under study and other microorganisms in the same genus. Such information is useful in developing strategies for bioproducts enhancement and industrial applications. In 2001, the first draft genome of a bioleaching microorganism was published, and it came from the well-studied bioleaching bacteria *At. ferrooxidans*. This published draft genome has been used to construct or predict important metabolic abilities. Sulfur assimilation, quorum sensing, extracellular polysaccharides, and iron and sulfur oxidation are all examples [17,174]. The value of genetic studies lies in the indirect improvement of the bioleaching process based on a better understanding of the organisms' genetics and physiology [175]. Presently, over 55 bacterial and 36 archaeal draft genomes of bioleaching microorganisms have been published [14]. This has allowed for more accurate predictions of the genetic and metabolic potential of bioleaching microorganisms, such as models for energy metabolism in *A. caldus* and *A. thiooxidans*, and models for overall metabolism in *Sulfobacillus* spp. and *Ferrovum* spp. [176–179]. However, one limitation of these predictions is that the majority of whole-genome sequencing projects published have only been performed on bioleaching microorganisms that have been kept in the laboratory for several years, if not decades [166]. Concerns have been raised about the evolutionary ability of these isolates, as well as the possibility of genetic changes, such as mutations and gene loss, which may have resulted in genome rearrangement [166]. Furthermore, most of these organisms' robustness has not been tested in significant field application and/or commercial bioleaching work. Relevantly, both metabolic and functional predictions derived from such an old stock of microorganisms, such as metal resistance, growth parameters, and metabolite production, may influence the outcomes of bioleaching application [180]. As a result, it is critical that research be tailored to the specific environments in which the biomining process will take place and that organisms derived from within should then be used for feasibility and field studies leading to applications. Nonetheless, the importance of functional gene predictions, particularly in bioleaching genomics, is critical for progress toward process improvement [181,182].

5.2. Recent Developments in Bioleaching Bench Applications

5.2.1. Impact of Flotation Reagents on Bioleaching Microorganisms

Metals of interest are typically concentrated in mineral processing using flotation reagents, such as collectors, depressants, and frothers, to increase recovery. Flotation reagents separate metals of interest from gangue by increasing the aeration of these metals in water and facilitating the attachment of air bubbles required for the metal of interest's levitation. The presence of these flotation reagents, on the other hand, presents a set of challenges for microorganisms involved in bioleaching. Furthermore, the results differ depending on the perspectives of the various researchers. In some cases, flotation reagents absorbed into concentrates or tailings may have an impact on the activity of bioleaching microorganisms, either a positive or a negative one. So far, contradictory results from various studies indicate that residual flotation reagents in concentrates and tailings may inhibit or reduce the oxidation ability and growth of bioleaching microorganisms or may contribute to improving these microorganisms' bioleaching ability [60,183–186].

Few studies have been conducted with the goal of understanding the potential effects of interactions between bacteria and flotation reagents. Dehghan and Dianati [184] investigated the effects of lead-zinc flotation reagents on the activity of bioleaching bacteria using three organic reagents (including xanthate collectors and gasoline) and two inorganic salts. Their findings indicated that the presence of flotation reagents had a positive effect on the activity of the bacteria (individually or as consortia), with zinc bioleaching being greater in the presence of flotation reagents than in the absence of flotation reagents. This beneficial effect was attributed to the addition of salts, which served as carbon and energy sources for the bioleaching bacteria. Jafari et al. [187], Christel et al. [138], Zhang et al. [188], and Zeng et al. [189] found that flotation reagents positively influenced the activity of bioleaching bacteria. Dong and Lin [185] conducted another study in which they compared the effects of five different flotation reagents on the bioleaching of copper from chalcopyrite. Copper extraction was reduced by 45% by butylamine, 36% by ethyl xanthate, 20% by isoamyl xanthate, 13% by isopropyl xanthate, and 11% by butyl xanthate when compared to the control. Guo et al. [190], Pacholewska et al. [191], Dopson et al. [192], Okibe and Johnson, [193], and Tuovinen [194] all found similar results. However, the authors' differing results raise the question of how the effects of flotation reagents can differ and have different effects on the bioleaching process. This lack of agreement highlights the likely dissimilarities that result from environmental uniqueness, implying that organisms from different environments may be behaving differently, supporting our assertions that studies for different mine areas should be conducted, and organisms from that region may likely work best for the particular ore.

To summarize, answers may be found in understanding the mechanisms of attachment used by microorganisms to make contact with and interact directly with the metal in order for leaching activity to occur. To attach to the metal, bioleaching microorganisms must produce EPS, which is dependent on the ions on the mineral surface. *Acidithiobacillus ferrooxidans*, for example, attaches to pyrite via an electrochemical interaction with the negatively charged pyrite surface mediated by exopolymer-complexed iron (III) ions [138,195]. The residual flotation reagents on the concentrate and tailing can change the properties of the mineral surface, which can either positively or negatively influence the bioleaching process [185,187,194]. Before commercialization decisions are made, a credible strategy for gaining this understanding will involve the execution of small-scale trial runs that integrate optimization with a focus on the flotation type and specific consortia involvement to determine the effect it will have on the organisms involved [60,185,193].

5.2.2. Impact of Ferrous/Ferric Ions Concentration on Bioleaching Experiments

The oxidation of Fe^{2+} to Fe^{3+} in the presence of oxygen is the primary mechanism used by bioleaching microorganisms for metal dissolution. This Fe^{2+} is the main source of energy, whereas the biologically regenerated Fe^{3+} initiates metal dissolution [161,196]. These reactions occur in a series of steps and are mutually dependent. The rate of metal dissolution is determined by the concentration of Fe^{3+}, which is determined by microbial oxidation of Fe^{2+}. As a consequence, the initial concentration of Fe^{2+} prior to the start of the bioleaching process is critical [197,198]. Therefore, before being used for bioleaching, the prepared inoculum must reach a significant concentration of Fe^{2+}. When bioleaching experiments using an inoculum strength containing 9 g/L of Fe^{2+} solution were used, Xiang et al. [198] observed a maximum of 95% copper leached in 5 days compared to 12 days without this specific concentration of Fe2+. However, as a precaution, it was suggested that a suitable concentration of Fe^{2+} should be at least 6.25 g/L. Dong et al. [196] obtained similar results by increasing copper leaching from five different copper sulfide minerals early in the leaching process by adding adequate amounts of ferrous ions.

Ferric irons were found to have beneficial effects on *At. ferrooxidans* involved in bioleaching in a study conducted by Chen et al. [17]. The bacterial tolerance to nickel in the presence of ferric irons was the basis for this study. They discovered that 600 mM nickel inhibited the growth and sulfur metabolism of *At. ferrooxidans* completely. However, when traces of ferric iron (between 0.5 and 5 mM) were added, *At. ferrooxidans* were able to tolerate a high concentration of nickel, and as the ferric iron concentration was increased, metal resistance genes and sulfur metabolism were significantly enhanced. When trace amounts of ferric irons were used, this effect was found to be beneficial. However, once the concentration was increased above 5 mM, no further beneficial effects on the growth of *At. ferrooxidans* were observed.

5.2.3. Catalyst Application

The reason for adding a catalyst is to try to speed up electron transfer by lowering the activation energy and thus increasing the rate of reaction. Furthermore, metal ions, such as Ag^{+}, Hg^{2+}, Bi^{3+}, Cu^{2+}, and Co^{2+}, have been used and found to have superior catalytic properties to non-metallic catalysts [15,25]. Non-metallic catalysts have a low bioleaching yield and require a lot of catalyst to achieve significant process efficiency [15,87]. Guo et al. [23] investigated the catalytic effects of silver ions (Ag^{+}) and copper ions (Cu^{2+}) on the dissolution of realgar (As_2S_2) and discovered that the leaching efficiency was increased to more than 86% compared to 4% for As_2S_2 leaching without Ag^{+}. In this case, the catalytic effect of Cu^{2+} was not as strong as that of Ag^{+}. This study also discovered that these catalysts can be used for either chemical or bioleaching methods to improve metal dissolution. Zeng et al. [22] investigated the catalytic effects of copper ions on the bioleaching of cobalt from lithium batteries by *At. ferrooxidans* in another study. Their findings revealed that almost all cobalt (>99%) was solubilized in the presence of 75 g/L copper ions in 6 days, while only 43.1% was solubilized in 10 days without the catalyst. Metal ion catalyst research is primarily conducted in the laboratory, with industrial applications appearing to be scarce. It is possible that elucidating the organism's up- and down-regulatory mechanisms during these reactions will provide information on how to improve and manipulate the efficient dissolution rate. Furthermore, improved bioleaching kinetics will lead to increased process efficiency and economics in commercial applications [15].

5.2.4. Metabolomics Application in Bioleaching

The petroleum-refining processes of hydroprocessing and fluid catalytic cracking (FCC) generate massive amounts of spent catalysts containing toxic and valuable metals [199]. These spent catalysts are typically disposed of in approved dumpsites that pose a risk to the environment [200]. These wastes are being investigated as a secondary source of valuable metals, such as Ni, V, Mo, Co, W, Al, and others, in order to solve environmental issues while also meeting current metal demand. Metal extraction from spent catalysts can be accomplished via a hydrometallurgical or bio-hydrometallurgical process. Mouna and Baral [201] used a bio-hydrometallurgical approach to leach lanthanum from spent fluid catalytic cracking catalyst (SFCCC) using the fungus *Aspergillus niger* in a study published in 2019. Lanthanum recovery efficiency of 63% was observed at 1% pulp density but decreased as pulp density increased due to the SFCCC's inhibition effect on *A. niger* activity.

Hydrochloric acid had a high recovery efficiency of 68% to chemical leaching. In this study, bio-hydrometallurgy (a greener process) was recommended because the leaching efficiencies of the two were comparable. However, because these metals have low recovery rates (less than 50%), efforts must be made to develop technologies that allow for high recovery rates [199,202].

5.3. Metabolomics Application in Bioleaching

At the present time, readily available high-throughput genomic technologies are hastening progress in understanding the diversity and genetic make-up of bioleaching microorganisms isolated from extreme environments, albeit they are most often decades-old laboratory isolates [178,203]. Consequently, the expansion of metabolomics applications in tandem with genomics aim to become routine preliminary inclusions in both bioprospecting and inoculum development research for bioleaching. The identification and quantification of low-molecular-weight compounds known as metabolites produced at a specific point in time from a sample understudy (metabolome) is defined as metabolomics [204–207]. Metabolomics is distinct, in that the measurement and quantification of metabolites provides information about an organism's functional status [208]. Simply put, the RNA or transcriptome contains potential genes with unknown functions. Rather than just a prediction, metabolomics provides an immediate answer as to what organisms are capable of producing under specific conditions or premised by catabolite induction. Identifying a larger array of these small molecules released during bioleaching, which may be associated with bioleaching effectiveness, according to Brisson et al. [209], could potentially contribute to a better understanding of the bioleaching process. This is supported by metabolomics studies conducted by the same researchers who investigated and tested metabolites exuded by a *Paecilomyces* fungus as potential contributors to the bioleaching of rare earth elements from monazite [209]. They chose eight of the identified organic metabolites for abiotic testing, and two of the identified metabolites, citric and citramalic acid, significantly increased leaching. Thus, metabolomics is regarded as a significant advancement that will improve the economic feasibility of industrializing bioleaching processes.

6. Research Outlooks

This review identified several gaps that researchers need to address in terms of low-grade processing and its environmental impact. One of the gray areas identified was the critical need for technological development of pyrometallurgical processes to address the process's high energy requirements, which make product recovery from low-grade processing costly, unsustainable, and harmful to the environment.

Other gray areas were identified in biomining, which is expected to play a significant role globally in the efficient extraction and processing of low-grade ores on an industrial scale while causing minimal environmental damage. The incubation period, which is determined by the lag phase of bacteria growth, is one area in need of investigation. The lengthy period has hampered the adoption of bioleaching on an industrial scale, making commercial metal sulfide processing impossible. There is also a need for research, ranging from identifying suitable microorganisms to comprehending the molecular factors that are critical to their survival in harsh environments. The need for such research demonstrates that understanding microorganisms is a fertile field of study.

This review buttresses that little research has focused on polymerase chain reaction (PCR) and quantification/real-time polymerase chain reaction (qPCR/RT-PCR) to understand dynamics in toxicity, tolerance, and adaptation in bioleaching processes. Another unexploited area in terms of research is the complexity and unique interspecific interactions, as well as the diverse chemical profiles and ingenuity that come with the evolution of different organisms, particularly within consortia and signal transduction. It should also be noted that knowledge of an important step regulated by cell-to-cell communication mechanisms known as "quorum sensing" (QS), as seen in Gram-negative bacteria's role in bioleaching, is lacking.

Though much is known about the electron transfer that contributes to the survival of *At. ferrooxidans* replication, studies that look at other bioleaching organisms besides model organisms are scarce. However, the general lack of agreement among researchers on the effects of flotation reagents on the bioleaching process may be due to dissimilarities caused by environmental uniqueness, implying that organisms from different environments behave differently. As a result, studies concerning a specific mine area must be conducted using organisms from that region. Finally, it was discovered that most studies on metal ion catalysts are conducted on a laboratory scale, with little or no application on an industrial scale.

7. Concluding Remarks

Significant progress has been made in understanding the environmental impact of mining activities. However, the importance of mining in a variety of applications, such as the generation of metal for domestic and industrial purposes or income, implies that the mining process will continue as long as there is a demand that must be met. As a result, this review highlighted the current challenges in processing low-grade ores, technological advancement, and the subsequent environmental implications. This review's major conclusions are highlighted below

1. As a consequence of the global rate of decline in high-grade ores, the possibility of low-grade ore reserves serving as major future metal sources is high. Furthermore, mine tailings represent an easy-to-process source of metals with high profit potential. The nature, mining, and processing of low-grade ores present challenges that would otherwise impede profitability. The continuous technological evolution of available processes, on the other hand, has the potential to mitigate this effect.
2. The environmental impacts of mining processes are a major challenge for the mining industry, regardless of ore grade. This review focuses on changes implemented in terms of the types of chemicals used or energy-saving methods aimed at meeting the required environmental standards as outlined by various governments of mining countries. Some of these advancements in traditional methods have resulted in a preference for hydrometallurgy over pyrometallurgy. Although hydrometallurgy presents greater product recovery challenges than pyrometallurgy, the latter faces challenges due to the process's exhaustion of innovative possibilities and its inherent high carbon footprint. Meeting and resolving the problem of developing effective methods for processing low-quality-grade ores remains at the heart of the modern mining debate.
3. Although bioleaching has its own set of challenges and continues to use many of the primary energy-intensive units found in conventional mining, such as milling and crushing, it currently meets the majority of environmental safety requirements but falls short in terms of process efficiency in terms of reaction time for metal recovery. As a result, its industrial application and commercial acceptance are limited. Consequently, bioleaching modification influenced by re-defined optimization methods based on the understanding of the genetic make-up and metabolites produced by microorganisms involved may be the key to process improvement. Such knowledge may elucidate microbial functioning under specific conditions, such as high metal concentration, increasing its feasibility and acceptability for industrial application. Current advances in molecular techniques and omics technologies provide effective tools for enhancing our understanding of these extremophiles and will be critical in unravelling their mystery.

Finally, continuous technological advancement of traditional methods has the potential to improve metal recoveries from low-grade ores while reducing the environmental pollution. Nonetheless, despite the challenges that bioleaching faces, it remains an appealing technology in comparison to conventional methods in this current era. It may also be beneficial to intensify molecular studies on bioleaching microorganisms in order to accelerate the potential for bioleaching process improvement.

Author Contributions: Conceptualization, G.N.I. and R.N.; writing—original draft preparation, R.N., G.N.I. and N.C.; methodology, G.N.I.; software, R.N.; validation, G.N.I., N.C. and T.S.M.; formal analysis, R.N., G.N.I. and N.C.; investigation, R.N. and G.N.I.; resources, N.C. and T.S.M.; data curation, R.N.; visualization, R.N., G.N.I., N.C. and T.S.M.; supervision, G.N.I. and T.S.M.; project administration, T.S.M.; funding acquisition, N.C. and T.S.M. All authors have read and agreed to the published version of the manuscript.

Funding: This research was funded by the Department of Science and Innovation (DSI); Technology Innovation Agency (TIA) grant number DST/CON 0197/2017 and the NRF, Grant No: 12039.

Acknowledgments: The Institute for the Development of Energy for African Sustainability (IDEAS), University of South Africa (UNISA), is acknowledged. The opinions expressed and conclusions reached are those of the authors and not necessarily endorsed by the DSI, TIA, and UNISA.

Conflicts of Interest: The authors declare no conflict of interest.

References

1. Johnson, D.B. The evolution, current status, and future prospects of using biotechnologies in the mineral extraction and metal recovery sectors. *Minerals* **2018**, *8*, 343. [CrossRef]
2. Hermanus, M. Mining redesigned—Innovation and technology needs for the future—A South African perspective. *J. S. Afr. Inst. Min. Metall.* **2017**, *117*, 811–818. [CrossRef]
3. National Research Council. *Evolutionary and Revolutionary Technologies for Mining*; National Academies Press: Washington, DC, USA, 2002; ISBN 9780309073400.
4. Ahmadi, A.; Khezri, M.; Abdollahzadeh, A.A.; Askari, M. Bioleaching of copper, nickel and cobalt from the low grade sulfidic tailing of Golgohar Iron Mine, Iran. *Hydrometallurgy* **2015**, *154*, 1–8. [CrossRef]
5. Norgate, T.; Haque, N. Energy and greenhouse gas impacts of mining and mineral processing operations. *J. Clean. Prod.* **2010**, *18*, 266–274. [CrossRef]
6. Johnson, D.B. Development and application of biotechnologies in the metal mining industry. *Environ. Sci. Pollut. Res.* **2013**, *20*, 7768–7776. [CrossRef]
7. Dold, B. Sustainability in metal mining: From exploration, over processing to mine waste management. *Rev. Environ. Sci. Biotechnol.* **2008**, *7*, 275–285. [CrossRef]
8. Kulczycka, J.; Lelek, Ł.; Lewandowska, A.; Wirth, H.; Bergesen, J.D. Environmental Impacts of Energy-Efficient Pyrometallurgical Copper Smelting Technologies: The Consequences of Technological Changes from 2010 to 2050. *J. Ind. Ecol.* **2016**, *20*, 304–316. [CrossRef]
9. Northey, S.; Mohr, S.; Mudd, G.M.; Weng, Z.; Giurco, D. Modelling future copper ore grade decline based on a detailed assessment of copper resources and mining. *Resour. Conserv. Recycl.* **2014**, *83*, 190–201. [CrossRef]
10. Kržanović, D.; Conić, V.; Bugarin, D.; Jovanović, I.; Božić, D. Maximizing economic performance in the mining industry by applying bioleaching technology for extraction of polymetallic mineral deposits. *Minerals* **2019**, *9*, 400. [CrossRef]
11. Watling, H.R. Review of biohydrometallurgical metals extraction from polymetallic mineral resources. *Minerals* **2014**, *5*, 1–60. [CrossRef]
12. Jamieson, H.E. Geochemistry and mineralogy of solid mine waste: Essential knowledge for predicting environmental impact. *Elements* **2011**, *7*, 381–386. [CrossRef]
13. Ilyas, S.; Bhatti, H.N.; Bhatti, I.A.; Sheikh, M.A.; Ghauri, A. Bioleaching of metal ions from low grade sulphide ore: Process optimization by using orthogonal experimental array design. *Afr. J. Biotechnol.* **2010**, *9*, 2801–2810. [CrossRef]
14. Martinez, P.; Vera, M.; Bobadilla-Fazzini, R.A. Omics on bioleaching: Current and future impacts. *Appl. Microbiol. Biotechnol.* **2015**, *99*, 8337–8350. [CrossRef] [PubMed]
15. Pathak, A.; Morrison, L.; Healy, M.G. Catalytic potential of selected metal ions for bioleaching, and potential techno-economic and environmental issues: A critical review. *Bioresour. Technol.* **2017**, *229*, 211–221. [CrossRef]
16. Borja, D.; Nguyen, K.A.; Silva, R.A.; Park, J.H.; Gupta, V.; Han, Y.; Lee, Y.; Kim, H. Experiences and Future Challenges of Bioleaching Research in South Korea. *Minerals* **2016**, *6*, 128. [CrossRef]
17. Chen, A.; Hao, X.; Xiao, Y.; Zou, K.; Liu, H.; Liu, X.; Yin, H.; Qiu, G.; Liang, Y. Responses of Acidithiobacillus thiooxidans A01 to Individual and Joint Nickel (Ni^{2+}) and Ferric (Fe^{3+}). *Minerals* **2019**, *9*, 82. [CrossRef]
18. Macomber, L.; Hausinger, R.P. Mechanisms of nickel toxicity in microorganisms. *Metallomics* **2011**, *3*, 1153–1162. [CrossRef]
19. Masaki, Y.; Hirajima, T.; Sasaki, K.; Miki, H.; Okibe, N. Microbiological Redox Potential Control to Improve the Efficiency of Chalcopyrite Bioleaching. *Geomicrobiol. J.* **2018**, *35*, 648–656. [CrossRef]
20. Tavakoli, H.Z.; Abdollahy, M.; Ahmadi, S.; Darban, A.K. Enhancing recovery of uranium column bioleaching by process optimization and kinetic modeling. *Trans. Nonferrous Met. Soc. China* **2017**, *27*, 2691–2703. [CrossRef]
21. Amiri, F.; Mousavi, S.M.; Yaghmaei, S.; Barati, M. Bioleaching kinetics of a spent refinery catalyst using Aspergillus niger at optimal conditions. *Biochem. Eng. J.* **2012**, *67*, 208–217. [CrossRef]

22. Zeng, G.; Deng, X.; Luo, S.; Luo, X.; Zou, J. A copper-catalyzed bioleaching process for enhancement of cobalt dissolution from spent lithium-ion batteries. *J. Hazard. Mater.* **2012**, *199–200*, 164–169. [CrossRef] [PubMed]
23. Guo, P.; Zhang, G.; Cao, J.; Li, Y.; Fang, Z.; Yang, C. Catalytic effect of Ag^+ and Cu^{2+} on leaching realgar (As_2S_2). *Hydrometallurgy* **2011**, *106*, 99–103. [CrossRef]
24. Pradhan, D.; Mishra, D.; Kim, D.J.; Ahn, J.G.; Chaudhury, G.R.; Lee, S.W. Bioleaching kinetics and multivariate analysis of spent petroleum catalyst dissolution using two acidophiles. *J. Hazard. Mater.* **2010**, *175*, 267–273. [CrossRef]
25. Liu, W.; Yang, H.-Y.; Song, Y.; Tong, L.-L. Catalytic effects of activated carbon and surfactants on bioleaching of cobalt ore. *Hydrometallurgy* **2015**, *152*, 69–75. [CrossRef]
26. Priester, M.; Ericsson, M.; Dolega, P.; Löf, O. Mineral grades: An important indicator for environmental impact of mineral exploitation. *Miner. Econ.* **2019**, *32*, 49–73. [CrossRef]
27. Mozaffari, S. Measurement While Drilling System in Aitik Mine. Master's Thesis, School of Applied Geosciences and Mining, Lulea University of Technology, Lulea, Sweden, 2007; p. 30.
28. Kontinen, A.; Hanski, E. The Talvivaara Black Shale-Hosted Ni-Zn-Cu-Co Deposit in Eastern Finland. In *Mineral Deposits of Finland*; Elsevier Inc.: Amsterdam, The Netherlands, 2015; ISBN 9780124104761.
29. Riekkola-Vanhanen, M. Talvivaara mining company—From a project to a mine. *Miner. Eng.* **2013**, *48*, 2–9. [CrossRef]
30. Lutandula, M.S.; Maloba, B. Recovery of cobalt and copper through reprocessing of tailings from flotation of oxidised ores. *J. Environ. Chem. Eng.* **2013**, *1*, 1085–1090. [CrossRef]
31. Falagán, C.; Grail, B.M.; Johnson, D.B. New approaches for extracting and recovering metals from mine tailings. *Miner. Eng.* **2017**, *106*, 71–78. [CrossRef]
32. Andrews, W.J.; Moreno, C.J.G.; Nairn, R.W. Potential recovery of aluminum, titanium, lead, and zinc from tailings in the abandoned Picher mining district of Oklahoma. *Miner. Econ.* **2013**, *26*, 61–69. [CrossRef]
33. Mudd, G.M.; Weng, Z.; Jowitt, S.M.; Turnbull, I.D.; Graedel, T. Quantifying the recoverable resources of by-product metals: The case of cobalt. *Ore Geol. Rev.* **2013**, *55*, 87–98. [CrossRef]
34. Mazalto, M. La réforme du secteur minier en République démocratique du Congo: Enjeux de gouvernance et perspectives de reconstruction. *Afr. Contemp.* **2008**, *227*, 53–80. [CrossRef]
35. Van Zyl, D.; Shields, D.; Agioutantis, Z.; Joyce, S. Waste Not, Want Not—Rethinking the Tailings and Mine Waste Issue. *AusIMM Bull.* December 2016. Available online: https://www.ausimmbulletin.com/feature/waste-not-want-not-rethinking-the-tailings-and-mine-waste-issue/ (accessed on 22 March 2022).
36. Davou, L.C.; Mangset, W.E. Evaluation of Radiation Hazard Indices and Excess Lifetime Cancer Risk due to Natural Radioactivity in Mined Tailings in Some Locations in JOS Plateau State Nigeria. *IOSR J. Appl. Phys.* **2015**, *7*, 67–72. [CrossRef]
37. Van Jaarsveld, J.G.S.; Lukey, G.C.; Van Deventer, J.S.J.; Graham, A. The Stabilisation of Mine Tailings by Reactive Geopolymerisation (gold). In *Proceedings of the MINPREX 2000 International Congress on Mineral and Processing and Extractive Metallurgy, Melbourne, Australia, 11–13 September 2000*; Griffiths, P., Spry, A., Eds.; The Australian Institute of Mining and Metallurgy: Carlton, Australia, 2000; pp. 363–371.
38. Karaca, O.; Cameselle, C.; Reddy, K. Mine tailing disposal sites: Contamination problems, remedial options and phytocaps for sustainable remediation. *Rev. Environ. Sci. Biotechnol.* **2018**, *17*, 205–228. [CrossRef]
39. Rathna, R.; Nakkeeran, E. Biological treatment for the recovery of minerals from low-grade ores. In *Current Developments in Biotechnology and Bioengineering*; Elsevier B.V.: Amsterdam, The Netherlands, 2020; pp. 437–458. ISBN 9780444643216.
40. Spooren, J.; Binnemans, K.; Björkmalm, J.; Breemersch, K.; Dams, Y.; Folens, K.; González-Moya, M.; Horckmans, L.; Komnitsas, K.; Kurylak, W.; et al. Near-zero-waste processing of low-grade, complex primary ores and secondary raw materials in Europe: Technology development trends. *Resour. Conserv. Recycl.* **2020**, *160*, 104919. [CrossRef]
41. Tabelin, C.B.; Park, I.; Phengsaart, T.; Jeon, S.; Villacorte-Tabelin, M.; Alonzo, D.; Yoo, K.; Ito, M.; Hiroyoshi, N. Copper and critical metals production from porphyry ores and E-wastes: A review of resource availability, processing/recycling challenges, socio-environmental aspects, and sustainability issues. *Resour. Conserv. Recycl.* **2021**, *170*, 105610. [CrossRef]
42. Mochizuki, Y.; Tsubouchi, N. Removal of gangue components from low-grade iron ore by hydrothermal treatment. *Hydrometallurgy* **2019**, *190*, 105159. [CrossRef]
43. Mokmeli, M. Pre feasibility study in hydrometallurgical treatment of low-grade chalcopyrite ores from Sarcheshmeh copper mine. *Hydrometallurgy* **2020**, *191*, 105215. [CrossRef]
44. Asghari, I.; Mousavi, S.M.; Amiri, F.; Tavassoli, S. Bioleaching of spent refinery catalysts: A review. *J. Ind. Eng. Chem.* **2013**, *19*, 1069–1081. [CrossRef]
45. Malekian, H.; Salehi, M.; Biria, D. Investigation of platinum recovery from a spent refinery catalyst with a hybrid of oxalic acid produced by Aspergillus niger and mineral acids. *Waste Manag.* **2019**, *85*, 264–271. [CrossRef]
46. Srichandan, H.; Mohapatra, R.K.; Parhi, P.K.; Mishra, S. Bioleaching approach for extraction of metal values from secondary solid wastes: A critical review. *Hydrometallurgy* **2019**, *189*, 105122. [CrossRef]
47. Shah, S.S.; Palmieri, M.C.; Sponchiado, S.R.P.; Bevilaqua, D. Environmentally sustainable and cost-effective bioleaching of aluminum from low-grade bauxite ore using marine-derived Aspergillus niger. *Hydrometallurgy* **2020**, *195*, 105368. [CrossRef]
48. Birinci, M.; Gök, R. Characterization and flotation of low-grade boehmitic bauxite ore from Seydişehir (Konya, Turkey). *Miner. Eng.* **2021**, *161*, 106714. [CrossRef]

49. Gibson, B.; Wonyen, D.G.; Chelgani, S.C. A review of pretreatment of diasporic bauxite ores by flotation separation. *Miner. Eng.* **2017**, *114*, 64–73. [CrossRef]
50. Jiang, Y.-R.; Li, W.; Feng, R. Preparation and performance of 4-alkyl-4,4-bis(hydroxycarbamoyl) carboxylic acid for flotation separation of diaspore against aluminosilicates. *Miner. Eng.* **2011**, *24*, 1571–1579. [CrossRef]
51. Xia, L.-Y.; Zhong, H.; Liu, G.-Y. Flotation techniques for separation of diaspore from bauxite using Gemini collector and starch depressant. *Trans. Nonferrous Met. Soc. China Engl. Ed.* **2010**, *20*, 495–501. [CrossRef]
52. Smith, P. The processing of high silica bauxites—Review of existing and potential processes. *Hydrometallurgy* **2009**, *98*, 162–176. [CrossRef]
53. Petersen, J. Heap leaching as a key technology for recovery of values from low-grade ores—A brief overview. *Hydrometallurgy* **2016**, *165*, 206–212. [CrossRef]
54. Ghorbani, Y.; Franzidis, J.-P.; Petersen, J. Heap leaching technology—Current state, innovations and future directions: A review. *Miner. Process. Extr. Metall. Rev.* **2015**, *37*, 73–119. [CrossRef]
55. Calvo, G.; Palacios, J.-L.; Valero, A. The influence of ore grade decline on energy consumption and GhG emissions: The case of gold. *Environ. Dev.* **2022**, *41*, 100683. [CrossRef]
56. Sudagar, A.J.; Andrejkovičová, S.; Rocha, F.; Patinha, C.; Soares, M.R.; Velosa, A.L.; da Silva, E.F. Combined Influence of Low-Grade Metakaolins and Natural Zeolite on Compressive Strength and Heavy Metal Adsorption of Geopolymers. *Minerals* **2021**, *11*, 486. [CrossRef]
57. Park, I.; Kanazawa, Y.; Sato, N.; Galtchandmani, P.; Jha, M.K.; Tabelin, C.B.; Jeon, S.; Ito, M.; Hiroyoshi, N. Beneficiation of Low-Grade Rare Earth Ore from Khalzan Buregtei Deposit (Mongolia) by Magnetic Separation. *Minerals* **2021**, *11*, 1432. [CrossRef]
58. Alves Dias, P.; Bobba, S.; Carrara, S.; Plazzotta, B.; European Commission; Joint Research Centre. *The Role of Rare Earth Elements in Wind Energy and Electric Mobility: An Analysis of Future Supply/Demand Balances*; Publication Office of the European Union: Luxembourg, 2020; ISBN 9789276270164.
59. Tian, M.; Gao, Z.; Han, H.; Sun, W.; Hu, Y. Improved flotation separation of cassiterite from calcite using a mixture of lead (II) ion/benzohydroxamic acid as collector and carboxymethyl cellulose as depressant. *Miner. Eng.* **2017**, *113*, 68–70. [CrossRef]
60. Jafari, M.; Shafaei, S.Z.A.; Abdollahi, H.; Gharabaghi, M.; Chelgani, S.C. A comparative study on the effect of flotation reagents on growth and iron oxidation activities of *Leptospirillum ferrooxidans* and *Acidithiobacillus ferrooxidans*. *Minerals* **2016**, *7*, 2. [CrossRef]
61. Komnitsas, K.; Petrakis, E.; Bartzas, G.; Karmali, V. Column leaching of low-grade saprolitic laterites and valorization of leaching residues. *Sci. Total Environ.* **2019**, *665*, 347–357. [CrossRef] [PubMed]
62. Norgate, T.; Jahanshahi, S. Energy and greenhouse gas implications of deteriorating quality ore reserves. In Proceedings of the 5th Australian Conference on Life Cycle Assessment: Achieving Business Benefits from Managing Life Cycle impacts, Melbourne, Australia, 22–24 November 2006.
63. Norgate, T.; Jahanshahi, S. Low grade ores—Smelt, leach or concentrate? *Miner. Eng.* **2010**, *23*, 65–73. [CrossRef]
64. Chetty, D.; Gutzmer, J. Quantitative mineralogy to address energy consumption in smelting of ores from the Kalahari Manganese Field, South Africa. In Proceedings of the Infacon XV: International Ferro-Alloys Congress, Southern African Institute of Mining and Metallurgy, Cape Town, South Africa, 25–28 February 2018.
65. Jin, P.; Jiang, Z.; Bao, C.; Hao, S.; Zhang, X. The energy consumption and carbon emission of the integrated steel mill with oxygen blast furnace. *Resour. Conserv. Recycl.* **2017**, *117*, 58–65. [CrossRef]
66. Kaushik, P.; Fruehan, R.J. Mixed burden softening and melting phenomena in blast furnace operation Part 3—Mechanism of burden interaction and melt exudation phenomenon. *Ironmak. Steelmak.* **2007**, *34*, 10–22. [CrossRef]
67. Fisher, K.G. Refining of Gold At the Rand Refinery. In *The Extractive Metallurgy of Gold in South Africa*; Stanley, G.G., Ed.; Monograph Series M7; Chamber of Mines of South Africa; The South African Institute of Mining and Metallurgy: Johannesburg, South Africa, 1987; Volume 2, pp. 615–653.
68. Agarwal, J.C.; Beecher, N.; Davies, D.S.; Hubred, G.L.; Kakaria, V.K.; Kust, R.N. Processing of ocean nodules: A technical and economic review. *JOM* **1976**, *28*, 24–31. [CrossRef]
69. Soderholm, K.; Soderholm, P.; Pettersson, M.; SvaHn, N.; WiKlund, R.; Helenius, H. *Environmental Regulation and Mining Sector Competitiveness*; Luleå Tekniska Universitet: Luleå, Sweden, 2014.
70. Steinberg, W.S.; Geyser, W.; Nell, J. The history and development of the pyrometallurgical processes at Evraz Highveld steel & vanadium. *J. S. Afr. Inst. Min. Metall.* **2011**, *111*, 705–710.
71. Wang, P.; Li, J.X.; Zhou, L.Y.; Long, H.M. Theoretical and experimental investigation of oxygen blast furnace process with high injection of hydrogenous fuel. *Ironmak. Steelmak.* **2013**, *40*, 312–317. [CrossRef]
72. Takahashi, K.; Nouchi, T.; Sato, M.; Ariyama, T. Perspective on Progressive Development of Oxygen Blast Furnace for Energy Saving. *Tetsu-to-Hagane/J. Iron Steel Inst. Jpn.* **2016**, *102*, 365–374. [CrossRef]
73. Abbruzzese, C.; Duarte, M.; Paponetti, B.; Toro, L. Biological and chemical processing of low-grade manganese ores. *Miner. Eng.* **1990**, *3*, 307–318. [CrossRef]
74. Jiang, T.; Yang, Y.; Huang, Z.; Zhang, B.; Qiu, G. Leaching kinetics of pyrolusite from manganese–silver ores in the presence of hydrogen peroxide. *Hydrometallurgy* **2004**, *72*, 129–138. [CrossRef]
75. Welham, N. Activation of the carbothermic reduction of manganese ore. *Int. J. Miner. Process.* **2002**, *67*, 187–198. [CrossRef]
76. Cheng, Z.; Zhu, G.; Zhao, Y. Study in reduction-roast leaching manganese from low-grade manganese dioxide ores using cornstalk as reductant. *Hydrometallurgy* **2009**, *96*, 176–179. [CrossRef]

77. Rao, S.; Yang, T.; Zhang, D.; Liu, W.; Chen, L.; Hao, Z.; Xiao, Q.; Wen, J. Leaching of low grade zinc oxide ores in $NH_4Cl–NH_3$ solutions with nitrilotriacetic acid as complexing agents. *Hydrometallurgy* **2015**, *158*, 101–106. [CrossRef]
78. Zhang, S.; Zhang, B.; Liu, B.; Sun, S. A review of Mn-containing oxide catalysts for low temperature selective catalytic reduction of NO_x with NH_3: Reaction mechanism and catalyst deactivation. *RSC Adv.* **2017**, *7*, 26226–26242. [CrossRef]
79. Sinha, M.K.; Purcell, W. Reducing agents in the leaching of manganese ores: A comprehensive review. *Hydrometallurgy* **2019**, *187*, 168–186. [CrossRef]
80. Nazari, A.M.; Radzinski, R.; Ghahreman, A. Review of arsenic metallurgy: Treatment of arsenical minerals and the immobilization of arsenic. *Hydrometallurgy* **2017**, *174*, 258–281. [CrossRef]
81. Marsden, J. Energy Efficiency and Copper Hydrometallurgy. In Proceedings of the Sixth International Symposium, Phoenix, AZ, USA, 17–20 August 2008; pp. 29–42.
82. Zheng, F.; Chen, F.; Guo, Y.; Jiang, T.; Travyanov, A.Y.; Qiu, G. Kinetics of Hydrochloric Acid Leaching of Titanium from Titanium-Bearing Electric Furnace Slag. *JOM* **2016**, *68*, 1476–1484. [CrossRef]
83. Guo, Q.; Qu, J.; Han, B.; Zhang, P.; Song, Y.; Qi, T. Innovative technology for processing saprolitic laterite ores by hydrochloric acid atmospheric pressure leaching. *Miner. Eng.* **2015**, *71*, 1–6. [CrossRef]
84. Langová, Š.; Leško, J.; Matýsek, D. Selective leaching of zinc from zinc ferrite with hydrochloric acid. *Hydrometallurgy* **2009**, *95*, 179–182. [CrossRef]
85. Abdel-Aal, E.A. Kinetics of sulfuric acid leaching of low-grade zinc silicate ore. *Hydrometallurgy* **2000**, *55*, 247–254. [CrossRef]
86. El Ouni, A.; Brusseau, M.L. *Rosemont Copper Mine. Conceptual Site Model for Assessing Contaminant Transport and Pathways*; Soil, Water, and Environmental Science Department, The Contaminant Transport Lab, University of Arizona: Tucson, AZ, USA, 2016.
87. Zhang, R.; Wei, D.; Shen, Y.; Liu, W.; Lu, T.; Han, C. Catalytic effect of polyethylene glycol on sulfur oxidation in chalcopyrite bioleaching by *Acidithiobacillus ferrooxidans*. *Miner. Eng.* **2016**, *95*, 74–78. [CrossRef]
88. Santos, F.M.F.; Pina, P.S.; Porcaro, R.; Oliveira, V.A.; Silva, C.A.; Leão, V.A. The kinetics of zinc silicate leaching in sodium hydroxide. *Hydrometallurgy* **2010**, *102*, 43–49. [CrossRef]
89. Ju, S.; Motang, T.; Shenghai, Y.; Yingnian, L. Dissolution kinetics of smithsonite ore in ammonium chloride solution. *Hydrometallurgy* **2005**, *80*, 67–74. [CrossRef]
90. Abkhoshk, E.; Jorjani, E.; Al-Harahsheh, M.S.; Rashchi, F.; Naazeri, M. Review of the hydrometallurgical processing of non-sulfide zinc ores. *Hydrometallurgy* **2014**, *149*, 153–167. [CrossRef]
91. Liu, Y.; Zhang, Y.; Chen, F.; Zhang, Y. The alkaline leaching of molybdenite flotation tailings associated with galena. *Hydrometallurgy* **2012**, *129–130*, 30–34. [CrossRef]
92. Ding, Z.; Yin, Z.; Hu, H.; Chen, Q. Dissolution kinetics of zinc silicate (hemimorphite) in ammoniacal solution. *Hydrometallurgy* **2010**, *104*, 201–206. [CrossRef]
93. Chen, A.; Zhao, Z.W.; Jia, X.; Long, S.; Huo, G.; Chen, X. Alkaline leaching Zn and its concomitant metals from refractory hemimorphite zinc oxide ore. *Hydrometallurgy* **2009**, *97*, 228–232. [CrossRef]
94. Earnest, C.M.; Gann, K.; Stong, B. Improved Quantification of Gibbsite in Bauxite Ores by Thermogravimetric Methods (TGA and DTG). *Adv. Appl. Chem. Biochem.* **2018**, *1*, 9–17. [CrossRef]
95. Guan, C.; Chen, L.; Zheng, Y.; Sun, W.; Zheng, Y. Hydrochloric acid leaching for upgrading flotation concentrate from a low-grade bauxite ore. *Physicochem. Probl. Miner. Process.* **2017**, *53*, 1038–1046. [CrossRef]
96. Safarian, J. Extraction of Iron and Ferrosilicon Alloys from Low-Grade Bauxite Ores. In *Extraction 2018*; Springer: Cham, Switzerland, 2018; pp. 825–837.
97. Sukla, L.B.; Pattanaik, A.; Pradhan, D. Advances in Beneficiation of Low-Grade Bauxite. In *Light Metals 2019*; Chesonis, C., Ed.; Springer International Publishing: Cham, Switzerland, 2019; pp. 3–10.
98. McDonald, R.G.; Whittington, B.I. Atmospheric acid leaching of nickel laterites review: Part I. Sulphuric acid technologies. *Hydrometallurgy* **2008**, *91*, 35–55. [CrossRef]
99. Astuti, W.; Hirajima, T.; Sasaki, K.; Okibe, N. Comparison of effectiveness of citric acid and other acids in leaching of low-grade Indonesian saprolitic ores. *Miner. Eng.* **2016**, *85*, 1–16. [CrossRef]
100. Chen, D.; Zhao, H.; Hu, G.; Qi, T.; Yu, H.; Zhang, G.; Wang, L.; Wang, W. An extraction process to recover vanadium from low-grade vanadium-bearing titanomagnetite. *J. Hazard. Mater.* **2015**, *294*, 35–40. [CrossRef]
101. Gok, O.; Anderson, C.G. Dissolution of low-grade chalcopyrite concentrate in acidified nitrite electrolyte. *Hydrometallurgy* **2013**, *134–135*, 40–46. [CrossRef]
102. Tian, X.; Wen, X.; Yang, C.; Liang, Y.; Pi, Z.; Wang, Y. Reductive leaching of manganese from low-grade manganese dioxide ores using corncob as reductant in sulfuric acid solution. *Hydrometallurgy* **2010**, *100*, 157–160. [CrossRef]
103. Liu, W.; Xu, H.; Yang, X.; Shi, X. Extraction of molybdenum from low-grade Ni–Mo ore in sodium hypochlorite solution under mechanical activation. *Miner. Eng.* **2011**, *24*, 1580–1585. [CrossRef]
104. Chen, T.; Lei, C.; Yan, B.; Xiao, X. Metal recovery from the copper sulfide tailing with leaching and fractional precipitation technology. *Hydrometallurgy* **2014**, *147–148*, 178–182. [CrossRef]
105. Muravyov, M.I.; Bulaev, A.G.; Kondrat'Eva, T.F. Complex treatment of mining and metallurgical wastes for recovery of base metals. *Miner. Eng.* **2014**, *64*, 63–66. [CrossRef]
106. Bian, Z.; Miao, X.; Lei, S.; Chen, S.-E.; Wang, W.; Struthers, S. The Challenges of Reusing Mining and Mineral-Processing Wastes. *Science* **2012**, *337*, 702–703. [CrossRef] [PubMed]

107. Acheampong, M.A.; Meulepas, R.J.; Lens, P.N. Removal of heavy metals and cyanide from gold mine wastewater. *J. Chem. Technol. Biotechnol.* **2010**, *85*, 590–613. [CrossRef]
108. Akcil, A.; Koldas, S. Acid Mine Drainage (AMD): Causes, treatment and case studies. *J. Clean. Prod.* **2006**, *14*, 1139–1145. [CrossRef]
109. Lottermoser, B. *Mine Wastes*; Springer: Berlin/Heidelberg, Germany, 2003; Volume 7, ISBN 4186530777.
110. Johnson, D. Biodiversity and ecology of acidophilic microorganisms. *FEMS Microbiol. Ecol.* **1998**, *27*, 307–317. [CrossRef]
111. McCarthy, T.S. The impact of acid mine drainage in South Africa. *S. Afr. J. Sci.* **2011**, *107*, 1–7. [CrossRef]
112. Michaud, M.L.; Plante, B.; Bussière, B.; Benzaazoua, M.; Leroux, J. Development of a modified kinetic test using EDTA and citric acid for the prediction of contaminated neutral drainage. *J. Geochem. Explor.* **2017**, *181*, 58–68. [CrossRef]
113. Argane, R.; El Adnani, M.; Benzaazoua, M.; Bouzahzah, H.; Khalil, A.; Hakkou, R.; Taha, Y. Geochemical behavior and environmental risks related to the use of abandoned base-metal tailings as construction material in the upper-Moulouya district, Morocco. *Environ. Sci. Pollut. Res.* **2016**, *23*, 598–611. [CrossRef] [PubMed]
114. Cardoso, P.G.; Pereira, E.; Duarte, A.C.; Azeiteiro, U.M. Temporal characterization of mercury accumulation at different trophic levels and implications for metal biomagnification along a coastal food web. *Mar. Pollut. Bull.* **2014**, *87*, 39–47. [CrossRef]
115. Yarsan, E.; Yipel, M. The Important Terms of Marine Pollution "Biomarkers and Biomonitoring, Bioaccumulation, Bioconcentration, Biomagnification". *J. Mol. Biomarkers Diagn.* **2013**, *S1*, 1–4. [CrossRef]
116. Cárdenas, J.P.; Quatrini, R.; Holmes, D.S. Genomic and metagenomic challenges and opportunities for bioleaching: A mini-review. *Res. Microbiol.* **2016**, *167*, 529–538. [CrossRef]
117. Moodley, I.; Sheridan, C.M.; Kappelmeyer, U.; Akcil, A. Environmentally sustainable acid mine drainage remediation: Research developments with a focus on waste/by-products. *Miner. Eng.* **2018**, *126*, 207–220. [CrossRef]
118. Issaad, M.; Boutaleb, A.; Kolli, O.; Edahbi, M.; Benzaazoua, M.; Hakkou, R. Environmental characterization of mine waste at the Pb–Zn Sidi Kamber abandoned mine (NE Algeria). *Rendiconti Lincei* **2019**, *30*, 427–441. [CrossRef]
119. Sajjad, W.; Zheng, G.; Zhang, G.; Ma, X.; Xu, W.; Ali, B.; Rafiq, M. Diversity of Prokaryotic Communities Indigenous to Acid Mine Drainage and Related Rocks from Baiyin Open-Pit Copper Mine Stope, China. *Geomicrobiol. J.* **2018**, *35*, 580–600. [CrossRef]
120. Plaza-Cazón, J.; Benítez, L.; Murray, J.; Kirschbaum, P.; Donati, E. Influence of Extremophiles on the Generation of Acid Mine Drainage at the Abandoned Pan de azúcar mine (Argentina). *Microorganisms* **2021**, *9*, 281. [CrossRef] [PubMed]
121. Favas, P.J.C.; Sarkar, S.K.; Rakshit, D.; Venkatachalam, P.; Prasad, M.N.V. Acid Mine Drainages from Abandoned Mines: Hydrochemistry, Environmental Impact, Resource Recovery, and Prevention of Pollution. In *Environmental Materials and Waste*; Academic Press: Cambridge, MA, USA, 2016; pp. 413–462.
122. Török, Z.; Modoi, O.C.; Roba, C.; Ozunu, A. Environmental risks due to heavy metal pollution of water resulted from mining wastes in NW Romania. *Environ. Eng. Manag. J.* **2014**, *13*, 2325–2336. [CrossRef]
123. Kolo, M.T.; Amin, Y.M.; Khandaker, M.U.; Abdullah, W.H.B. Radionuclide concentrations and excess lifetime cancer risk due to gamma radioactivity in tailing enriched soil around Maiganga coal mine, Northeast Nigeria. *Int. J. Radiat. Res.* **2017**, *15*, 71–80. [CrossRef]
124. Blight, G. Mine Waste: A Brief Overview of Origins, Quantities, and Methods of Storage. In *Waste*; Academic Press: Cambridge, MA, USA, 2011; pp. 77–88. [CrossRef]
125. Malatse, M.; Ndlovu, S. The viability of using the Witwatersrand gold mine tailings for brickmaking. *J. S. Afr. Inst. Min. Metall.* **2015**, *115*, 321–327. [CrossRef]
126. Beulah, M.; Sudhir, M.R.; Mohan, M.K.; Gayathri, G.; Jain, D. Mine Waste-Based Next Generation Bricks: A Case Study of Iron Ore Tailings, Red Mudand GGBS Utilization in Bricks. *Adv. Mater. Sci. Eng.* **2021**, *2021*, 9499613. [CrossRef]
127. Gou, M.; Zhou, L.; Then, N.W.Y. Utilization of tailings in cement and concrete: A review. *Sci. Eng. Compos. Mater.* **2019**, *26*, 449–464. [CrossRef]
128. Lyashenko, V.; Topolnij, F.; Dyatchin, V. Development of technologies and technical means for storage of waste processing of ore raw materials in the tailings dams. *Technol. Audit Prod. Reserv.* **2019**, *5*, 33–40. [CrossRef]
129. Snyders, C.A.; Akdogan, G.; Bradshaw, S.M.; van Vreden, J.H.; Smith, R. The development of a caustic pre-leaching step for the recovery of Au from a refractory ore tailings heap. *Miner. Eng.* **2018**, *121*, 23–30. [CrossRef]
130. Anjum, F.; Shahid, M.; Akcil, A. Biohydrometallurgy techniques of low grade ores: A review on black shale. *Hydrometallurgy* **2012**, *117–118*, 1–12. [CrossRef]
131. Mulligan, C.N.; Kamali, M. Bioleaching of copper and other metals from low-grade oxidized mining ores by *Aspergillus niger*. *J. Chem. Technol. Biotechnol.* **2003**, *78*, 497–503. [CrossRef]
132. Nguyen, V.K.; Lee, J.-U. A comparison of microbial leaching and chemical leaching of arsenic and heavy metals from mine tailings. *Biotechnol. Bioprocess Eng.* **2015**, *20*, 91–99. [CrossRef]
133. Bayat, B.; Sari, B. Comparative evaluation of microbial and chemical leaching processes for heavy metal removal from dewatered metal plating sludge. *J. Hazard. Mater.* **2010**, *174*, 763–769. [CrossRef]
134. Szolucha, M.M.; Chmielewski, A.G. A comparison of uranium recovery from low-grade ore by bioleaching and acid leaching. *Physicochem. Probl. Miner. Process.* **2017**, *53*, 136–149. [CrossRef]
135. Johnson, D.B. Biomining-biotechnologies for extracting and recovering metals from ores and waste materials. *Curr. Opin. Biotechnol.* **2014**, *30*, 24–31. [CrossRef]

136. Pradhan, D. Bioleaching of Cu and Zn from Complex Sulfide using an Isolated Iron Oxidizing Bacteria. *Lett. Appl. NanoBioSci.* **2020**, *10*, 1825–1832. [CrossRef]
137. Clark, M.; Batty, J.; van Buuren, C.; Dew, D.; Eamon, M. Biotechnology in minerals processing: Technological breakthroughs creating value. *Hydrometallurgy* **2006**, *83*, 3–9. [CrossRef]
138. Christel, S.; Dopson, M.; Vera, M.; Sand, W.; Herold, M.; Wilmes, P.; Buetti-Dinh, A.; Pivkin, I.; Trötschel, C.; Poetsch, A.; et al. Systems Biology of Acidophile Biofilms for Efficient Metal Extraction. *Adv. Mater. Res.* **2015**, *1130*, 312–315. [CrossRef]
139. Amin, M.M.; Elaassy, I.; El-Feky, M.G.; Kawady, N.A.; Talaat, M.S.; Sallam, A.S.M. Recovery of uranium from low-grade ore using microorganism isolated from uraniferous rock sample. *Sep. Sci. Technol.* **2018**, *53*, 2232–2237. [CrossRef]
140. Ghosh, S.; Bal, B.; Das, A.P. Enhancing Manganese Recovery from Low-Grade Ores by Using Mixed Culture of Indigenously Isolated Bacterial Strains. *Geomicrobiol. J.* **2018**, *35*, 242–246. [CrossRef]
141. Valix, M.; Usai, F.; Malik, R. Fungal bio-leaching of low grade laterite ores. *Miner. Eng.* **2001**, *14*, 197–203. [CrossRef]
142. Mulligan, C.N.; Kamali, M.; Gibbs, B.F. Bioleaching of heavy metals from a low-grade mining ore using Aspergillus niger. *J. Hazard. Mater.* **2004**, *110*, 77–84. [CrossRef] [PubMed]
143. Ghosh, A.; Pandey, B.D. Bioleaching of low grade granitic chalcopyrite ore by hyperthermophiles: Elucidation of kinetics-mechanism. *Metall. Res. Technol.* **2015**, *112*, 506. [CrossRef]
144. Ye, M.; Li, G.; Yan, P.; Ren, J.; Zheng, L.; Han, D.; Sun, S.; Huang, S.; Zhong, Y. Removal of metals from lead-zinc mine tailings using bioleaching and followed by sulfide precipitation. *Chemosphere* **2017**, *185*, 1189–1196. [CrossRef] [PubMed]
145. Park, J.; Han, Y.; Lee, E.; Choi, U.; Yoo, K.; Song, Y.; Kim, H. Bioleaching of highly concentrated arsenic mine tailings by *Acidithiobacillus ferrooxidans*. *Sep. Purif. Technol.* **2014**, *133*, 291–296. [CrossRef]
146. Ming, Z.; Ya-Na, L.; Shu-Fa, Z.; Juan, M.; Tie-You, D. Removal of Cu, Zn and Pb from mine tailings by bioleaching: Effects of initial pH. *Int. J. Environ. Stud.* **2012**, *69*, 616–624. [CrossRef]
147. Li, H.M.; Ke, J. Influence of Cu^{2+} and Mg^{2+} on the growth and activity of Ni^{2+} adapted Thiobacillus ferrooxidans. *Miner. Eng.* **2001**, *14*, 113–116. [CrossRef]
148. Olson, G.J.; Brierley, J.A.; Brierley, C.L. Bioleaching review part B: Progress in bioleaching: Applications of microbial processes by the minerals industries. *Appl. Microbiol. Biotechnol.* **2003**, *63*, 249–257. [CrossRef]
149. Ehrlich, H.L. Beginnings of rational bioleaching and highlights in the development of biohydrometallurgy: A brief history. *Eur. J. Miner. Process. Environ. Prot.* **2004**, *4*, 1303–1868.
150. Colmer, A.R.; Hinkle, M.E. The role of microorganisms in acid mine drainage: A preliminary report. *Science* **1947**, *106*, 253–256. [CrossRef] [PubMed]
151. González-Toril, E.; Gómez, F.; Malki, M.; Amils, R. 20 The Isolation and Study of Acidophilic Microorganisms. *Methods Microbiol.* **2006**, *35*, 471–510. [CrossRef]
152. Hiraishi, A.; Nagashima, K.V.P.; Matsuura, K.; Shimada, K.; Takaichi, S.; Wakao, N.; Katayama, Y. Phylogeny and photosynthetic features of Thiobacillus acidophilus and related acidophilic bacteria: Its transfer to the genus Acidiphilium as Acidiphilium acidophilum comb. nov. *Int. J. Syst. Bacteriol.* **1998**, *48*, 1389–1398. [CrossRef] [PubMed]
153. Clark, D.A.; Norris, P.R. *Acidimicrobium ferrooxidans* gen. nov., sp. nov.: Mixed-culture ferrous iron oxidation with *Sulfobacillus* species. *Microbiology* **1996**, *142*, 785–790. [CrossRef]
154. Vera, M.; Schippers, A.; Sand, W. Progress in bioleaching: Fundamentals and mechanisms of bacterial metal sulfide oxidation—Part A. *Appl. Microbiol. Biotechnol.* **2013**, *97*, 7529–7541. [CrossRef]
155. Adeleke, R.A. Getting rid of the unwanted: Highlights of developments and challenges of biobeneficiation of iron ore minerals—A review. *J. Ind. Microbiol. Biotechnol.* **2014**, *41*, 1731–1741. [CrossRef]
156. Gikas, P. Kinetic responses of activated sludge to individual and joint nickel (Ni(II)) and cobalt (Co(II)): An isobolographic approach. *J. Hazard. Mater.* **2007**, *143*, 246–256. [CrossRef]
157. Ramos-Zúñiga, J.; Gallardo, S.; Martínez-Bussenius, C.; Norambuena, R.; Navarro, C.A.; Paradela, A.; Jerez, C.A. Response of the biomining *Acidithiobacillus ferrooxidans* to high cadmium concentrations. *J. Proteom.* **2019**, *198*, 132–144. [CrossRef]
158. Sampson, M.; Phillips, C. Influence of base metals on the oxidising ability of acidophilic bacteria during the oxidation of ferrous sulfate and mineral sulfide concentrates, using mesophiles and moderate thermophiles. *Miner. Eng.* **2001**, *14*, 317–340. [CrossRef]
159. Chen, B.-Y.; Liu, H.-L.; Chen, Y.-W.; Cheng, Y.-C. Dose-response assessment of metal toxicity upon indigenous *Thiobacillus thiooxidans* BC1. *Process Biochem.* **2004**, *39*, 737–748. [CrossRef]
160. Pradhan, D.; Ahn, J.-G.; Kim, D.-J.; Lee, S.-W. Effect of Ni^{2+}, V^{4+} and Mo^{6+} concentration on iron oxidation by *Acidithiobacillus ferrooxidans*. *Korean J. Chem. Eng.* **2009**, *26*, 736–741. [CrossRef]
161. Siezen, R.J.; Wilson, G. Bioleaching genomics. *Microb. Biotechnol.* **2009**, *2*, 297–303. [CrossRef] [PubMed]
162. Rouch, D.A.; Lee, B.T.O.; Morby, A.P. Understanding cellular responses to toxic agents: A model for mechanism-choice in bacterial metal resistance. *J. Ind. Microbiol. Biotechnol.* **1995**, *14*, 132–141. [CrossRef]
163. Elzeky, M.; Attia, Y. Effect of bacterial adaptation on kinetics and mechanisms of bioleaching ferrous sulfides. *Chem. Eng. J. Biochem. Eng. J.* **1995**, *56*, B115–B124. [CrossRef]
164. Kim, D.-J.; Pradhan, D.; Ahn, J.-G.; Lee, S.-W. Enhancement of metals dissolution from spent refinery catalysts using adapted bacteria culture—Effects of pH and Fe(II). *Hydrometallurgy* **2010**, *103*, 136–143. [CrossRef]
165. Xia, L.; Liu, X.; Zeng, J.; Yin, C.; Gao, J.; Liu, J.; Qiu, G. Mechanism of enhanced bioleaching efficiency of *Acidithiobacillus ferrooxidans* after adaptation with chalcopyrite. *Hydrometallurgy* **2008**, *92*, 95–101. [CrossRef]

166. Cardenas, J.P.; Valdés, J.; Quatrini, R.; Duarte, F.; Holmes, D.S. Lessons from the genomes of extremely acidophilic bacteria and archaea with special emphasis on bioleaching microorganisms. *Appl. Microbiol. Biotechnol.* **2010**, *88*, 605–620. [CrossRef]
167. Kassim, M.A.; Hashim, N.H.F.; Yusof, N.A. Archaea Domain as Biocatalyst in Environmental Biotechnology and Industrial Applications: A Review. *J. Adv. Microbiol.* **2017**, *5*, 1–21. [CrossRef]
168. Nguyen, V.K.; Lee, M.H.; Park, H.J.; Lee, J.-U. Bioleaching of arsenic and heavy metals from mine tailings by pure and mixed cultures of Acidithiobacillus spp. *J. Ind. Eng. Chem.* **2015**, *21*, 451–458. [CrossRef]
169. Remonsellez, F.; Galleguillos, F.; Moreno-Paz, M.; Parro, V.; Acosta, M.; Demergasso, C. Dynamic of active microorganisms inhabiting a bioleaching industrial heap of low-grade copper sulfide ore monitored by real-time PCR and oligonucleotide prokaryotic acidophile microarray. *Microb. Biotechnol.* **2009**, *2*, 613–624. [CrossRef] [PubMed]
170. Wang, Y.; Su, L.; Zhang, L.; Zeng, W.; Wu, J.; Wan, L.; Qiu, G.; Chen, X.; Zhou, H. Bioleaching of chalcopyrite by defined mixed moderately thermophilic consortium including a marine acidophilic halotolerant bacterium. *Bioresour. Technol.* **2012**, *121*, 348–354. [CrossRef] [PubMed]
171. Zhang, R.-B.; Wei, M.-M.; Ji, H.-G.; Chen, X.-H.; Qiu, G.-Z.; Zhou, H.-B. Application of real-time PCR to monitor population dynamics of defined mixed cultures of moderate thermophiles involved in bioleaching of chalcopyrite. *Appl. Microbiol. Biotechnol.* **2009**, *81*, 1161–1168. [CrossRef]
172. Liljeqvist, M.; Rzhepishevska, O.I.; Dopson, M. Gene Identification and Substrate Regulation Provide Insights into Sulfur Accumulation during Bioleaching with the Psychrotolerant Acidophile Acidithiobacillus ferrivorans. *Appl. Environ. Microbiol.* **2013**, *79*, 951–957. [CrossRef] [PubMed]
173. Yarzábal, A.; Appia-Ayme, C.; Ratouchniak, J.; Bonnefoy, V. Regulation of the expression of the *Acidithiobacillus ferrooxidans* rus operon encoding two cytochromes c, a cytochrome oxidase and rusticyanin. *Microbiology* **2004**, *150*, 2113–2123. [CrossRef] [PubMed]
174. Valdés, J.; Cárdenas, J.; Quatrini, R.; Esparza, M.; Osorio, H.; Duarte, F.; Lefimil, C.; Sepulveda, R.; Jedlicki, E.; Holmes, D. Comparative genomics begins to unravel the ecophysiology of bioleaching. *Hydrometallurgy* **2010**, *104*, 471–476. [CrossRef]
175. Gericke, M. Review of the role of microbiology in the design and operation of heap bioleaching processes. *J. S. Afr. Inst. Min. Metall.* **2012**, *112*, 1005–1012.
176. Ullrich, S.R.; Poehlein, A.; Tischler, J.S.; González, C.; Ossandon, F.J.; Daniel, R.; Holmes, D.; Schlömann, M.; Mühling, M. Genome Analysis of the Biotechnologically Relevant Acidophilic Iron Oxidising Strain JA12 Indicates Phylogenetic and Metabolic Diversity within the Novel Genus "Ferrovum". *PLoS ONE* **2016**, *11*, e0146832. [CrossRef]
177. Travisany, D.; Cortés, M.P.; Latorre, M.; Di Genova, A.; Budinich, M.; Bobadilla-Fazzini, R.A.; Parada, P.; González, M.; Maass, A. A new genome of Acidithiobacillus thiooxidans provides insights into adaptation to a bioleaching environment. *Res. Microbiol.* **2014**, *165*, 743–752. [CrossRef]
178. Justice, N.B.; Norman, A.; Brown, C.T.; Singh, A.; Thomas, B.C.; Banfield, J.F. Comparison of environmental and isolate Sulfobacillus genomes reveals diverse carbon, sulfur, nitrogen, and hydrogen metabolisms. *BMC Genom.* **2014**, *15*, 1–17. [CrossRef]
179. Chen, L.; Ren, Y.; Lin, J.; Liu, X.; Pang, X.; Lin, J. Acidithiobacillus caldus Sulfur Oxidation Model Based on Transcriptome Analysis between the Wild Type and Sulfur Oxygenase Reductase Defective Mutant. *PLoS ONE* **2012**, *7*, e39470. [CrossRef] [PubMed]
180. Latorre, M.; Cortés, M.P.; Travisany, D.; Di Genova, A.; Budinich, M.; Reyes-Jara, A.; Hödar, C.; González, M.; Parada, P.; Bobadilla-Fazzini, R.A.; et al. The bioleaching potential of a bacterial consortium. *Bioresour. Technol.* **2016**, *218*, 659–666. [CrossRef] [PubMed]
181. Ter Kuile, B.H.; Westerhoff, H. Transcriptome meets metabolome: Hierarchical and metabolic regulation of the glycolytic pathway. *FEBS Lett.* **2001**, *500*, 169–171. [CrossRef]
182. Millard, P.; Smallbone, K.; Mendes, P. Metabolic regulation is sufficient for global and robust coordination of glucose uptake, catabolism, energy production and growth in Escherichia coli. *PLOS Comput. Biol.* **2017**, *13*, e1005396. [CrossRef] [PubMed]
183. Jafari, M.; Chelgani, S.C.; Shafaie, S.; Abdollahi, H.; Hadavandi, E. Study effects of conventional flotation reagents on bioleaching of zinc sulfide. *J. Ind. Eng. Chem.* **2019**, *78*, 364–371. [CrossRef]
184. Dehghan, R.; Dianati, M. The effects of Pb-Zn flotation reagents on the bioleaching process by mesophilic bacteria. *Int. J. Miner. Process.* **2015**, *143*, 80–86. [CrossRef]
185. Dong, Y.; Lin, H. Influences of flotation reagents on bioleaching of chalcopyrite by *Acidthiobacillus ferrooxidans*. *Miner. Eng.* **2012**, *32*, 27–29. [CrossRef]
186. Ageeva, S.N.; Kondrat'Eva, T.F.; Karavaiko, G.I. Phenotypic Characteristics of *Thiobacillus ferrooxidans* Strains. *Microbiology* **2001**, *70*, 186–194. [CrossRef]
187. Jafari, M.; Shafaei, S.Z.; Abdollahi, H.; Gharabaghi, M.; Chelgani, S.C. Effect of Flotation Reagents on the Activity ofL. Ferrooxidans. *Miner. Process. Extr. Metall. Rev.* **2018**, *39*, 34–43. [CrossRef]
188. Zhang, C.-G.; Xia, J.-L.; Zhang, R.-Y.; Peng, A.-A.; Nie, Z.-Y.; Qiu, G.-Z. Comparative study on effects of Tween-80 and sodium isobutyl-xanthate on growth and sulfur-oxidizing activities of Acidithiobacillus albertensis BY-05. *Trans. Nonferrous Met. Soc. China* **2008**, *18*, 1003–1007. [CrossRef]

189. Zeng, G.-M.; Shi, J.-G.; Yuan, X.-Z.; Liu, J.; Zhang, Z.-B.; Huang, G.-H.; Li, J.-B.; Xi, B.-D.; Liu, H.-L. Effects of Tween 80 and rhamnolipid on the extracellular enzymes of Penicillium simplicissimum isolated from compost. *Enzym. Microb. Technol.* **2006**, *39*, 1451–1456. [CrossRef]
190. Guo, Z.; Yao, J.; Wang, F.; Yuan, Z.; Bararunyeretse, P.; Zhao, Y. Effect of three typical sulfide mineral flotation collectors on soil microbial activity. *Environ. Sci. Pollut. Res.* **2016**, *23*, 7425–7436. [CrossRef] [PubMed]
191. Pacholewska, M.; Cwalina, B.; Steindor, K. The influence of flotation reagents on sulfur-oxidizing bacteria Acidithiobacillus thiooxidans. *Physicochem. Probl. Miner. Process.* **2008**, *42*, 37–46.
192. Dopson, M.; Sundkvist, J.-E.; Lindström, E.B. Toxicity of metal extraction and flotation chemicals to Sulfolobus metallicus and chalcopyrite bioleaching. *Hydrometallurgy* **2006**, *81*, 205–213. [CrossRef]
193. Okibe, N.; Johnson, D.B. Toxicity of flotation reagents to moderately thermophilic bioleaching microorganisms. *Biotechnol. Lett.* **2002**, *24*, 2011–2016. [CrossRef]
194. Tuovinen, O.H. Inhibition of *Thiobacillus ferrooxidans* by mineral flotation reagents. *Eur. J. Appl. Microbiol. Biotechnol.* **1978**, *5*, 301–304. [CrossRef]
195. Gehrke, T.; Telegdi, J.; Thierry, D.; Sand, W. Importance of Extracellular Polymeric Substances from *Thiobacillus ferrooxidans* for Bioleaching. *Appl. Environ. Microbiol.* **1998**, *64*, 2743–2747. [CrossRef]
196. Dong, Y.; Lin, H.; Xu, X.; Zhou, S. Bioleaching of different copper sulfides by *Acidithiobacillus ferrooxidans* and its adsorption on minerals. *Hydrometallurgy* **2013**, *140*, 42–47. [CrossRef]
197. Hassanshahian, M.; Ghoebani, S. Isolation and Characterization of Iron and Sulfur-Oxidizing Bacteria from Maiduk Copper Mine at Shahrbabk Province in Iran. *Geomicrobiol. J.* **2018**, *35*, 261–265. [CrossRef]
198. Xiang, Y.; Wu, P.; Zhu, N.; Zhang, T.; Liu, W.; Wu, J.; Li, P. Bioleaching of copper from waste printed circuit boards by bacterial consortium enriched from acid mine drainage. *J. Hazard. Mater.* **2010**, *184*, 812–818. [CrossRef]
199. Pathak, A.; Kothari, R.; Vinoba, M.; Habibi, N.; Tyagi, V. Fungal bioleaching of metals from refinery spent catalysts: A critical review of current research, challenges, and future directions. *J. Environ. Manag.* **2021**, *280*, 111789. [CrossRef] [PubMed]
200. Al-Salem, S.M.; Constantinou, A.; Leeke, G.A.; Hafeez, S.; Safdar, T.; Karam, H.J.; Al-Qassimi, M.; Al-Dhafeeri, A.T.; Manos, G.; Arena, U. A review of the valorization and management of industrial spent catalyst waste in the context of sustainable practice: The case of the State of Kuwait in parallel to European industry. *Waste Manag. Res. J.* **2019**, *37*, 1127–1141. [CrossRef] [PubMed]
201. Mouna, H.M.; Baral, S.S. A bio-hydrometallurgical approach towards leaching of lanthanum from the spent fluid catalytic cracking catalyst using *Aspergillus niger*. *Hydrometallurgy* **2019**, *184*, 175–182. [CrossRef]
202. Sibley, S.F. *Overview of Flow Studies for Recycling Metal Commodities in the United States*; US Department of the Interior, US Geological Survey: Reston, VA, USA, 2011.
203. Martínez, P.; Gálvez, S.; Ohtsuka, N.; Budinich, M.; Cortés, M.P.; Serpell, C.; Nakahigashi, K.; Hirayama, A.; Tomita, M.; Soga, T.; et al. Metabolomic study of Chilean biomining bacteria *Acidithiobacillus ferrooxidans* strain Wenelen and *Acidithiobacillus thiooxidans* strain Licanantay. *Metabolomics* **2012**, *9*, 247–257. [CrossRef] [PubMed]
204. Johnson, C.H.; Ivanisevic, J.; Siuzdak, G. Metabolomics: Beyond biomarkers and towards mechanisms. *Nat. Rev. Mol. Cell Biol.* **2016**, *17*, 451–459. [CrossRef] [PubMed]
205. Rawat, M.; Rangarajan, S. Omics approaches for elucidating molecular mechanisms of microbial bioremediation. In *Smart Bioremediation Technologies*; Elsevier Inc.: Amsterdam, The Netherlands, 2019; ISBN 9780128183076.
206. Rochfort, S. Metabolomics reviewed: A new "omics" platform technology for systems biology and implications for natural products research. *J. Nat. Prod.* **2005**, *68*, 1813–1820. [CrossRef] [PubMed]
207. Viant, M.R. Recent developments in environmental metabolomics. *Mol. Biosyst.* **2008**, *4*, 980–986. [CrossRef]
208. Pandey, A.; Tripathi, P.H.; Tripathi, A.H.; Pandey, S.C.; Gangola, S. Omics technology to study bioremediation and respective enzymes. In *Smart Bioremediation Technologies*; Elsevier Inc.: Amsterdam, The Netherlands, 2019; ISBN 9780128183076.
209. Brisson, V.L.; Zhuang, W.Q.; Alvarez-Cohen, L.; Brisson, V.L.; Zhuang, W.Q.; Alvarez-Cohen, L. Metabolomic Analysis Reveals Contributions of Citric and Citramalic Acids to Rare Earth Bioleaching by a *Paecilomyces* Fungus. *Front. Microbiol.* **2020**, *10*, 3008. [CrossRef]

minerals

MDPI

Article

Mineralogical Characteristics of Pegmatite Tailings and Beneficiation Assessment of Pollucite in Recovering Cesium

Yong Chen [1,2,3], Jiankang Wen [1,2,*], Yongsheng Song [1,2], Wenjuan Li [1,2], Shuang Liu [1,2,*] and Ying Liu [1,2]

1 National Engineering Research Center for Environmentally Friendly Metallurgy in Producing Premium Non-ferrous Metals, GRINM Co., Ltd., Beijing 101407, China; chenyong@grinm.com (Y.C.); songyongsheng@grinm.com (Y.S.); liwenjuan@grinm.com (W.L.); rddly@grinm.com (Y.L.)
2 GRINM Resources and Environment Tech. Co., Ltd., Beijing 101407, China
3 General Research Institute of Nonferrous Metals, Beijing 100088, China
* Correspondence: wjk@grinm.com (J.W.); liushuang@grinm.com (S.L.)

Abstract: The technological mineralogical characteristics of cesium-containing minerals in tailings were examined by means of chemical analysis, the energy spectrum analysis method, and MLA (mineral liberation analyzer) to determine the element content, phase analysis, associated mineral components, degree of liberation, particle size, etc. The results showed that the samples mainly contained spodumene, quartz, feldspar, mica, and other minerals. Pollucite was the main cesium-containing mineral in the sample, which had a cesium oxide content that was as high as 34.58%. The mineral content of pollucite in the sample was relatively low—only 1.23%. The pollucite monomer content and the amount of rich intergrowth was 85.25%, and the metal distribution of cesium in the +0.074 mm sample was as high as 87.06%. Spodumene was the main mineral associated with pollucite. The beneficiation evaluation of this tailing sample was conducted using a combined process that integrated desliming, magnetic separation, and froth flotation, and a pollucite concentrate containing 4.45% Cs_2O was obtained with a 63.71 recovery rate. This indicates that little pollucite was removed by means of desliming and magnetic separation before froth flotation recovery, but during the froth flotation stage in spodumene and feldspar, a large pollucite loss was observed. Therefore, to improve pollucite recovery, a pollucite-specific adsorption reagent should be synthesized.

Keywords: feasibility; pollucite; mineralogical characteristics; tailing; froth flotation

Citation: Chen, Y.; Wen, J.; Song, Y.; Li, W.; Liu, S.; Liu, Y. Mineralogical Characteristics of Pegmatite Tailings and Beneficiation Assessment of Pollucite in Recovering Cesium. *Minerals* **2022**, *12*, 541. https://doi.org/10.3390/min12050541

Academic Editors: Shenghua Yin, Leiming Wang and Hyunjung Kim

Received: 31 March 2022
Accepted: 26 April 2022
Published: 27 April 2022

1. Introduction

Cesium is an important rare metal element that has great application prospects in ion rocket engines, in the information industry, in medicine, as a catalyst, and in other contexts [1–3]. At present, the main source of cesium extraction is pollucite, but with the depletion of high-grade cesium resources [4], the clean and efficient utilization of low-grade cesium resources and secondary resources has attracted attention.

World-class TANCO pegmatite, which is located in Bernic Lake, is the most significant rare metal deposit in Manitoba. This highly fractionated lithium–cesium–tantalum(LCT)-type pegmatite contains several distinct mineralogical zones that have historically been mined for Li, Ta, Cs, Rb, and Be ores. Currently, the TANCO mine produces pollucite for the manufacture of cesium formate, which is used for high-density and low-viscosity drilling as well as in the completion fluid that is primarily used in oil and gas wells. The development of the West TMA of the TANCO Mine was approved in 1991 through alterations made to Environment Act License No. 973. The placement of tailings in the West TMA started in 1991 and has continued to the present, with approximately three million tons of tailings having been deposited there to date.

The occurrence state and property state of minerals are the key factors for determining the ore washability, and it is very important to accurately determine the occurrence state of useful minerals and the parameters of the useful minerals in any ore [5,6]. Therefore, the key to the development and utilization of low-grade secondary tailings lies in the mineralogical study of these kinds of tailings using a systematic process to determine the mineral composition of the ore, the dissemination characteristics of the minerals, and the occurrence state of useful minerals in the tailings. Its purpose is to reveal the key factors of valuable mineral recovery to formulate the best recovery process suitable for ore properties.

There are two types of tailings in the TANCO mine; the east part of the mine contains tailings from the smelting and chemical industry, and the west part of the mine contains tailings left by ore dressing. In this paper, a systematic process mineralogy study of the TANCO west tailings was carried out to determine the types of minerals as well as the occurrence and dissemination characteristics of the cesium in the tailings, and an automatic mineral analysis system (Mineral Liberation Analyzer, MLA) was used to determine and analyze the minerals that were found to contain cesium. According to the process mineralogical characteristics of cesium-bearing minerals in the sample and the test pollucite beneficiation results, an evaluation determining pollucite recovery from the sample was carried out. The purpose of this study was to provide suggestions for the rational secondary development and utilization of tailings to recover cesium-containing minerals.

2. Materials and Methods

2.1. Samples

The ore sample was taken from the comprehensive ore sample (1000 kg) of TANCO west tailings. The ore sample was dried naturally, mixed, and divided to obtain the sample (200 kg) required for the test, and the sample was reserved for preparation. The silt particles in the sample were fine, and the overall size of the particles was less than 0.6 mm. Some of the mica flakes had widths that were more than 1.0 mm. Mica, quartz, feldspar, and other minerals were able to be initially identified by the naked eye.

2.2. Methods

The chemical composition of the sample was determined. The major elements were analyzed by means of X-ray fluorescence (XRF) using a sequential EDX-7000 spectrophotometer (Shimazu, Kyoto, Japan). Minor elements were measured by inductively coupled plasma–mass spectroscopy (ICP–MS) (PlasmaQuant MS Q, Analytik Jena AG, Jena, Germany).

Mineralogical characterization was carried out by various methods, including mineral liberation analysis (MLA), X-ray diffraction analysis (XRD), and optical microscopy analysis.

MLA system is an automatic mineral analyzer that is mainly used to accurately test the mineralogical characteristics and parameters of the process, such as mineral composition and content, particle size distribution, degree of liberation, mineral occurrence, the dissemination relationship, etc. The MLA system is composed of a scanning electron microscope (Jeol JSM6510, Japan Electronics Co., Ltd., Tokyo, Japan), an energy dispersive spectrometer (Genesis XM2 EDS, EDAX Inc., Berwyn, IL, USA), and MLA software (Suit2.9, JKTeck, Brisbane, Australia). MLA measurements were performed at 20 kV with a 1.5-micron pixel resolution using the XBSE method. The working principle of MLA is to use backscattered images to distinguish between different phases and to combine these images with the energy dispersive spectrometer to collect and analyze relevant information and then use image analysis technology to calculate and process the obtained data.

XRD patterns were recorded using a Rigaku D/MaxrA rotating anode X-ray diffractometer equipped with a Cu Kα tube and an Ni filter (λ = 0.1542 nm) at a scanning rate of 0.0167°/s.

Optical microscopy analysis was performed on a ZEISS Axioskop 40 microscope.

Samples were taken to determine the feed sample particle size distribution of the ore using a laser particle size instrument (Bettersize 2000, Dandong Baxter Instrument Co., Ltd., Dandong, China).

To prepare the ore for froth flotation, the sample (1000 g) was screened by a sieve with a mesh size of 0.15 mm. The oversize particles were ground to −150 μm and were deslimed afterward. After desliming, magnetic separation was conducted to purify the froth flotation feed by removing the Fe minerals. After magnetic separation, froth flotation was adopted. Froth flotation experiments were conducted using a 1.5 L stirred froth flotation cell. The slurry was stirred at 660 rpm. The solid content in the cell was then adjusted to 33% via the addition of approximately 1.2 L of tap water. In the stage of mica flotation, the pH value of the pulp was 7. After mica flotation, the pH value of the pulp was controlled at 9 with NaOH (AR, Macklin). The collector (details was shown in Table 1) and frother (MIBC, Macklin) were added at the selected dosages. The pH was readjusted before the addition of the reagent to ensure that froth flotation occurred at the target pH during the conditioning period, which was set at 3 min.

Table 1. Collector in different froth flotation states.

Froth Flotation Stage	Collector	Reagent Grade	Reagent Company
Mica	3030C	-	Solvi
Spodumene and Feldspar	Oxidized paraffin soap 731 Odium oleate	Industrial grade	Bei Zhen City Lvfeng Chemical plant, Jinzhou, China
Pollucite	Ether amine T-49	Industrial grade	Yunnan Yuntong Tiefeng ore dressing Pharmaceutical Co., Ltd., Chuxiong, China

Wet screening was carried out using an XSB-88 standard vibration sieve machine.

A wet magnetic separator (XCSQ 50 × 70) manufactured by Wuhan Exploration Machinery Co., Ltd. was used to run the tests. The magnetic field intensity was 1 Tesla, and the magnetic concentration was 20%.

A hydrocyclone (JAX-75) was used for desliming, the pulp concentration was 12% (solid–liquid weight ratio), and the working pressure was 0.4 Mpa.

3. Results and Discussion

3.1. Chemical Composition and Analysis

The sample was analyzed by XRF analysis and multi-element analysis, and the analysis and test results are shown in Tables 2 and 3, respectively. The main chemical components in the sample were SiO_2, Al_2O_3, Na_2O, and K_2O. The Li_2O, Rb_2O, and Cs_2O contents in the sample were 1.16%, 0.34%, and 0.51%, respectively, which are close to the mining grade of pegmatitic granite, and the mining index was able to be reached at the current metal price. Lithium, cesium, and rubidium were the main target elements for recovery in the sample.

Table 2. Semi-quantitative analysis results of XRF.

Element	Rb	Cs	Ta	Ga	Ge	Cu	Zn	Sr
Content, %	0.158	0.286	0.01	0.008	0.001	0.005	0.007	0.003
Element	Zr	Ti	Cr	Fe	Mn	Si	Al	Ca
Content, %	0.001	0.017	0.006	0.373	0.072	33.53	7.133	0.222
Element	Mg	K	Na	P	S	F		
Content, %	0.032	1.6	2.91	0.64	0.01	0.32		

Table 3. Results of chemical multi-element analysis.

Constituent	Li_2O	Rb_2O	Cs_2O	Ta_2O_5	Nb_2O_5	FeO	MnO	F
Content, %	1.16	0.34	0.51	0.037	0.006	0.43	0.069	0.21
Constituent	Al_2O_3	CaO	MgO	K_2O	Na_2O	SiO_2	P	Ig
Content, %	13.52	0.34	0.063	1.93	3.64	74.51	0.48	2.755

Table 4 shows that cesium mainly exists as an independent mineral of pollucite, comprising 83.33% of the total elemental cesium. About 13% of the cesium in the sample was distributed in mica. Therefore, cesium can be fully recovered by recovering pollucite and mica.

Table 4. Results of chemical cesium phase analysis.

Cesium Phase	Cs_2O in Mica	Cs_2O in Pollucite	Cs_2O in Others	Total
Content, %	0.067	0.424	0.018	0.509
Proportion of total cesium, %	13.16	83.30	3.54	100.00

3.2. Mineral Composition and Content of Sample

The XRD results for the samples are shown in Figure 1, and the relative weight contents of the main minerals in the sample are shown in Figure 2. Through microscopic identification, XRD, and MLA, the main minerals in the sample were determined to be quartz, feldspar (mainly albite, followed by orthoclase), spodumene, and mica, followed by hornblende, pollucite, phosphoponite, and apatite, among others. The results show that the main lithium-bearing minerals were spodumene and mica, followed by phosphoponite; the main rubidium-bearing mineral was determined to be mica, followed by feldspar; and the main cesium-bearing mineral was pollucite, followed by mica.

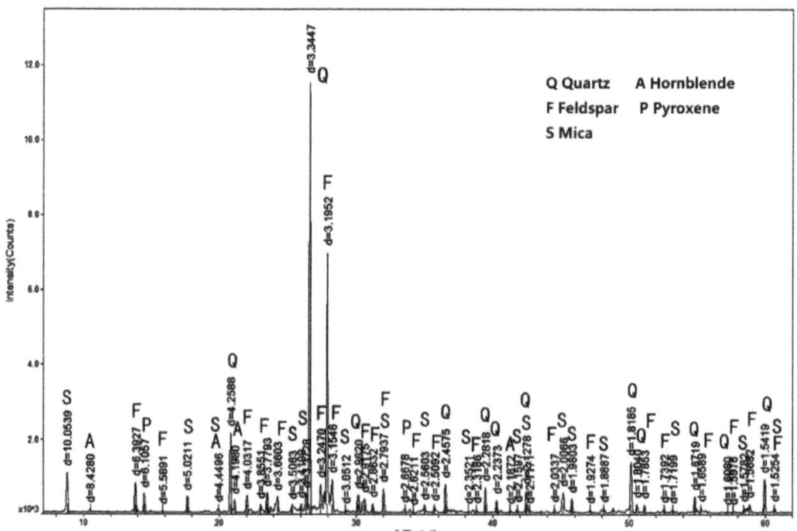

Figure 1. XRD analysis results of the sample.

3.3. Size Characteristics of Main Cesium-Bearing Minerals

The results of the size characteristics of pollucite and mica are shown in Figure 3. In general, pollucite and mica had coarse particles that were less than 0.6 mm in size. In a +0.075 mm sample, mica accounted for 57.74%, and pollucite accounted for more than 70%. In +0.038 mm samples, mica accounted for 83.82%, and pollucite accounted for more than 90%. The metal distribution of cesium in the +0.074 mm sample was as high as 87.06%. Both cesium-containing minerals had coarse particles and demonstrated reasonable froth flotation recovery.

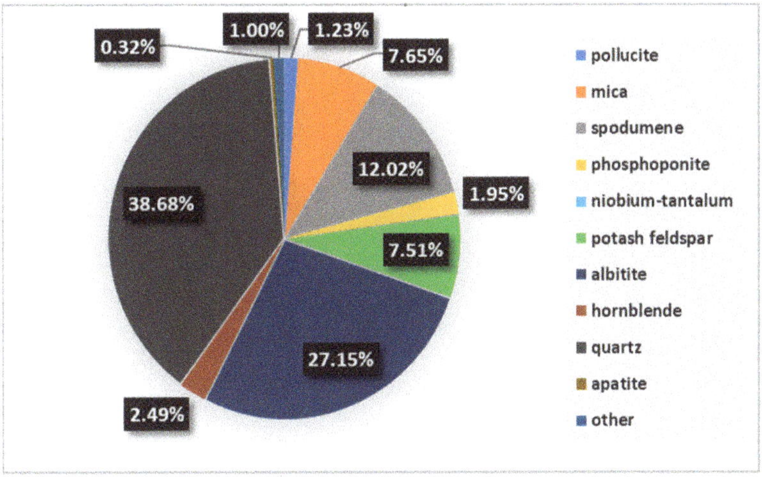

Figure 2. Results determining the mineral composition and content.

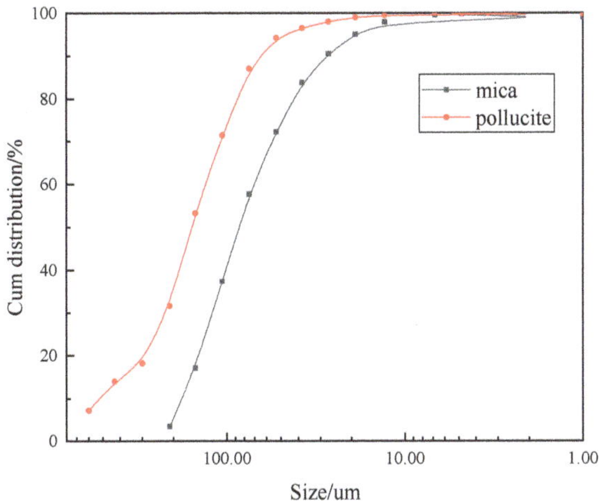

Figure 3. Cumulative distribution curve of particle size of the main cesium-containing minerals.

3.4. Degree of Liberation of Cesium-Bearing Minerals

The results determining the degree of liberation in pollucite and mica are shown in Table 5. The degrees of liberation for mica and pollucite were high. Mica had a higher dissociation degree than pollucite. The sum of the pollucite monomer and rich intergrowth (>3/4) was 85.25%, and that of mica was more than 90%. Judging from the dissemination feature of the minerals, the sample can be separated without grinding.

Table 5. Liberation degrees of pollucite and mica.

Mineral	Monomer, %	Intergrowth, %			
		>3/4	3/4~1/2	1/2~1/4	<1/4
pollucite	76.33	8.92	10.77	2.57	1.41
mica	84.38	6.45	4.86	2.45	1.86

3.5. Cesium Content in Cesium-Bearing Minerals

The results of the energy spectrum micro-domain composition analysis determining the main cesium-bearing minerals are shown in Tables 6 and 7 and Figure 4. The average Cs_2O content in mica was only 0.31%, while the average Cs_2O content in pollucite was 34.58%. Because of the low cesium oxide content in mica, mica is not the main mineral for cesium recovery. There was a relatively high cesium oxide content in pollucite, meaning that a high-grade cesium concentrate can be obtained via pollucite recovery.

Table 6. Energy spectrum micro-domain composition analysis for pollucite.

Number	Content, %				
	Cs_2O	Rb_2O	Na_2O	Al_2O_3	SiO_2
1	34.39	0.00	2.23	17.16	44.73
2	33.56	0.00	2.63	17.27	45.04
4	34.46	0.00	2.62	16.81	44.61
5	33.24	0.00	2.66	17.44	45.15
6	34.15	0.88	2.20	16.63	44.65
7	33.88	0.55	2.35	16.82	44.89
8	33.74	0.00	2.67	16.99	45.10
9	36.86	0.00	1.55	15.29	44.81
Average	34.28	0.18	2.36	16.80	44.87

Table 7. Energy spectrum micro-domain composition analysis of mica.

Number	Content, %				
	Cs_2O	Rb_2O	K_2O	FeO	MnO
1	0.12	2.55	11.09	0.37	0.05
2	0.16	2.69	10.09	0.81	0.08
3	0.59	4.08	9.19	0.25	0.88
4	0.08	3.39	9.92	0.58	0.06
5	0.05	2.24	10.70	0.96	0.22
6	0.72	3.10	10.38	0.99	0.37
7	0.16	2.40	10.48	1.26	0.05
8	0.15	2.28	10.99	1.29	0.05
9	0.30	3.54	9.11	2.43	0.11
Average	0.25	2.84	10.36	0.81	0.22

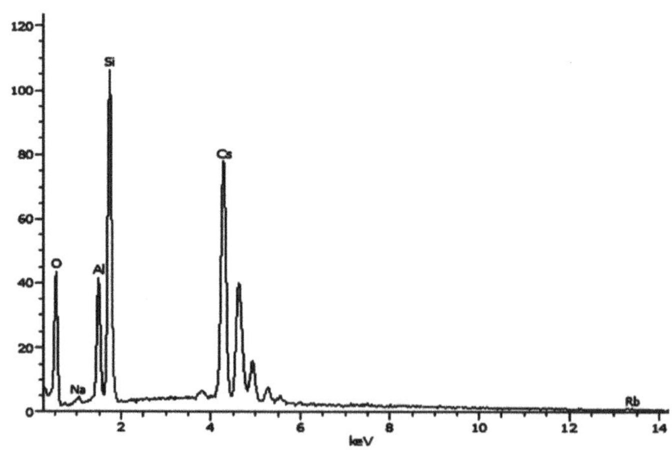

Figure 4. X-ray energy spectrum composition diagram for pollucite.

3.6. Occurrence of Cesium-Bearing Minerals

The embedded characteristics results showing the relationship between cesium-bearing minerals and other minerals are shown in Figures 5–7. Pollucite was an independent mineral of cesium in the samples and existed in both granular and irregular fine-grained aggregate forms. Pollucite was found in a wide range of sizes, generally between 30 μm and 300 μm. Most pollucite was single or shows rich intergrowth, and a small amount of pollucite was embedded in other minerals. Undissociated pollucite was closely intercalated with spodumene, and spodumene is commonly found in intercalated minerals.

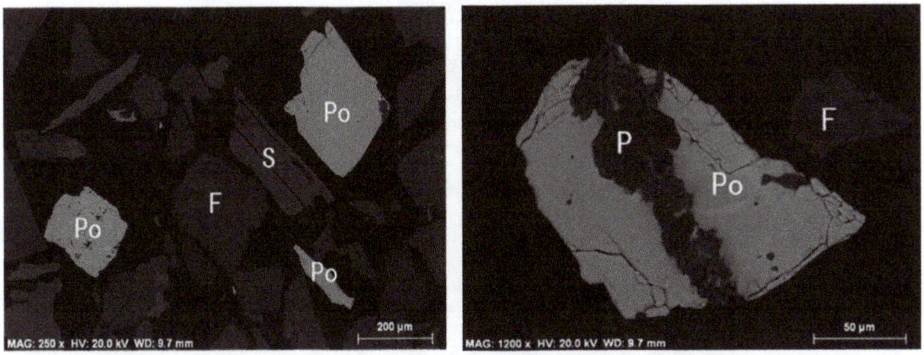

(a) (b)

Figure 5. Characteristics of pollucite (Po), albite (F), spodumene (P), and mica (S): (**a**) pollucite occurs as a monomer; (**b**) intergrowth of coarse-grained pollucite and spodumene.

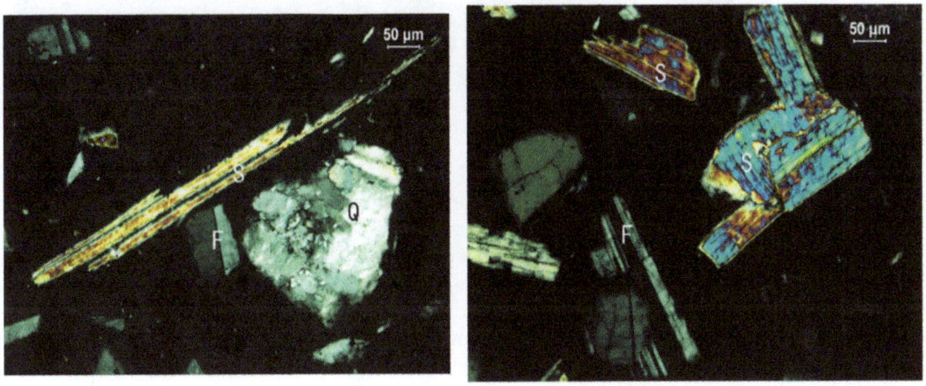

(a) (b)

Figure 6. Orthogonal polarization diagram of mica (S), albite (F), and quartz (Q): (**a**) mica exists in the form of thin strips; (**b**) mica exists in the form of flakes.

Mica is mainly composed of muscovite (lepidolite) and occasionally biotite. Mica usually has a flaky or scaly shape, and flakes are usually less than 0.03 mm thick. Mica can have different particle sizes, with muscovite with a sheet width of more than 1.0 mm being able to be seen with the naked eye, but fine and scaly mica can be microscopic and less than 0.02 mm in size. Coarse-grained mica is mostly found in the monomer state, in which fine feldspar and other minerals are sometimes disseminated; fine lepidolite is often closely interlocked with feldspar, quartz, and other minerals.

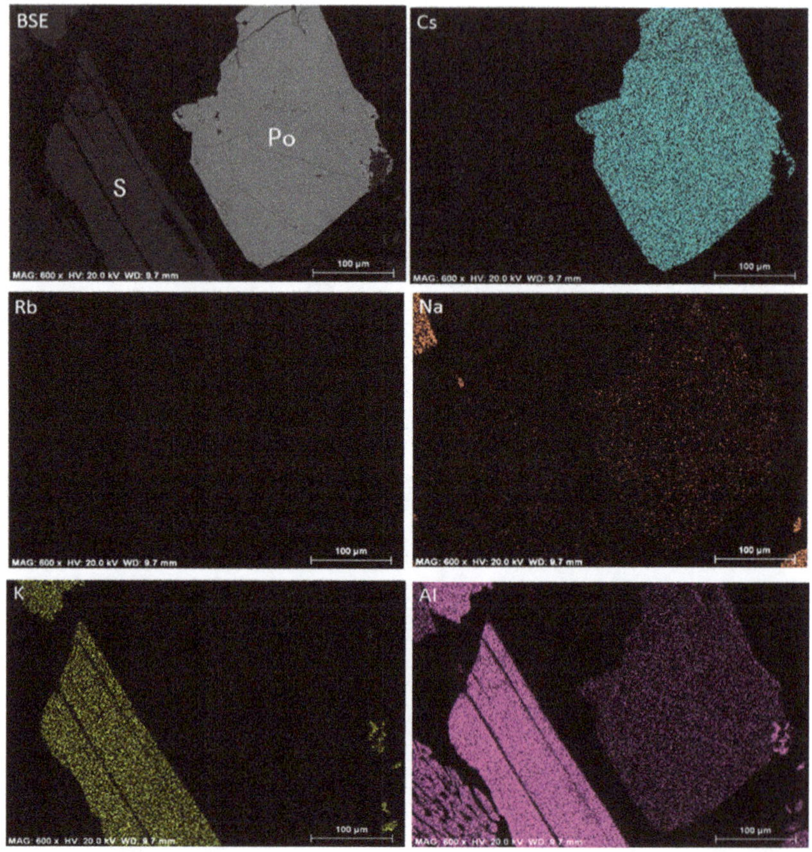

Figure 7. SEM–EDS spectrum of pollucite (Po) and mica (S).

3.7. Pollucite Beneficiation

In the sample, Cs_2O was present at a concentration of 0.51%, and cesium oxide mainly existed in the form of pollucite, with only a small portion existing in the form of mica. Pollucite has a high degree of liberation. The occurrence state of cesium indicates that most cesium can be effectively enriched by froth flotation, and cesium's maximum possible recovery is 80.26%.

In the sample, the cesium oxide content in pollucite was higher, and the particle size of the pollucite in the sample was in a range that is suitable for froth flotation and has a high liberation degree, indicating that this portion of the pollucite can be floated without grinding. However, the coarse pollucite fraction with a low froth flotation efficiency still needs to be ground. During the grinding and froth flotation process, minerals are often contaminated by multivalent cations, which, in turn, affect mineral floatability. The research shows that Ca^{2+}, Mg^{2+}, Al^{3+}, and other cations can activate silicate minerals in specific conditions.

There were some Fe minerals (hematite, limonite, etc.) in the samples, and the Fe minerals needed to be removed by magnetic separation before the pollucite was recovered by means of froth flotation. This is because these Fe minerals will compete with pollucite minerals during the froth flotation process.

In general, it is easy for minerals to react with adequate reagents and under adequate froth flotation conditions without slime, and these minerals have good floatability [7,8]. However, in nature, pegmatite ore is often weathered and easily covered by slime particles, which has a greater impact on froth flotation. During the froth flotation process, a large amount of slime adheres to the surface of useful minerals and gangue minerals, reducing the floatability of useful minerals and improving the floatability of gangue minerals, thus reducing the differences in floatability between them.

When the pulp contains slime and metal ions, it can seriously affect the floatability difference between pollucite and the gangue minerals [9–11]. Therefore, pretreatment should be carried out before the pollucite floatation to remove any impurities and slime on the mineral surface to create favorable conditions for the subsequent froth flotation process.

According to the above analysis and the mineralogical characteristics of the samples, pollucite beneficiation tests were carried out in accordance with Figure 8. During the process, particles larger than 0.15 mm in size were screened out first because spodumene and pollucite intergrowth are common in the +0.15 mm size fraction. Desliming was also essential. Although there was little slime in the sample, the slime had great interference with the ability of silicate minerals to float, so it is suggested to remove the slime before commencing froth flotation. At the same time, the Fe minerals in the sample had a great interference on the froth flotation of silicate minerals, and these could be removed by means of magnetic separation. During the froth flotation stage, the reagent is important, specifically the selectivity of the collector. Because the pollucite, albite, spodumene, and quartz in the samples are silicate minerals, the surfaces of these minerals are very similar. The test results are shown in Table 8. Pollucite concentrate with a Cs_2O grade of 4.45% and a Cs_2O recovery rate of 63.71% was obtained. There was a large gap (about 20 percentage points for pollucite recovery) between the test index and maximum pollucite recovery. Pollucite loss was the largest during the spodumene and feldspar froth flotation stage, which may be because spodumene, feldspar, and pollucite are silicate minerals and because these minerals have similar surface properties. Additionally, the collectors that were used during the floatation process, sodium oleate and 731, have limited selectivity, resulting in the spodumene and feldspar also catching some of the pollucite during the froth flotation stage. In order to obtain a higher pollucite recovery rate, other special pollucite activators or collectors should be explored in the future.

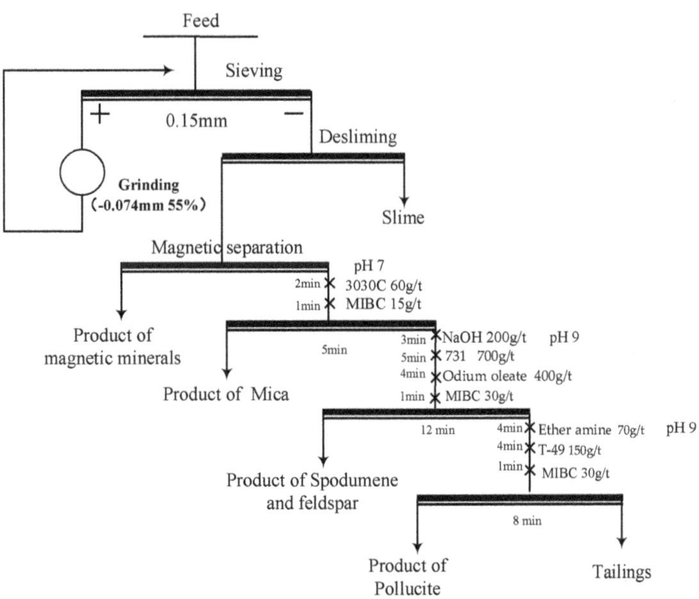

Figure 8. Beneficiation flowsheet for pollucite.

Table 8. Pollucite beneficiation results.

Product	Yield, %	Grade (Cs_2O), %	Recovery (Cs_2O), %
Slime	8.92	0.22	3.88
Fe minerals	3.04	0.21	1.26
Mica	7.21	0.32	4.56
Spodumene and feldspar	43.06	0.22	18.73
Pollucite	7.24	4.45	63.71
Tailings	30.53	0.13	7.85
Feed	100.00	0.51	100.00

4. Conclusions

The cesium oxide content in the sample was 0.5%, which is higher than that of common granite raw ore and allows for comprehensive utilization. The cesium in the sample mainly existed in the form of pollucite and pollucite with a higher liberation degree (the monomer and rich intergrowth was 85.25%), and this is within the appropriate particle size range that is suitable for froth flotation. The maximum cesium recovery rate was determined to be 80.26% according to the process mineralogy analysis.

The analysis of the results of the beneficiation test shows that it is feasible to recover pollucite from tailings. During the pollucite recovery process, it is important to consider the influence of Fe minerals, mud, and metal ions on pollucite recovery when using froth flotation. Before recovering pollucite using froth flotation, it is necessary to eliminate the interference of these factors to the maximum extent possible to create a good froth flotation environment for pollucite floatation. However, pollucite, feldspar, spodumene, and mica are silicate minerals with similar surface properties, and presently, there is no special collector or activator for pollucite froth flotation, indicating that pollucite recovery during the froth flotation process needs to be further improved.

Author Contributions: Y.C. and Y.S. conceived of and designed the experiments; Y.C., J.W., Y.S., W.L. and S.L. performed the experiments; Y.C. and Y.L. analyzed the data; Y.C. and W.L. wrote the paper. All authors have read and agreed to the published version of the manuscript.

Funding: This research was funded by SINOMINE (Hong Kong) Rare Metal Resource Co., Ltd. grant number ZH2020J006.

Acknowledgments: The authors acknowledge SINOMINE (Hong Kong) Rare Metal Resource Co., Ltd. for the samples and financial support, and we express our sincere appreciation to the reviewers for their insightful comments, which have greatly aided us in improving the quality of the paper.

Conflicts of Interest: The authors declare no conflict of interest.

References

1. Koch, E.C. Special materials in pyrotechnics, Part II: Application of cesium and rubidium compounds in pyrotechnics. *J. Pyrotech.* **2002**, *15*, 9–24.
2. Li, J.P.; Xu, S.H. Eye-forming metals, cesium and rubidium. *Chem. World* **2015**, *85*, 108–117.
3. Huang, W.F.; Li, X.D. Application of cesium and research status of extraction and separation technology. *Rare Met. Cem. Carbide* **2003**, *31*, 18–20.
4. U.S. Geological Survey. *Mineral Commodity Summaries*; U.S. Geological Survey: Reston, VA, USA, 2022.
5. Jia, M.X.; Ying, P.; Fu, Q. Discussion on certain problems on the utilization of some kinds of ore from the view of process mineralogy. *Nonferrous Met. Miner. Process. Sect.* **2015**, *2*, 1–4.
6. Dong, D.; Cheng, H.W.; Guo, B.W.; He, L.J. Present situation and prospect of spodumene beneficiation technology. *Prot. Util. Miner. Resour.* **2018**, *4*, 130–134.
7. Yu, F.S.; Wang, Y.H.; Wang, J.M.; Xie, Z.F. First-principle investigation on mechanism of Ca ion activating flotation of spodumene. *Rare Met.* **2014**, *33*, 358–362. [CrossRef]
8. Li, X.D.; Huang, W.F.; Wen, J.L.; Chen, Y. Study on process mineralogy and mineral processing technology of spodumene ore. *Bull. Chin. Ceram. Soc.* **2014**, *33*, 1207–1211.
9. Liu, J.; Gong, G.C.; Han, Y.X.; Zhu, Y. New insights into the adsorption of oleate on cassiterite: A DFT study. *Minerals* **2017**, *7*, 236. [CrossRef]

10. Xu, L.H.; Peng, T.F.; Tian, J.; Lu, Z.; Hu, Y.; Sun, W. Anisotropic surface physicochemical properties of spodumene and albite crystals: Implications for flotation separation. *Appl. Surf. Sci.* **2017**, *426*, 1005–1015. [CrossRef]
11. Chen, J.H.; Chen, Y.; Li, Y.Q. Effect of vacancy defects on electronic properties and activation of sphalerite(110)surface by first principles. *Trans. Nonferrous Met. Soc. China* **2010**, *20*, 502–506. [CrossRef]

minerals

MDPI

Article

Process Mineralogy of the Tailings from Llallagua: Towards a Sustainable Activity

Pura Alfonso [1,*], Miguel Ruiz [2], Rubén Néstor Zambrana [3], Miquel Sendrós [1], Maite Garcia-Valles [4], Hernan Anticoi [5], Nor Sidki-Rius [1] and Antonio Salas [2]

[1] Departament d'Enginyeria Minera, Industrial i TIC, Universitat Politècnica de Catalunya Barcelona Tech, Av. Bases de Manresa 61-63, 08242 Manresa, Barcelona, Spain; msendros@upc.edu (M.S.); nor.sidki@upc.edu (N.S.-R.)

[2] Facultad Nacional de Ingeniería, Universidad Técnica de Oruro, Ciudadela Universitaria, Oruro 0401, Bolivia; miguelruiz@mecanica.edu.bo (M.R.); antoniosalas@metalurgia.edu.bo (A.S.)

[3] Tecnología, Campus Universitario María Barzola, Universidad Nacional Siglo XX, Llallagua 0502, Bolivia; rubennzm@gmail.com

[4] Departament de Mineralogia, Petrologia i Geologia Aplicada, Universitat de Barcelona, Carrer Martí i Franquès, s/n, 08028 Barcelona, Spain; maitegarciavalles@ub.edu

[5] Grupo de Investigación de Ingeniería Cartográfica y Explotación de Minas, Escuela Politécnica de Ingeniería de Minas y Energía, Universidad de Cantabria, Bvd Ronda Rufino Peón, 254, Tanos, 39300 Torrelavega, Spain; hernan.anticoi@upc.edu

* Correspondence: maria.pura.alfonso@upc.edu

Abstract: There are significant tin reserves in the dumps and tailings from Llallagua. Currently, this waste is being processed using gravity concentration or a combination of gravity concentration with a final stage of froth flotation. A process mineralogy study of the tailings and their products after processing in Llallagua was carried out to determine the failings of the processing system in order to contribute to designing an improved new processing scheme. The mineralogy of the feed tailings, concentrate, and final tailings was determined by X-ray diffraction, scanning electron microscopy, and mineral liberation analysis. The tailings were composed of quartz, tourmaline, illite, K-feldspar, plagioclase, cassiterite, rutile, zircon, and monazite. The concentrate essentially contains cassiterite (57.4 wt.%), tourmaline, quartz, hematite, rutile and rare earth minerals, mainly monazite and minor amounts of xenotime and florencite. The concentrate contained 52–60 wt.% of SnO_2 and 0.9–1.3 wt.% REE. The final tailings contained 0.23–0.37 wt.% SnO_2 and 0.02 wt.% of Rare Earth Elements (REE). Only 57.6 wt.% of cassiterite from the concentrate was liberated. The non-liberated cassiterite was mainly associated with quartz, tourmaline, and rutile. The average grain size of monazite was 45 μm and 57.5 wt.% of this was liberated. In other cases, it occurs in mixed particles associated with tourmaline, quartz, cassiterite, and muscovite. To improve the sustainability of this mining activity, the concentrate grade and the metal recovery must be improved. Reducing the particle size reduction of the processed tailings would increase the beneficiation process rates. In addition, the recovery of the REE present in the concentrate as a by-product should be investigated.

Keywords: process mineralogy; tailings; cassiterite; REE; mineral liberation; quantitative mineralogy

Citation: Alfonso, P.; Ruiz, M.; Zambrana, R.N.; Sendrós, M.; Garcia-Valles, M.; Anticoi, H.; Sidki-Rius, N.; Salas, A. Process Mineralogy of the Tailings from Llallagua: Towards a Sustainable Activity. *Minerals* **2022**, *12*, 214. https://doi.org/10.3390/min12020214

Academic Editors: Shenghua Yin and Leiming Wang

Received: 14 December 2021
Accepted: 5 February 2022
Published: 7 February 2022

1. Introduction

At present, much attention is being paid to the inefficiency of mining activities, which produces large amounts of environmental pollution. Approximately 5 to 7 billion tons of mine tailings are generated each year [1]. Their environmental impact is considerable since, in addition to occupying large areas, they can emit potentially toxic elements [2]. For this reason, mining should be carried out in a more sustainable way. Although there are different concepts of what is considered to be sustainable mining [3], it is generally described as mining that minimizes environmental impacts, operates with a social license, produces economic gains, and focuses on safety and resource efficiency [4]. Therefore, sustainable mining must optimize resources in addition to preventing environmental pollution [5,6].

The mining activities related to the Siglo XX mine, located in Llallagua (Bolivia), are a clear example of a low sustainable mining. As a result of the mining operations, the tin reserves from Llallagua are found both in an ore deposit and in the waste produced from the extraction and processing of the primary ores.

The Llallagua tin deposit began to be exploited on a large scale at the beginning of the 20th century with the extraction of high-grade tin ores. This mine, also called La Salvadora, was considered the richest tin mine in the world in the last century. There are no contemporary estimates, but the Llallagua tin production reached more than 500,000 tons between the beginning of the 20th century and 1930s [7], and low-grade resources of approximately 1,000,000 t of tin [7,8] have been reported. Initially, certain veins contained more than 25% cassiterite, but currently, veins contain less SnO_2 [9]. Additionally, 20.7 million tons of dumps and tailings were produced as a result of the Llallagua mine, with an average of 0.3% SnO_2 [10]. According to Villalpando [11], 300,000 t of tin reserves remain in placers, dumps, and tailings in Bolivia.

The residues from Llallagua are reprocessed, and even the newly generated tailings are tin-rich, i.e., of economic interest. This makes this mining area a clear example of an unsustainable operation. A great deal of energy is expended to mine the tin ores with a consequent low recovery rate, generating tailings that can be reprocessed. The environmental costs of generating and handling such large quantities of materials, which are rich in potentially toxic elements, are great. In addition to causing economic losses, low efficiency processes have an impact on environmental pollution, since a low recovery rate indicates that metals remain in the final tailings and, consequently, end up in the environment. The pollution created by the Llallagua waste is significant and, for example, the Catavi river receives As [12].

Currently, the Llallagua mining activities represent an important contribution to the Bolivian economy. The mine is mainly operated through mining cooperatives, the most important being the Multiactiva Catavi Ltd. Cooperative (Llallagua, Bolivia), which has been operating since 1986. Approximately 38% of the population in the area is exclusively dedicated to mining activities.

The aim of this work was to provide an example of unsustainable mining activities and show how processing mineralogy can contribute to identifying weaknesses in processing systems and help to improve them. To this end, process mineralogy was carried out using tailings from the Catavi plant of the Llallagua mine. The study could be used to revise and modify the current system of operation, making it more efficient, improving the recovery rates, and making it more environmentally friendly.

Another important motivation to conduct an exhaustive characterization of processing materials in the Llallagua mine is that they contain abundant rare earth-rich minerals, mainly monazite [13–16]. Rare earth elements (REE) are considered to be essential raw materials, especially in key fields, such as the development of clean technologies [17] Zhou. The extreme dependence on China, which is the main provider of these elements, makes them critical for many countries [18,19]. Thus, it is necessary to extract them from different sources, such as tailings [20], where they can be found in relative abundance and recovered as a by-product from certain mining processes [21].

2. Materials and Methods

2.1. Materials

The investigated tailings are the result of the processing of the run-off-mine materials from the Llallagua ore deposit, in the department of Potosí, Bolivia. This deposit belongs to a major tin district located in the Central Andean tin belt (Figure 1a). The exploited deposit is of the porphyry type and it occurs in association with the quartz-diritic to rhyodacitic La Salvadora stock of Lower Miocene age, intruded in Silurian sedimentary rocks [22,23]. Most parts of this stock are composed of hydrothermal breccia [24] and the mineralization is cassiterite, which occurs in veins and is disseminated in the stock [22,24,25]. The mineralization occurred in two stages: in the first stage, cassiterite pyrrhotite, tourmaline, bismuthinite, fluorapatite, and monazite were formed; in the second, late sulfides were superimposed to the early formed mineralization [22,26].

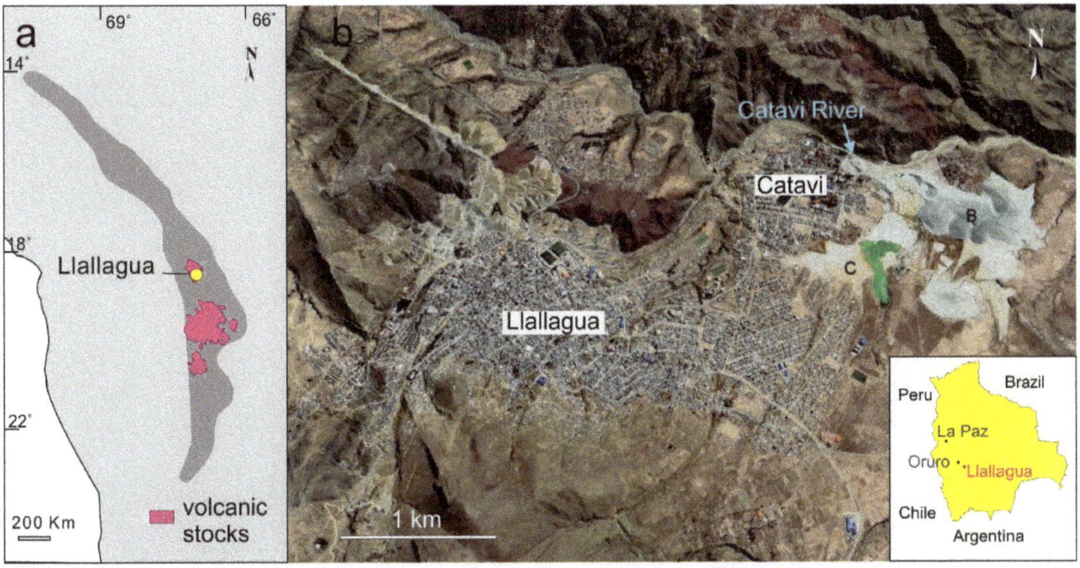

Figure 1. (**a**) Location of Llallagua in the Central Andean tin belt. Data from [27]; (**b**) View of Llallagua city and the mining wastes: A, dumps; B, sand tailings; and C, fine tailings.

Monazite can be very abundant as it is associated with the mineralization of cassiterite veins [15]. The stock is highly altered and almost all of the original feldspars have been transformed. Alteration includes sericitization and tourmalinization [22].

The dumps and tailings produced as a consequence of the exploitation of the Llallagua deposit cover a vast area close to the city (Figure 1b), with an estimation of about 18 Mt of tailings and 1.2 of dumps [10].

The investigated tailings are processed in the C-4 treatment plant located in Catavi, close to Llallagua. The supply of this plant corresponds to three types of materials: (1) those that come directly from the La Salvadora mine. (2) large dumps formed by the removal of waste material in order to reach the mineralized zone; and (3) tailings constituted by sands and fine materials resulted from previous processing of the ores. The plant processes about 250 t/day, obtaining about 1 t of concentrate with 51–57% SnO_2 [10].

At the plant (Figures 2 and 3), the tailing sands are initially introduced into ball mills for size reduction. Then, they are transported to a hydraulic classifier. From here, the +250 µm material is returned to the ball mill and the −250 m material is fed through several shaking tables, separating the high-density minerals from the lighter minerals that constitute the new tailings. Finally, the first concentrate follows two different routes: If it is rich in sulfides, it is subjected to flotation, where these minerals are removed and a final concentrate is obtained. In other cases, to increase the grade of the concentrate, it passes from the shaking tables to a hydrocyclone; from there, it is further ground in a ball mill and finally passed through another shaking table to obtain the final concentrate.

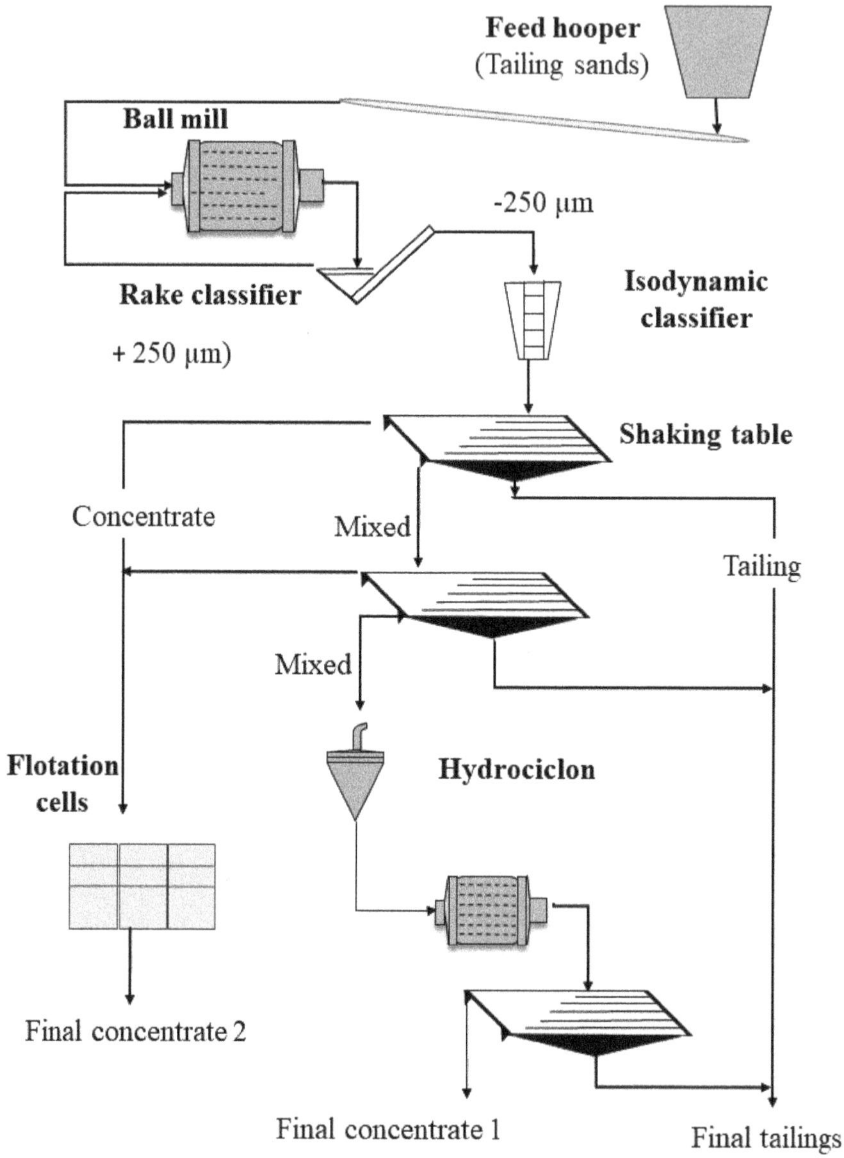

Figure 2. Simplified flux diagram followed by the samples used in this study in the Catavi processing plant.

The residual solid material is sent to waste piles located in the vicinity of the plant, while the liquid is removed using a gutter that empties directly into the river.

The samples used in this study were as follows: the sandy tailings that constitute the feed material of the processing plant; concentrate 1, which had been subjected to flotation; concentrate 2, which had not been floated; an intermediate concentrate; and the final tailings.

Figure 3. View of the Catavi processing plant: (**a**) feed tailings; (**b**) ball mill for the comminution; (**c**), view of one of the three floors with the material being processed in the shaking tables; (**d**), detail of the gravity separation in a shaking table.

2.2. Methods

The chemical composition of major elements and traces was determined. Major elements were analyzed by X-ray fluorescence (XRF) using a sequential X-Ray PW2400 spectrophotometer (Philips, Amsterdam, The Netherlands) located at the Centres Científics i Tecnològics de la Universitat de Barcelona. Traces and Sn were determined at the ALS Laboratories. Minor elements were measured using ICP-MS from the acid digestion of fused glass beads. Sn was determined by XRF. Fusion was obtained using lithium or sodium borate.

The particle size distribution of the tailings and concentrate was determined using a LS 13 320 Particle Size Analyzer (Beckman Coulter, Brea, CA, USA). Before being measured, the samples were treated with sodium pyrophosphate and mechanically agitated for 24 h in order to achieve complete disaggregation.

Mineralogical characterization was carried out by X-ray powder diffraction (XRD), optical and scanning electron microscopy. The XRD spectra were measured from powdered samples in a Bragg–Brentano PANAnalytical X'Pert Diffractometer (graphite monochromator, automatic gap, Cu Kα radiation at λ = 1.54061 Å, powered at 45 kV–40 mA, and a scanning range of 4–100° with a 0.017°2θ step scan and a measuring time of 50 s). Identification and Rietveld semiquantitative evaluation of phases were conducted with the PANanalytical X'Pert HighScore software (PANalytical, Almelo, The Netherlands).

Scanning electron microscopy with energy-dispersive spectral analysis (SEM–EDS) was performed with a Hitachi TM-1000 tabletop electron microscope (EDX, High-Technologies Corporation, Tokyo, Japan).

Automated mineral liberation analysis (MLA) was used to study the particle liberation characteristics. Samples were prepared in thick sections and, to avoid the segregation of particles according to their densities [28], the round thick sections were cut into two vertical slices. Analyses were carried out at the University of Tasmania using a FEI MLA650 (FEI, Hillsboro, OR, USA) environmental scanning electron microscope equipped with a Bruker Quantax Esprit 1.9 EDS system with two XFlash 5030 SDD detectors (Bruker, Berlin, Germany). MLA measurements were performed at 20 kV with a 1.5-micron pixel resolution using the XBSE method, which collects a range of BSE images at a specified resolution, segments the images into different mineral grains based on BSE contrast and textural features, and collects a single ED spectrum in the center of each identified mineral grain. A Mineral Liberation Analysis (MLA) software package v3.1 was used.

3. Results

3.1. Chemical Composition

The chemical composition of major components of dumps and tailings was previously reported [29]. A comparison of the chemistry of the feed tailings, concentrates, and final tailings from the Catavi plant is presented in Table 1. The contents of major elements were similar in the feed and final tailings, i.e., highly rich in SiO_2 and Al_2O_3, with the most abundant oxide from the concentrate being SnO_2. Fe, Ti, and P_2O_5 also were increased in the concentrated fraction.

Table 1. Chemical composition, in wt.%, of the tailings and processing products from the C-4 processing plant.

Material	SiO_2	Al_2O_3	Fe_2O_3	TiO_2	MnO	CaO	K_2O	MgO	Na_2O	P_2O_5	SnO_2	LOI *
Feed	77.75	11.19	2.58	0.44	0.01	0.07	1.43	1.09	0.37	0.19	0.40	1.79
Concentrate	11.42	4.64	5.55	3.52	0.03	0.00	0.22	0.61	0.24	0.72	52.43	3.03
Final tailings	79.20	10.52	2.42	0.43	0.01	0.07	1.20	1.09	0.38	0.15	0.29	1.78

* LOI: loss on ignition.

The Sn content of the feed tailings varied between 0.40 and 1.7 wt.% [30]. After the concentration process using the shaking tables, the concentrate displayed 41.3% and 59.0 wt.% SnO_2. The final tailings still contained relatively high amounts of Sn, between 0.23 and 0.37 wt.% SnO_2. These contents are still of economic interest. The average grade for greisen-type tin deposits is 0.3% Sn [31].

The concentrate contained significant amounts of W, i.e., from 4990 to 6700 ppm, with 70 ppm remaining in the final tailings. The Pb content was 1560 ppm and Zn and Ag occurred in small amounts, i.e., 90 and 38 ppm, respectively.

There were also minor amounts of radioactive elements present, i.e., U in a range from 32 to 41 ppm and Th in a range from 66 and 80 ppm. Both contents were low in the final tailings, i.e., 3 and 6 ppm, respectively.

Indium occurs in relatively high concentrations in certain deposits of the Central Andean tin belt [32,33]. However, in Llallagua, the contents obtained were negligible, reaching only 0.03 ppm in the concentrate. Moreover, Nb and Ta were found in significant amounts in certain tin deposits, since they can replace Sn in the cassiterite structure [34]. However, in the case of the materials from Llallagua, the contents were low, i.e., 17 ppm and 12 ppm, respectively.

Other elements that are currently of great interest are germanium and gallium. In the materials from Llallagua, both were higher in the concentrate than in the final tailings. The concentrate had up to 9 ppm Ge and between 49 and 63 ppm Ga. The latter can be of interest as a by-product as this represents double the average content of Ga in the crust (10–20 ppm) and is similar to the amount found in bauxite ores, from where it is extracted as a by-product [35].

The concentrate contained a significant REE content, ranging between 8800 and more than 12,000 ppm of total REE (Table 2). The most abundant elements were cerium, lanthanum, and neodymium (Figure 4). On the other hand, the total rare earths were only 170 ppm in the final tailings. This indicates that most have been removed to the concentrate, leaving about 2% in the light fraction. An intermediate concentrate was also analyzed and, as expected, it contained fewer rare earths than the final concentrate, which indicates that, at this stage, concentration had not been totally achieved. As expected, the intermediate concentrate had a lower rare earth content than the final concentrate.

Table 2. REE elements in the tailings and processing products from the C-4 plant, in ppm.

Element	Final Tailings	Intermediate Concentrate	Final Concentrate 1	Final Concentrate 2
ppm	LG-2	LG-4	LG-35	LG-41
Ce	74.5	3370	5670	4140
Dy	3.51	56.3	123	83.9
Er	1.72	19.1	47.6	30.1
Eu	1.09	31.5	56	40.3
Gd	3.86	109.5	210	144.5
Ho	0.63	8.5	18.9	13.05
La	37.2	1640	2860	2160
Lu	0.34	1.4	4.16	2.93
Pr	8.86	358	667	431
Nd	40.3	1250	2210	1520
Sm	5.48	185.5	366	251
Tb	0.6	13.3	23.3	16.85
Tm	0.25	2.1	5.81	3.75
Yb	1.74	11.3	32.7	21.8
Σ RRE	170.5	7056.5	12,294.5	8859.18

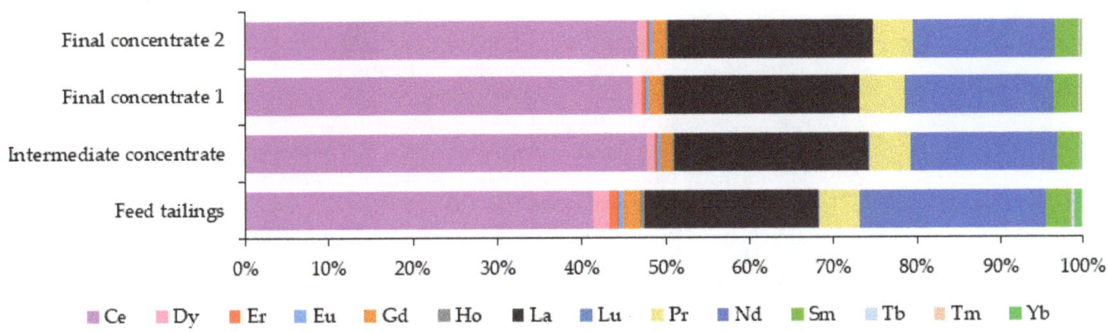

Figure 4. Ratios among the REE contents in the feed tailings and in the processing plant products.

3.2. Mineralogy

3.2.1. Major Minerals: XRD

The feed tailings were mainly composed of quartz and tourmaline of the dravite and schorl types, with minor amounts of micas. The micas were mainly sericite and illite. Muscovite, biotite, and highly altered feldspars were observed under the microscope, but they went undetected in the majority of XRD diffractograms. In the feed tailings, cassiterite was the most important metal oxide, although it was not abundant. In the final tailings, it was not present in sufficient amounts to be determined by XRD. After processing the material with shaking tables, the minerals of the feed tailings were grouped according to their densities, i.e., into a heavy fraction, or concentrate, and a light fraction, or final tailings (Figure 5).

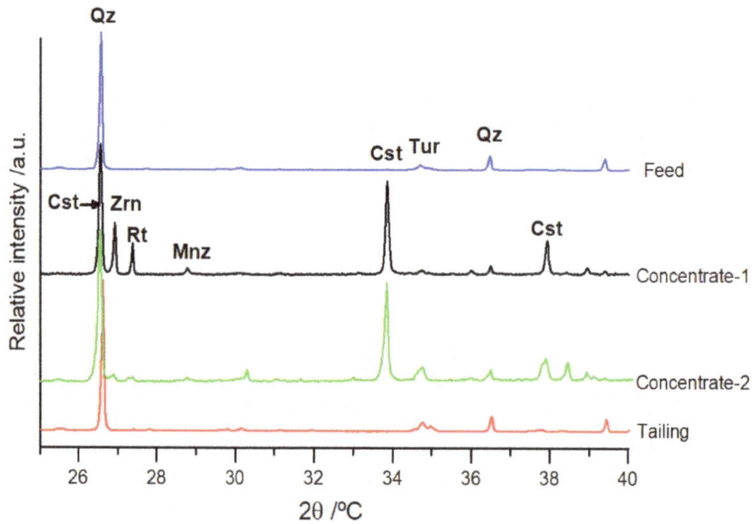

Figure 5. XRD diffractograms of materials from the Catavi processing plant. Qz: quartz; Tur: tourmaline; Cst: cassiterite; Zrn: zircon; Rt: rutile; Mnz: monazite.

Quartz was much more abundant in the final tailings. Tourmaline was also more abundant in the tailings, but with a less significant difference. The average density of tourmaline is 3.10 g/cm^3, whereas that of quartz is 2.6 g/cm^3. The floatability rates indicate that it is more difficult using gravity methods to separate tourmaline into the light fraction. Cassiterite was the predominant mineral in the concentrate. The semiquantitative determination using the Rietveld method provided a cassiterite content of 41–46 wt.%. A low intensity monazite peak can be observed in the concentrate diagram. The rutile and zircon contents were higher in the concentrates, although they can vary considerably according to the route followed during the processing, being up to 9 wt.% rutile and 9 wt.% zircon if the feed tailings were not floated (concentrate 1).

Most of the feed material remained in the final tailings; therefore, they demonstrated a mineralogical composition similar to that of the feed, with lower amounts of heavy minerals, such as cassiterite, rutile, zircon, and monazite.

3.2.2. Modal Mineralogy of the Concentrate

The mineralogy of the concentrate was also determined by MLA, which provided the modal estimation (Table 3). Cassiterite is the most abundant mineral in the concentrate, with 57.41 wt.%. This quantification is in accordance with the results of the chemical analyses; however, it is a contrast with the results obtained from XRD, in which the cassiterite contents were underestimated, probably because there is a superposition of the most intense peak reflections of cassiterite (at d_{hkl} = 3.353 Å) and quartz (at d_{hkl} = 3.345 Å). Zircon and rutile were also enriched in the concentrate. The MLA of concentrate 2 reported 0.69 wt.% of zircon. This is in accordance with the chemical composition results, i.e., 0.41 wt.% of Zr and 99 ppm of Hf, which corresponds to approximately 0.82 wt.% of zircon. In addition to Zr and Hf, this mineral probably contained a significant part of the U and Th determined in the chemical analysis of the concentrate. This mineral was frequently observed in the SEM as euhedral crystals with a homogeneous appearance or a slight zonation. The SEM observation and MLA map shows that in the concentrate, cassiterite only represented half of the components and most cassiterite particles were not completely liberated and contained some proportion of other minerals (Figure 6).

Table 3. Modal mineralogy provided by the MLA of concentrate 2 from Catavi.

Mineral	Formula	wt %
Cassiterite	SnO_2	57.41
Quartz	SiO_2	7.13
Plagioclase	$(Na,Ca)(Si,Al)_4O_8$	0.02
Tourmaline	$Na(Mg,Fe)_3Al_6(BO_3)3Si_6O_{18}(OH)_4$	11.22
Muscovite	$KAl_3Si_3O_{10}(OH)_2$	1.41
Biotite	$KMg_{2.5}Fe^{2+}{}_{0.5}AlSi_3O10(OH)1.75F_{0.25}$	0.32
Chlorite	$(Mg,Fe)_3(Si,Al)_4O_{10}(OH)_2(Mg,Fe)_3(OH)_6$	0.09
Kaolinite	$Al_2Si_2O_5(OH)_4$	0.23
Rutile	TiO_2	4.60
Ilmenite	$Fe^{2+}TiO_3$	0.04
Fe oxides	Fe_2O_3	10.44
Pyrrhotite	FeS	1.28
Pyrite	FeS_2	0.37
Arsenopyrite	FeAsS	0.05
Stannite	SnS	0.01
Galena	PbS	0.01
Crandallite	$CaAl_3(PO_4)_2(OH)_5H_2O$	0.10
Florencite	$CeAl_3(PO_4)_2(OH)_6$	0.16
Plumbogummite	$PbAl_3(PO_4)_2(OH)_5(H_2O)$	0.26
Fe-Pb Phosphate	$(Pb,Fe)_3(PO_4)_2$	0.33
Scorodite	$FeAsO_4.2(H_2O)$	0.15
Jarosite	$KFe_3(SO_4)2(OH)_6$	1.30
Barite	$Ba(SO_4)$	0.14
Xenotime	$Y(PO_4)$	0.04
Monazite	$(La,Ce,Nd,Th)(PO_4)$	1.62
Zircon	$ZrSiO_4$	0.69
Wolframite	$(Fe,Mn)(WO_4)$	0.47
Total		100.00

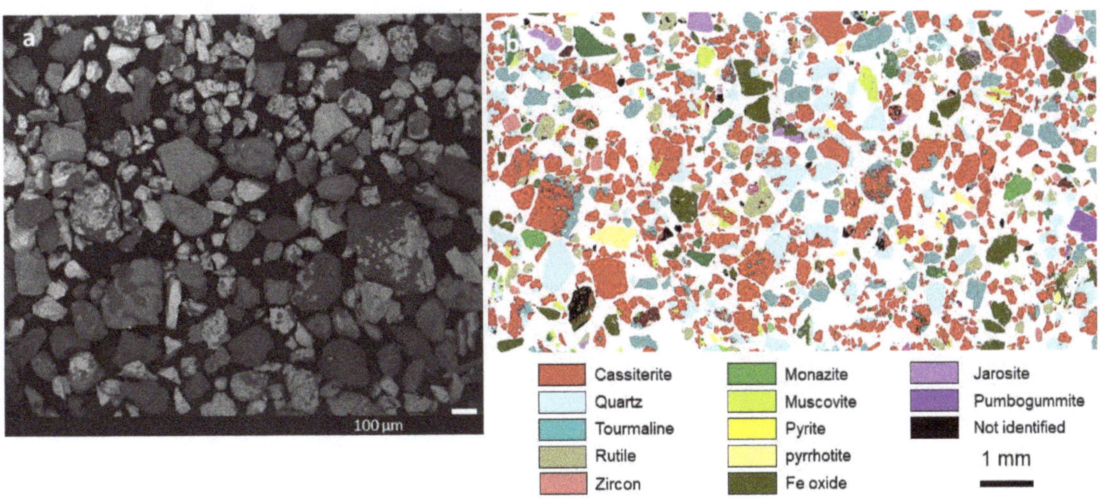

Figure 6. (**a**) SEM images of the concentrate 2 from Catavi; the lightest color is cassiterite. (**b**) False colour MLA map of the concentrate 2.

Iron oxides and hydroxides represented 10.44 wt.% and rutile 4.6 wt.%. Sulfides were not abundant, and pyrite pyrrhotite, arsenopyrite, galena, sphalerite and stannite occurred in minor amounts.

The REE-bearing phosphates were monazite and florencite. Monazite normally appears in idiomorphic crystals, with sizes of less than 100 μm, liberated or associated with tourmaline or quartz. Florencite was occasionally reported from Llallagua [15,26]. This is a crandallite-group REE-bearing phosphate that usually forms from the alteration of monazite [36]. Crandallite and plumbogummite were also determined by the MLA.

Sulfates, such as barite and jarosite, were also present in low amounts. Their presence has been confirmed by SEM observations.

3.3. Particle Size Distribution

The particle size distribution (PSD) data are used to predict the liberation size of ores and to determine the grinding parameters required to obtain an efficient recovery [37]. The PSD of the feed, concentrate, and final tailings was determined by laser analysis and the PSD of concentrate was additionally obtained by MLA. Both methods produced similar distribution curves. This indicates that the sample analyzed by MLA was representative of the different particle sizes. The MLA was expected to give a low particles size due to the stereological error associated with this technique [38,39]. However, in the C-4 plant concentrate, this lower size was not observed.

The particle size distribution shows that, in the feed and products of the Catavi plant, the number of finer particles was low (Figure 7a), i.e., only 5% of the population was less than 100 μm in the feed sands. After milling, the size reduction ratio in this finer phase was negligible. The feed tailings were the coarsest material and they were comminuted in the tumbling mills and the gravity separation produced a finer concentrate and coarser final tailings. This is because quartz, which is the main component of the tailings, is harder than cassiterite, which is the main product of the concentrate, with a Bond ball work index of 14–15 kWh/t in quartz compared to 10–12 kWh/t in cassiterite [40]. On the other hand, this could also be because cassiterite grains are originally smaller than quartz grains. In the concentrate, the grain size determined by MLA shows that the size of quartz was larger than that of cassiterite, monazite, and tourmaline (Figure 7b). Quartz grains reached up to 425 μm in diameter, accounting for 80 wt.% of the total of −145 μm. The cassiterite grain size was −343 μm, and 80 wt.% of the total was −124 μm. Monazite grains were −203 μm in size, with 80 wt.% less than 106 μm.

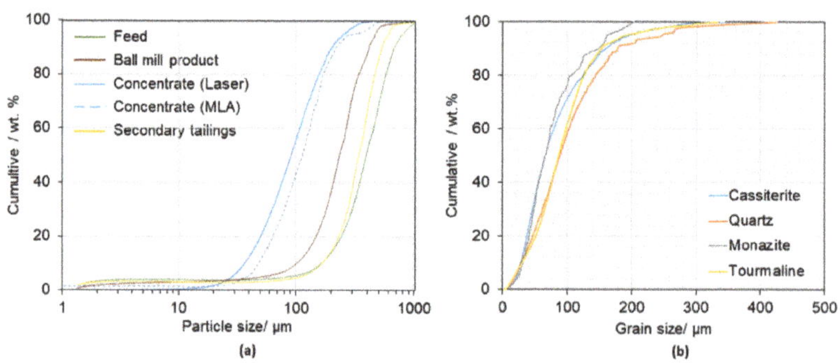

Figure 7. (**a**) Particle size of feed, concentrate and tailings from the Catavi plant, (**b**) Grain size of the minerals from the concentrate.

3.4. Mineral Liberation

The liberation characteristics of the concentrate were determined from the data provided by the MLA. The degree of liberation and the association among minerals could be established. These characteristics were determined in the minerals that constitute or may constitute ores, that is, cassiterite and minerals that contain rare earths, which in this case, were monazite, xenotime, and florencite.

The total number of particles analyzed in the MLA of the concentrate was 49,386 and 85,378 grains. Of these, there were 24,794 grains of cassiterite, 1190 of monazite, 88 of xenotime and 779 of florencite. The liberation characteristics of cassiterite and monazite in the concentrate of the Catavi plant are summarized in Table 4.

Table 4. Characteristics of ore-bearing particles from concentrate 2 of the C-4 plant.

	Cassiterite	Monazite
Total ore (wt %)	57.41	1.62
Number of particles	24,794	1190
Particles contain%	50.20	2.41
Number of liberated particles	19,708	586
Mass liberated respect the total ore %	65.74	63.83
Mass liberated respect the total concentrate %	37.74	1.03

3.4.1. Cassiterite

Cassiterite represented 57.41 wt.% of the total mass of the concentrate and 50.2% of the concentrated particles contained cassiterite. More than half of the cassiterite in the concentrate was liberated, i.e., 65.74 wt.% of the total cassiterite ore, which represents 37.74 wt.% of the total mass of the concentrate. Moreover, 24.28 wt.% of the non-liberated cassiterite grains occurred in binary particles and 9.99 wt.% occurred in ternary particles. As regards being associated with other minerals as binary and ternary particles, cassiterite was mainly in contact with quartz, tourmaline, rutile, and muscovite. This was already demonstrated in the SEM observations (Figures 8 and 9). In addition, the MLA allowed us to quantify these observations and determine the minor associations (Table 3).

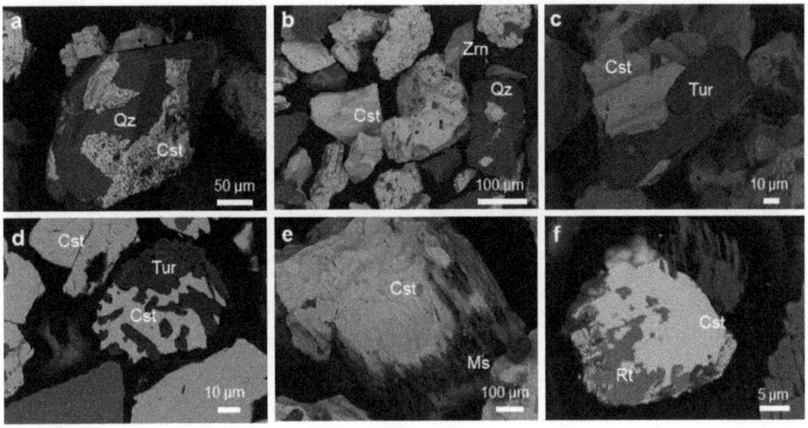

Figure 8. Images of the mineral association of binary particles of cassiterite in the concentrate from Catavi: (**a**,**b**) with quartz, (**c**,**d**) with tourmaline, (**e**) with muscovite, and (**f**) with rutile. Qz: quartz; Cst: cassiterite; Tur: tourmaline; Ms: muscovite; Rt: rutile.

3.4.2. Monazite

Monazite represented 1.62 wt.% of the total concentrate. In addition, 2.41% of the concentrated particles contained monazite and 63.83 wt.% of the total monazite was liberated, which represented 1.03 wt.% of the concentrate. The non-liberated monazite was mainly associated with cassiterite, tourmaline, and quartz, followed by rutile and zircon (Figure 9).

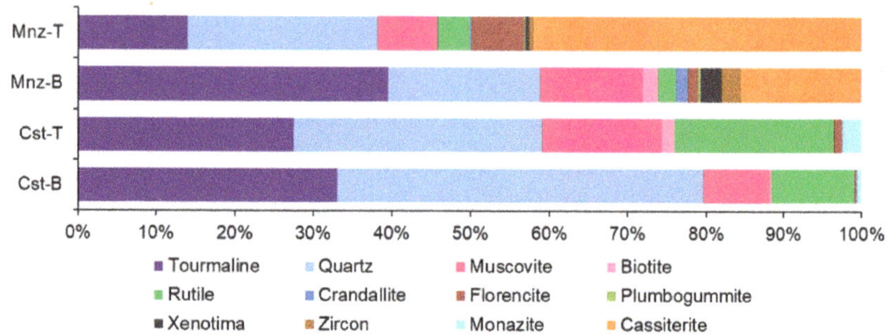

Figure 9. Mineral association of the non-liberated ore: Cst-B cassiterite in binary particles, Cst-T cassiterite in ternary particles, Mnz-B monazite in binary particles and Mnz-T monazite in ternary particles.

4. Discussion

The grinding process moderately reduced the size of the sands that constituted the feed tailings, which exhibited an average particle size of 540 μm. On the other hand, the average size of the cassiterite and monazite grains was 66 μm. Therefore, after grinding, one part of the ore remained unliberated. The analysis of cassiterite liberation by grade ranges is shown in Figure 10. A significant amount of cassiterite was found in mixed particles, with 65.73% of it being liberated.

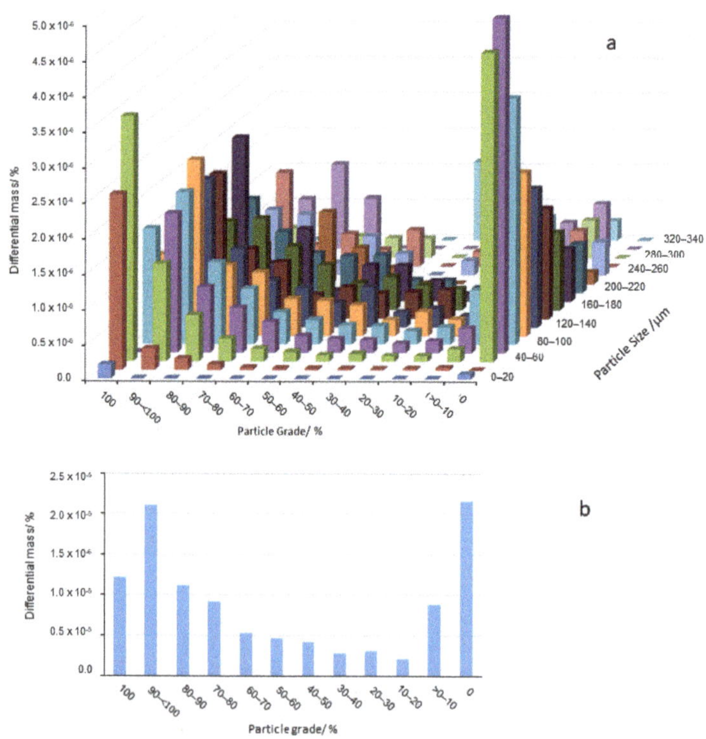

Figure 10. Distribution of cassiterite in the concentrated materials: (**a**) according to particle size and cassiterite grade classes; (**b**) according to the grade.

The distribution of the concentrate into the different particle sizes shows that the concentrate contained a low number of particles of less than 40 μm. Moreover, it shows that most particles of this class can be considered liberated cassiterite. Particles larger than +80 μm were mainly mixed. The most abundant content of cassiterite-free particles in the concentrate was in the size classes of +60 to +180 μm. This could be indicative of a loss of ore in the finest fractions during processing. This behavior was also reported in other cases, with the lack of fine cassiterite particles in the concentrate being attributed to an aqua flow phenomenon [40,41].

In the case of monazite, the distribution of the concentrate into the different particle sizes shows that there are few particles of liberated monazite less than 40 μm in size, with the most abundant content of liberated monazite particles occurring in the size class between 60 and 80 μm. The lower content of high-grade liberated monazite particles in the finer class size could be due to the fact that they are associated with cassiterite (Figure 11).

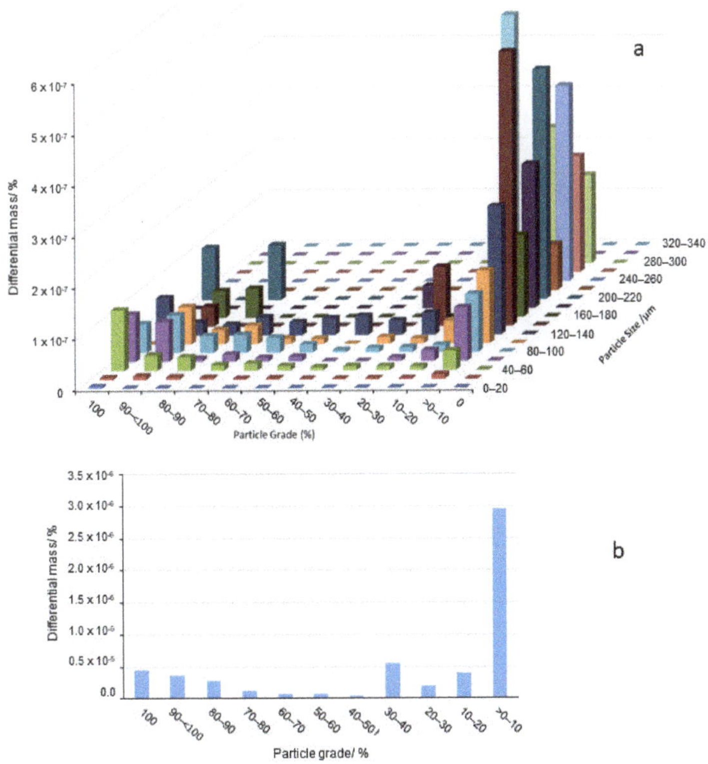

Figure 11. Distribution of monazite in the concentrated materials: (**a**) according to particle size and cassiterite grade classes; (**b**) according to the grade. The zero grade particles are not shown.

Therefore, particle size reduction should be necessary to increase the liberation and, consequently, the recovery of cassiterite and increase its grade in the concentrate. A high amount of energy is consumed during the comminution process, which means that when a significant size reduction is required, a large expense is also required [42]. An alternative could be the use of high-pressure grinding rolls (HPGR) before milling in the ball mill. This may represent an economical solution since HPGR produce a high number of microfractures in the material, which facilitates the increase in the liberation with a moderate size reduction [43]. Additionally, the grinding media charge and the increase in residence times should also be considered [44].

The REE content in the concentrate from Llallagua feed tailings (between 8859 and 12,294 ppm) is sufficient to be considered a suitable by-product of Sn production. The higher amount corresponds to samples that were not floated. In this case, the floated concentrate was also depleted in rutile and zircon as compared to the non-floated sample. Then, the flotation part of REE was floated with sulfide minerals, rutile, and zircon. The flotation of monazite-rich samples should be taken into account in the flow diagram, considering the inclusion of the flotation process or changing the depressants used, according to existing studies [45].

In other cases, lower REE concentrations in tailings were considered of interest, as was the case of the Kiirunavaara iron deposit [46] and in a copper flotation plant of Australia [47]. Other tin mining activities consider REE of interest for recovery as a by-product [48,49], such as thorium-rich monazite from Malaysia [50]. In contrast to this example, monazite from Llallagua has the advantage of being Th-poor [14–16].

5. Conclusions

The reprocessing of the sandy tailings from the Catavi plant produced final tailings with about 0.29 wt.% SnO_2, which is rich enough to be of economic interest. To make the mining activity of Llallagua more sustainable, Sn losses must be reduced. To this end, recovery needs to be improved. The average liberation size of cassiterite was 66μm, whereas that of the feed tailings was 540 μm. Finer particle size of the processed tailings is necessary to increase the liberation and recovery of Sn, and the cassiterite grade of the concentrate. It is evident that the efficiency of the mill is very low, and the comminution process should be improved. To this end, the control of the grinding media charge and the residence times should be studied. Using different combinations of equipment also demonstrated that other techniques may increase the reduction ratios.

In addition, there was a significant amount of REE in the concentrate, i.e., between 8859 and 12,294 ppm of total REE, which is of economic interest as a by-product. REE occurred mainly as monazite, and florencite, with minor amounts of xenotime. The density of these phosphate minerals is moderate, making their separation from the concentrate by gravitational separation methods difficult. Therefore, future studies should attempt to identify an optimal method to recover this ore should be recovered during the metallurgic process.

Author Contributions: Conceptualization, P.A. and M.R.; methodology, P.A.; Field work: P.A., M.R., R.N.Z., A.S.; software, P.A.; M.G.-V.; validation, P.A., H.A. and M.R.; formal analysis, M.G.-V., M.S.; investigation, P.A., M.S., H.A., N.S.-R.; resources, P.A., M.G.-V.; data curation, P.A.; writing—original draft preparation, P.A.; writing—review and editing, P.A., M.R., M.G.-V., H.A., N.S.-R.; supervision, P.A.; project administration, P.A.; funding acquisition, P.A. All authors have read and agreed to the published version of the manuscript.

Funding: This research was funded by the Centre de Cooperació al desenvolupament de la Universitat Politècnica de Cataluya, grant number 2018-U017 and 2019-B005.

Data Availability Statement: Not applicable.

Acknowledgments: We acknowledge at Multiactiva Catavi-Siglo XX cooperative the access to the processing plant and help in the sampling.

Conflicts of Interest: The authors declare no conflict of interest.

References

1. Edraki, M.; Baumgartl, T.; Manlapig, E.; Bradshaw, D.; Franks, D.M.; Moran, C.J. Designing mine tailings for better environmental, social and economic outcomes: A review of alternative approaches. *J. Clean. Prod.* **2004**, *84*, 411–420. [CrossRef]
2. Khan, M.M.; Mahajani, S.M.; Jadhav, G.N.; Vishwakarma, R.; Malgaonkar, V.; Mandre, S. Mechanical and thermal methods for reclamation of waste foundry sand. *J. Environ. Manag.* **2021**, *279*, 111628. [CrossRef] [PubMed]
3. Asr, E.T.; Kakaie, R.; Ataei, M.; Mohammadi, M.R.T. A review of studies on sustainable development in mining life cycle. *J. Clean. Prod.* **2019**, *229*, 213–231. [CrossRef]
4. Laurence, D. Establishing a sustainable mining operation: An overview. *J. Clean. Prod.* **2011**, *19*, 278–284. [CrossRef]
5. United Nations. The future We Want. In Proceedings of the United Nations Conference on Sustainable Development, Rio de Janeiro, Brazil, 20–22 June 2012.

6. Dold, B. Sustainability in metal mining: From exploration, over processing to mine waste management. *Rev. Environ. Sci. Biotechnol.* **2008**, *7*, 275–285. [CrossRef]
7. Ahlfeld, F. The tin deposits of Llallagua, Bolivia. *Econ. Geol.* **1936**, *31*, 219–221. [CrossRef]
8. Ahlfeld, F.; Schneider-Scherbina, A. Los yacimientos minerales y de hidrocarburos de Bolivia. *Bol. Dept. Nac. Geol.* **1964**, *5*, 1–388.
9. Hyrsl, J.; Petrov, A. Famous mineral localities: Llallagua, Bolivia. *Mineral. Rec.* **2006**, *37*, 17–163.
10. Zambrana, R.N. Retratamiento de Colas Finales Planta C-4 de Catavi y Propuesta Técnica Económica y Ambiental. Master's Thesis, Universitat Politècnica de Catalunya, Manresa, Spain, 2011.
11. Villalpando, B.A. The tin ore deposits of Bolivia. In *Geology of Tin Deposits in Asia and the Pacific*; Hutchinson, C.S., Ed.; Springer: Berlin/Heidelberg, Germany, 1988; pp. 201–215.
12. Romero, F.M.; Canet, C.; Alfonso, P.; Zambrana, R.N.; Soto, N. The role of cassiterite controlling arsenic mobility in an abandoned stanniferous tailings impoundment at Llallagua, Bolivia. *Sci. Total Environ.* **2014**, *481*, 100–107. [CrossRef]
13. Kempe, U.; Lehmann, B.; Wolf, D.; Rodionov, N.; Bombach, K.; Schwengfelder, U.; Dietrich, A. U–Pb SHRIMP geochronology of Th-poor, hydrothermal monazite: An example from the Llallagua tin-porphyry deposit, Bolivia. *Geochim. Cosmochim. Acta* **2008**, *72*, 4352–4366. [CrossRef]
14. Catlos, E.J.; Miller, N.R. Speculations Linking Monazite Compositions to Origin: Llallagua Tin Ore Deposit (Bolivia). *Resources* **2017**, *6*, 36. [CrossRef]
15. Betkowski, W.B.; Rakovan, J.; Harlov, D.E. Geochemical and textural characterization of phosphate accessory phases in the vein assemblage and metasomatically altered Llallagua tin porphyry. *Mineral. Petrol.* **2017**, *111*, 547–568. [CrossRef]
16. Zglinicki, K.; Szamałek, K.; Konopka, G. Monazite-bearing post processing wastes and their potential economic significance. *Miner. Resour. Manag.* **2020**, *36*, 37–58.
17. Zhou, B.; Li, Z.; Chen, C. Global potential of rare earth resources and rare earth demand from clean technologies. *Minerals* **2017**, *7*, 203. [CrossRef]
18. Massari, S.; Ruberti, M. Rare earth elements as critical raw materials: Focus on international markets and future strategies. *Resour. Policy* **2013**, *38*, 36–43. [CrossRef]
19. European Commission. Study on the Review of the List of Critical Raw Materials: 546 Executive Summary. European Commission, Directorate-General for Internal Market, 547 Industry, Entrepreneurship and SMEs. Available online: https://publications.europa.eu/en/publication548detail/-/publication/08fdab5f-9766-11e7-b92d-01aa75ed71a1/language-en (accessed on 15 November 2021).
20. Binnemans, K.; Jones, P.T.; Blanpain, B.; Van Gerven, T.; Pontikes, Y. Towards zero-waste valorisation of rare-earth-containing industrial process residues: A critical review. *J. Clean. Prod.* **2015**, *99*, 17–38. [CrossRef]
21. Pereira, L.; Birtel, S.; Möckel, R.; Michaux, B.; Silva, A.C.; Gutzmer, J. Constraining the economic potential of by-product recovery by using a geometallurgical approach: The example of rare earth element recovery at Catalão I, Brazil. *Econ. Geol.* **2019**, *114*, 1555–1568. [CrossRef]
22. Sillitoe, R.H.; Halls, C.; Grant, J.N. Porphyry tin deposits in Bolivia. *Econ. Geol.* **1975**, *70*, 913–927. [CrossRef]
23. Dietrich, A.; Lehmann, B.; Wallianos, A. Bulk Rock and melt inclusion geochemistry of Bolivian tin porphyry systems. *Econ. Geol.* **2000**, *95*, 313–326. [CrossRef]
24. Grant, J.N.; Halls, C.; Ávila Salinas, W.; Snelling, N.J. K-Ar ages of igneous rocks and mineralization in part of the Bolivian tin belt. *Econ. Geol.* **1979**, *74*, 838–851. [CrossRef]
25. Lehmann, B. Petrochemical factors governing the metallogeny of the Bolivian tin belt. In *Tectonics of the Southern Central Andes, Structure and Evolution of an Active Continental Margin*; Reutter, K.J., Scheuber, E., Wigger, P., Eds.; Springer: Berlin, Germany, 1994; pp. 317–326.
26. Grant, J.N.; Halls, C.; Avila, W.; Avila, G. Igneous geology and the evolution of hydrothermal systems in some subvolcanic tin deposits of Bolivia. *Geol. Soc. Lond. Spec. Publ.* **1977**, *7*, 117–126. [CrossRef]
27. Mlynarczyk, M.; Williams-Jones, A.E. The role of collisional tectonics in the metallogeny of the central Andean tin belt. *Earth Planet Sci. Lett.* **2005**, *240*, 656–667. [CrossRef]
28. Schulz, B.; Sandmann, D.; Gilbricht, S. SEM-Based Automated Mineralogy and Its Application in Geo- and Material Sciences. *Minerals* **2020**, *10*, 1004. [CrossRef]
29. Arancibia, J.R.H.; Alfonso, P.; García-Valles, M.; Martínez, S.; Canet, C.; Romero, F.M. Obtención de vidrio a partir de residuos de la minería del estaño en Bolivia. *Bol. Soc. Esp. Ceram. Vidr.* **2013**, *52*, 143–150. [CrossRef]
30. Menzie, W.D.; Reed, B.L.; Singer, D.A. Models of grades and tonnages of some lode tin deposits. In *Geology of Tin Deposits in Asia and the Pacific*; Springer: Berlin/Heidelberg, Germany, 1988; pp. 73–88.
31. Ishihara, S.; Murakami, H.; Marquez-Zavalia, M.F. Inferred indium resources of the Bolivian tin-polymetallic deposits. *Resour. Geol.* **2011**, *61*, 174–191. [CrossRef]
32. Jiménez-Franco, A.; Alfonso, P.; Canet, C.; Trujillo, J.E. Mineral chemistry of In-bearing minerals in the Santa Fe mining district, Bolivia. *Andean Geol.* **2018**, *45*, 410–432. [CrossRef]
33. Torró, L.; Melgarejo, J.C.; Gemmrich, L.; Mollinedo, D.; Cazorla, M.; Martínez, Á.; Artiaga, D.; Torres, B.; Alfonso, P.; Arce, O. Spatial and Temporal Controls on the Distribution of Indium in Xenothermal Vein-Deposits: The Huari Huari District, Potosí, Bolivia. *Minerals* **2019**, *9*, 304. [CrossRef]

34. Alfonso, P.; Garcia-Valles, M.; Llorens, T.; Tomasa, O.; Calvo, D.; Guasch, E.; Anticoi, H.; Oliva, J.; López Moro, J.; García Polonio, F. Nb-Ta mineralization from the rare element granite from Penouta, Galicia, Spain. *Mineral. Mag.* **2018**, *82*, S199–S222. [CrossRef]
35. Gray, F.; Kramer, D.A.; Bliss, J.D. *Gallium and Gallium Compounds*; Wiley-VCH Verlag GmbH & Co. KGaA: Weinheim, Germany, 2000; ISBN 978-3-527-30673-2.
36. Mordberg, L.E.; Stanley, C.J.; Germann, K. Rare earth element anomalies in crandallite group minerals from the Schugorsk bauxite deposit, Timan, Russia. *Eur. J. Mineral.* **2000**, *12*, 1229–1243. [CrossRef]
37. Evans, C.L.; Napier-Munn, T.J. Estimating error in measurements of mineral grain size distribution. *Miner. Eng.* **2013**, *52*, 198–203. [CrossRef]
38. Spencer, S.; Sutherland, D. Stereological correction of mineral liberation grade distributions estimated by single sectioning of particles. *Image Anal. Stereol.* **2000**, *19*, 175–182. [CrossRef]
39. Ueda, T.; Oki, T.; Koyanaka, S. Comparison of seven texture analysis indices for their applicability to stereological correction of mineral liberation assessment in binary particle systems. *Minerals* **2017**, *7*, 222. [CrossRef]
40. Gupta, A.; Yan, D. *Mineral Processing Design and Operations: An Introduction*; Elsevier: Amsterdam, The Netherlands, 2016; pp. 563–628.
41. Bru, K.; Sousa, R.; Leite, M.M.; Broadbent, C.; Stuart, G.; Pashkevich, D.; Martin, M.; Kern, M.; Parvaz, D.B. Pilot-scale investigation of two Electric Pulse Fragmentation (EPF) approaches for the mineral processing of a low-grade cassiterite schist ore. *Miner. Eng.* **2020**, *150*, 106270. [CrossRef]
42. Abouzeid, A.Z.M.; Fuerstenau, D.W. Grinding of mineral mixtures in high-pressure grinding rolls. *Int. J. Miner. Process.* **2009**, *93*, 59–65. [CrossRef]
43. Hamid, S.A.; Alfonso, P.; Anticoi, H.; Guasch, E.; Oliva, J.; Dosbaba, M.; Garcia-valles, M.; Chugunova, M. Quantitative mineralogical comparison between HPGR and ball mill products of a Sn-Ta ore. *Minerals* **2018**, *8*, 151. [CrossRef]
44. Genç, Ö. Analysis of grinding media effect on specific breakage rate function of particles in a full-scale open circuit three-compartment cement ball mill. *Miner. Eng.* **2015**, *81*, 10–17. [CrossRef]
45. Nduwa-Mushidi, J.; Anderson, C.G. Surface Chemistry and Flotation Behaviors of Monazite–Apatite–Ilmenite–Quartz–Rutile–Zircon with Octanohydroxamic Acid. *J. Sustain. Metall.* **2017**, *3*, 62–72. [CrossRef]
46. Wanhainen, C.; Pålsson, B.I.; Martinsson, O.; Lahaye, Y. Rare earth mineralogy in tailings from Kiirunavaara iron ore, northern Sweden: Implications for mineral processing. *Miner. Metall. Processing* **2017**, *34*, 189–200. [CrossRef]
47. Abaka-Wood, G.B.; Addai-Mensah, J.; Skinner, W. The Use of Mining Tailings as Analog of Rare Earth Elements Resources: Part 1–Characterization and Preliminary Separation. Miner. Process. *Extr. Metall. Rev.* **2021**, 1–15. [CrossRef]
48. Alves, F.E.; Neumann, R.; Ávila, C.A.; Ferreira, P.E.; Assumpção, C.D.S.; Carneiro, M.C.; Garcia, P.H. Mineralogical auditing of the Volta Grande Mine (SE Brazil) Sn-Ta-Nb-Li processing plant, aiming at REE recovery as by-products. *Appl. Earth Sci.* **2021**, *130*, 198–208. [CrossRef]
49. Adiputra, R.N.; Agustin, F.; Sulastri, A.; Abdullah, C.I.; Nugraha, I.; Andriansyah, R.; Hadiprayitno, M. The tin ore separation process and optimizing the rare earth mineral (monazite) as a by-product of tin mining in East Belitung Regency. *IOP Conf. Ser. Earth Environ. Sci.* **2020**, *413*, 012004, IOP Publishing. [CrossRef]
50. Udayakumar, S.; Noor, A.F.M.; Hamid, S.A.R.S.A.; Putra, T.A.R.; Anderson, C.G. Chemical and mineralogical characterization of Malaysian monazite concentrate. *Min. Metall. Explor.* **2020**, *37*, 415–431. [CrossRef]

minerals

MDPI

Article

Understanding the Effect of Stepwise Irrigation on Liquid Holdup and Hysteresis Behavior of Unsaturated Ore Heap

Leiming Wang [1,2,3,4], Shenghua Yin [1,2,*] and Bona Deng [3]

[1] Key Laboratory of the Ministry of Education of China for High-Efficient Mining and Safety of Metal Mines, University of Science and Technology Beijing, Beijing 100083, China; ustb_wlm@126.com
[2] School of Civil and Resources Engineering, University of Science and Technology Beijing, Beijing 100083, China
[3] Key Laboratory of Green Chemical Engineering Process of Ministry, Wuhan Institute of Technology, Wuhan 430205, China; bndeng@126.com
[4] Department of Materials Engineering, University of British Columbia, Vancouver, BC V6P 1Z4, Canada
* Correspondence: csuysh@126.com; Tel.: +86-13811668481

Abstract: Liquid is a crucial medium to contain soluble oxygen, valuable metal ions, and bacteria in unsaturated heap leaching. Liquid retention behavior is the first critical issue to be considered to efficiently extract low-grade minerals or wastes. In this study, the residual liquid holdup of an unsaturated packed bed was quantitatively discussed by liquid holdup (θ), residual liquid holdup ($\theta_{residual}$), relative liquid holdup (θ'), and relative porosity (n^*) using the designed measuring device. The detailed liquid holdup and the hysteresis behavior under stepwise irrigation are indicated and discussed herein. The results show that relative porosity of the packed bed was negatively related to particle size, and intra-particle porosity was more developed in the −4.0 + 2.0 mm packed bed. The higher liquid retention of the unsaturated packed bed could be obtained by using stepwise irrigation (incrementally improved from 0.001 to 0.1 mm/s) instead of uniform irrigation (0.1 mm/s). It could be explained in that some of the immobile liquid could not flow out of the unsaturated packed bed, and this historical irrigation could have accelerated formation of flow paths. The θ was sensitive to superficial flow rate (or irrigation rate) in that it obviously increased if a higher superficial flow rate (u) was introduced, however, the $\theta_{residual}$ was commonly affected by n^* and θ'. Moreover, the liquid hysteresis easily performed under stepwise irrigation condition, where θ and $\theta_{residual}$ were larger at u of the decreasing flow rate stage (DFRS) instead of u of the increasing flow rate stage (IFRS). These findings effectively quantify the liquid retention and the hysteresis behavior of ore heap, and the stepwise irrigation provides potential possibility to adjust liquid retention conditions.

Keywords: unsaturated packed bed; liquid holdup; hysteresis behavior; stepwise irrigation; heap leaching; flow behavior

Citation: Wang, L.; Yin, S.; Deng, B. Understanding the Effect of Stepwise Irrigation on Liquid Holdup and Hysteresis Behavior of Unsaturated Ore Heap. *Minerals* **2021**, *11*, 1180. https://doi.org/10.3390/min11111180

Academic Editor: Luis A. Cisternas

Received: 7 October 2021
Accepted: 22 October 2021
Published: 25 October 2021

1. Introduction

As an environmentally friendly, lower cost, and efficient recycling method of valuable metal resources, the solution mining method has been widely utilized in the recycling of lower grade minerals, mine tailings, and wastes [1–3]. It is commonly recognized that valuable metal ions continuously transfer between the reaction interface and the leaching solution, eventually obtaining pregnant leaching solution (PLS) and metal products [4]. In this leaching procedure, the leachate is regarded as the key medium of bacteria, soluble oxygen, and metallic ions [5–7]. The leachate could mainly include mobile and immobile liquids, and both simultaneously exist in this unsaturated ore heap. The liquid retention performance decisively affects solute transfer and industrial leaching operation [8–11]. Hence, to obtain a desirable leaching efficiency, the first step should be to deeply understand the liquid holdup and the hysteresis behavior in the unsaturated ore heap.

Relying on the current research findings, liquid retention in ore heaps is difficult to describe or predict, because it is highly affected by irrigation conditions (such as irrigation rate, irrigation methods, etc.) and heap conditions (such as inter-/intra-particle porosity, initial moisture content, etc.). On the one hand, the stratification and the segregation of raw crushed ore feeds [12–14] widely appear in the dumping procedure and heavily affect the flow behavior. In other words, the mobile liquid (shown as preferential flow, etc.) and the immobile liquid (shown as stagnant flow, etc.) co-exist in saturated and unsaturated conditions of industrial heaps [15–17]. In this heap, the fluid flow could be roughly divided into faster preferential flow (mainly driven by gravity) and slower liquid diffusion (mainly driven by capillarity forces). On the other hand, the flow paths are mainly dominated by developed conditions of pore structure, especially in ore heaps with lower porosity. The dual-pore structure includes inter-particle and intra-particle pores co-existing in unsaturated heaps [18,19]. Different from the heap packed by drain rocks or glass beads, the intra-particle pores of ore heap are well-developed. Thus, the liquid retention condition is an unresolved obstacle and is more complicated in ore heaps. However, heap irrigation in some industrial operations keep consistent for a long period without any stepwise adjustments, which easily accelerates the appearance of preferential flow, unsaturated leachate regions, heap compaction, and undesirable leaching rate [20].

Some recent studies revealed the possibility that liquid retention of ore heaps could be ameliorated by controlling irrigation rate (or superficial flow rate), heap porosity, initial moisture content, and so on [21]. Advanced scanning and characterizing methods were utilized to quantify these changes of liquid retention features in the unsaturated ore heap, which is mainly limited to transient, short-time conditions [22–24]. For instance, Fagan et al. (2014) evaluated the potential connections between liquid retention and dipper parameters of a bioleaching system using magnetic resonance imaging (MRI); based on MRI, the solute transfer and the liquid spreading with irrigation time were quantified [25]. Besides, the third-generation gamma transmission tomography system [26] and the UV fluorescence [27] were also utilized to reveal liquid spreading in an inter-particle scale [28]. However, even though these mentioned findings broke the study-scale bottleneck, these novel characterized methods are partially limited by their shortcomings such as high cost, transience, etc. Some new and lower cost methods such as stepwise irrigation were gradually applied in the valuable mass recycling, such as free phase light hydrocarbon [29], to obtain a higher recycle rate. In the unsaturated ore packed beds, McBride et al. (2015) explored the potential effect of heap porosity on liquid retention and preferential flow paths, and the results reveal that flexible irrigation could disturb and affect initial flow paths [30]. As a result, in the view of the current unsatisfactory liquid holdup behavior of unsaturated packed beds [31,32], adjusting the irrigation operation could promote the liquid retention and eventually obtain desirable extraction efficiency of ore heaps.

In this paper, the liquid holdup and the hysteresis behavior in an unsaturated ore heap under stepwise irrigation conditions are discussed. The liquid retained in the unsaturated ore packed bed was continuously and directly measured using an electronic balance, and the effect of stepwise irrigation on liquid holdup in the unsaturated packed bed was quantified by liquid holdup value (θ), residual liquid holdup value ($\theta_{residual}$), and relative porosity (n^*). The liquid hysteresis behavior especially was inferred in the drainage process and discussed under the effects of existing liquid holdup (historical irrigation). These results could provide theoretical support in enhanced recycling of waste or low grade minerals in unsaturated heaps.

2. Materials and Methods

2.1. Ore Samples and Its Pretreatment

The experimental ores in this study comprised one type of secondary copper sulfides, which were sampled from mines of Chile. Table 1 shows the detailed chemical composition of minerals and gangues obtained by X-ray powder-diffraction analyzer. The original ores were crushed by roller crusher and jaw crusher and further meshed into detailed intervals.

In this study, four particle size intervals including −19.0 + 13.5 mm, −11.2 + 9.5 mm, −8.0 + 5.6 mm, and −4.0 + 2.0 mm are respectively discussed. Their geometric mean diameters were 16.02 mm, 10.32 mm, 6.69 mm, and 2.83 mm, correspondingly. In this paper, the particle size distribution of crushed ore feeds is not considered to avoid the negative effect of particle segregation and stratification.

Table 1. Mass fractions of dominant ore minerals and gangues (wt.%).

Minerals	Mass Fraction (%)
Chalcopyrite ($CuFeS_2$)	0.6
Pyrite (FeS_2)	2.7
Illite ($K_{0.65}Al_{2.0}Al_{0.65}Si_{3.35}O_{10}(OH)_2$)	18.0
Plagioclase ($NaAlSi_3O_8$-$CaAl_2Si_2O_8$)	29.2
Quartz (SiO_2)	34.5
Others	15.6
Total	100

The solid density of the ore sample was 2.56 g/cm^3, and the irrigating liquid density was defined as 1.00 g/cm^3. To avoid the negative clogging effects caused by undesirable immigration of fine particles or powders on flow paths, all ore samples were processed by the following procedures: (1) ore samples were washed with deionized water to remove powders; (2) washed ore samples were totally dried to remove free water attached on ore surfaces in the experimental oven for 24 h; (3) dried ore samples were cooled down in lab conditions of 24 ± 2 °C for 0.5 h before formal stacking and column irrigation.

2.2. *Key Parameters of Liquid Retention*

2.2.1. Relative Porosity (n*)

To quantitatively describe the relationship of particle size and dual-pore structure (co-existence of intra-particle and inter-particle) of the ore packed bed, the relative porosity (n^*) was defined in Equation (1) as the ratio of the intra-particle porosity (ϕ_{intra}, %) and the inter-particle porosity (ϕ_{inter},%) of the ore packed bed. It obviously indicated that the intra-particle porosity played a dominant role when the n^* was over 0.5. The total/inter-particle porosity could be calculated by dry bulk density and wet bulk density. To better describe the influence of the irrigation process on heap porosity, the gray image and the image processing methods were utilized to describe the newly retained liquid. The proportion of black areas was the sum of the proportion of original ores and the proportion of newly retained liquid. The detailed relative porosity of the ore packed bed is shown in Table 2.

$$n^* = \frac{\phi_{intra}}{\phi_{inter}} \tag{1}$$

Table 2. Experimental design under different irrigation and heap porosity conditions.

Particle Size (Geometric Mean Diameter), mm	−19.0 + 13.5 (16.02)	−19.0 + 13.5 (16.02)	−11.2 + 9.5 (10.32)	−8.0 + 5.6 (6.69)	−4.0 + 2.0 (2.83)
Relative Porosity	0.143	0.143	0.159	0.176	0.196
Irrigation Method	Uniform	Stepwise	Stepwise	Stepwise	Stepwise
Superficial Flow Rate, mm/s	0.1	0.075~0.1	0.075~0.1	0.075~0.1	0.075~0.1
Moisture Content of Wetted Bed, %	/	0~10	0~10	0~10	0~10

2.2.2. Liquid Holdup (θ) and Residual Liquid Holdup ($\theta_{residual}$)

In the irrigation and the drainage processes, the volume of retained liquid in the packed bed could reach the steady state. In this paper, to quantitatively define the liquid retention features under the different packed conditions, the liquid holdup value (θ, %) and the residual liquid holdup value ($\theta_{residual}$, %) were respectively calculated by Equations (2) and (3). The θ and the $\theta_{residual}$ were utilized to describe the continuous changes or the residual steady state of the retained liquid in the unsaturated ore packed bed [33].

$$\theta = \frac{\int (v_{in} - v_{out})dt}{V} = \frac{v_{in}t - m_{out}/\rho}{V} \tag{2}$$

$$\theta_{residual} = \frac{V_{steady} - \int (v_{out})dt}{V} = \frac{\theta V - m_{out}/\rho}{V} = 1 - \frac{m_{out}}{V\rho} \tag{3}$$

where v_{in} is flow rate in, v_{out} is flow rate out, m_{out} is mass of liquid flow out, and ρ is liquid density.

2.2.3. Relative Unsaturated Liquid Holdup (θ')

Owing to the historical irrigation, there were some immobile liquids still reserved in the intra-particle and the inter-particle that could have affected formation of liquid bridge and flow paths [34]. To further reveal historical wetting conditions caused by stepwise irrigation, the relative unsaturated liquid holdup was defined. The net liquid holdup (θ_{net}) was the total liquid holdup minus the original liquid holdup of the wetted packed bed, as Equation (4) showed. The relative unsaturated liquid holdup (θ') was the ratio of net liquid holdup (θ_{net}) and total liquid holdup (θ) to quantify the contribution of inter- and intra-porosity in liquid holdup, as Equation (5) showed, where the m_{water} is weight of added liquid in the pre-wetting process.

$$\theta_{net} = \theta - \frac{m_{water}}{V} = \frac{(V_{in}t - m_{out}/\rho) - m_{water}}{V} \tag{4}$$

$$\theta' = \frac{\theta_{net}}{\theta} = 1 - \frac{m_{water}}{V_{in}t - M_{out}/\rho} \tag{5}$$

2.3. *Experimental Design and Irrigation Scheme*

To eliminate undesirable differences caused by column diameter, the author made data conversion and obtained superficial flow rates (0.001, 0.005, 0.01, 0.02, 0.05, and 0.10 mm/s). As Section 2.1 mentioned, four types of crushed ore feeds with different diameters were considered. The geometric mean diameters of crushed ore feeds were calculated as 16.02 mm, 10.32 mm, 6.69 mm, and 2.83 mm. In the stacking procedure of the ore heap, the crushed ore feeds with different geometric mean diameters were utilized independently.

By controlling the superficial flow rate (u), the stepwise irrigation was proposed to reveal the liquid holdup and the hysteresis behavior in the unsaturated packed ore. Specifically, the stepwise irrigation clearly defined that the irrigation procedure was started using the lowest u (0.001 mm/s) and stopped irrigation once liquid holdup (θ) reached the steady state. Then, the residual liquid holdup ($\theta_{residual}$) was obtained at the liquid drainage process; after that, we restarted the irrigation process using a higher u (0.005 mm/s) until the u was up to peak (0.1 mm/s). Finally, the irrigation procedure was restarted using a lower u (0.05 mm/s) until it reached the lowest value (0.001 mm/s).

During the irrigation and the drainage processes, the liquid holdup was measured per 15 s, and the residual steady state was confirmed if the residual liquid holdup was continuously unchanged for 15 min. The detailed experimental scheme under different dual-pore conditions is shown in Table 2.

The dual-pore condition of the unsaturated packed bed was described by relative porosity (n^*, %). Similar to the results of Zhang et al. (2018), it was clear that ore particle size was negatively proportional to relative porosity of the ore packed bed [35]. Thus, intra-particle porosity was more developed in the packed bed stacked by fine particles. Besides, to further reveal the effect of wetting conditions caused by stepwise irrigation on liquid retention, five types of the initial moisture content (0%, 1%, 3%, 5%, and 10%) of the unsaturated packed bed were selected. The wetting condition was prepared by thoroughly and uniformly mixing deionized water and dry ores in hydrophobic plastic bags before the stacking process.

2.4. Analytical Apparatus and Device

The liquid retention and the hysteresis behavior of the unsaturated ore packed bed can be mainly quantified by two methods [36]: (1) directly measuring the weight changes of liquid retention (used in this study); (2) indirectly calculating the flux differences of flow in and out. These two methods are in good agreement, and the former is easier to implement.

Figure 1 shows the structural composition of irrigation experimental device utilized in this paper, which could be mainly divided into three parts:

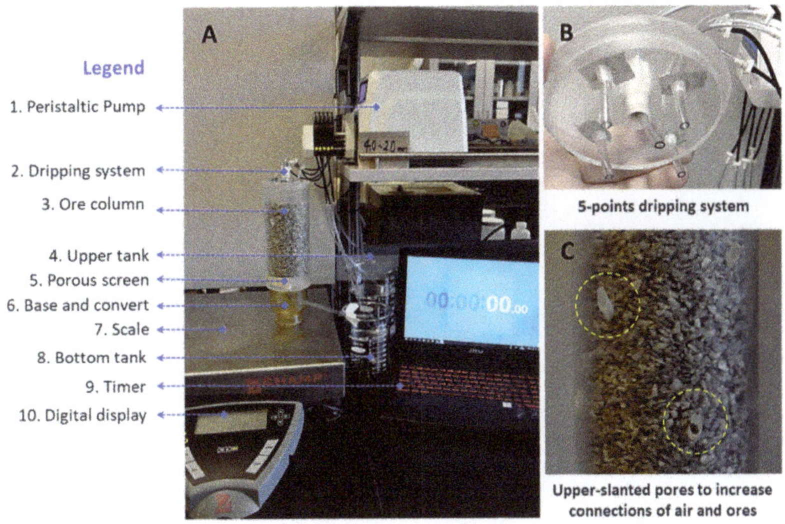

Figure 1. Experimental setup of stepwise liquid irrigation tests. (**A**): liquid holdup rate real-time characterization test bench; (**B**): 5 points dripping; (**C**): upper-slanted pores to eliminate pressure differences.

(1) Irrigation and column system: This included a liquid peristaltic pump (8CH/4RL Cole-Parmer, Montréal, QC, Canada), a hydrophobic plexiglass column (height was 150 mm, inner diameter was 90 mm) with upper-slanted holes (to eliminate the pressure differences in and out of the packed bed), a 5 point dripper (diameter of each tube was 3.5 mm to improve the irrigation uniformity and decrease the negative effects of segregated/stratified stacking on flow paths formation), and upper/bottom tanks (replaceable, maximum storage range was 2000 mL for each).

(2) Measuring system: This included electronic balance (Ohaus, Parsippany, NJ, USA); its sensitivity was 1 g, and the measuring range was 2000 g. This balance could continuously detect the changes of the column system due to the liquid injection and spreading in the unsaturated heaps.

(3) Recording and timer system: This mainly consisted of data storage and test laptop. Based on this device, the residual retained liquid of different ore packed beds could be continuously measured. The experimental data were carefully processed by the OriginPro

2016 64bt (OriginLab, Northampton, MA, USA), and gray image processing and porosity were calculated via MATLAB R2019b (The MathWorks, Inc., Natick, MA, USA).

3. Results and Discussion

3.1. Liquid Spreading Performance in the Irrigation and Drainage Procedure

It well known that liquid spreading in unsaturated ore heaps is basically divided into two existential statuses [27,37]: (1) preferential flow mainly exists in the macro-porous flow channels, which is driven by gravity forces and rapidly appears if the liquid breaks through the dry ore heap; (2) diffusion flow mainly exists in micro-porous flow channels, which is driven by capillary forces and slowly appears if the solute transfer and the liquid bridge are well developed. To better understand the effects of irrigation and drainage processes, liquid spreading evolutions (Figure 2A), liquid holdup, and corresponding porosity changes (Figure 2B) at a consistent u were discussed.

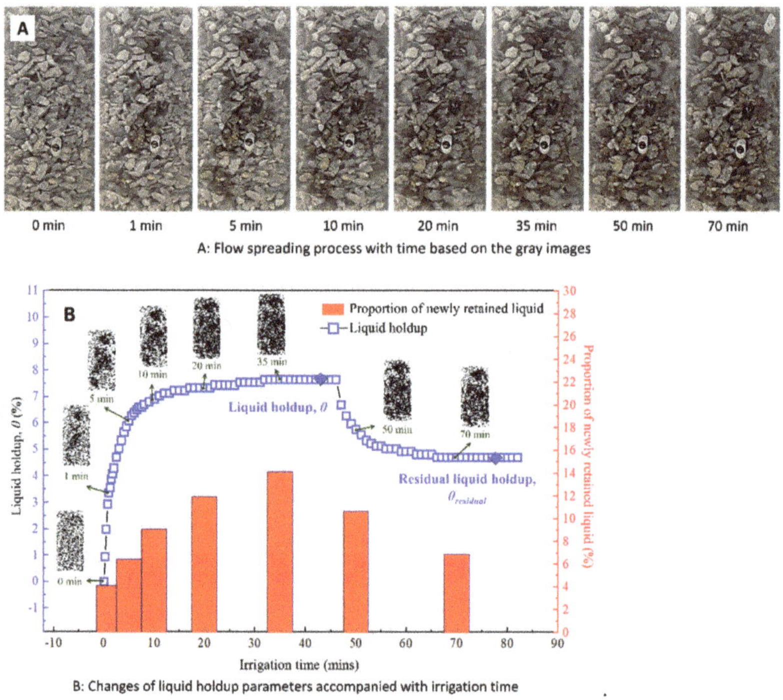

Figure 2. Liquid spreading process with irrigation time (**A**), quantification of liquid holdup and cross-section porosity in unsaturated ore packed bed (**B**).

It is clearly revealed by Figure 2A that crushed ore feeds in the column were gradually wetted accompanied by the liquid spreading process. This liquid spreading process could be explained via the key time notes that the liquid (leachate) longitudinally transferred, which broke through and rapidly flowed out of the ore heaps at 5 min; in other words, the preferential flow paths or channels tended to quickly form once the irrigation started. At 5–20 min, the increase rate of liquid holdup was significantly reduced. This indicates that the liquid gradually diffused and spread in the micro-scope pores/voids (also regarded as potential flow paths). In other word, the flow paths/channels gradually reached steady state, and the total liquid holdup tended to reach the peak value. In the 30–45 min period, the liquid holdup value roughly kept consistent, and the liquid retention of ore heaps reached the steady state. In the drainage process (after 50 min), the liquid holdup value (θ)

rapidly decreased once the irrigation operation stopped, and the θ gradually reached the residual liquid holdup at 70 min.

To further quantitively characterize this liquid spreading in irrigation and drainage processes, the liquid holdup and the proportion of newly retained liquid were calculated (Figure 2B). This showed that liquid holdup rapidly increased in the first 10 min and gradually reached the steady state liquid holdup (7.65%) at 31 min. This steady liquid holdup indicates that flow paths and immobile liquid (stagnant liquid regions) in the unsaturated ore heap reached the stable state [38]. At the steady state status of liquid holdup, the liquid holdup of the unsaturated ore heap could not fundamentally be improved if the heap porosity and the irrigation condition were consistent. In the drainage process, the liquid holdup decreased from 7.65% (θ) to 4.72% ($\theta_{residual}$). In other words, the residual liquid holdup (4.72%) could not reduce to 0%, and amounts of immobile liquid were still reserved in the pores/voids of ore heaps. Therefore, the liquid hysteresis behavior verified in this paper showed that the mobile liquid flowed out of the unsaturated ore heap, while the immobile liquid was still retained inside [39–41]. This liquid spreading was also quantified by the gray images showing that the proportion of newly retained liquid gradually increased with liquid irrigation, reached the peak value (14.11%) at the peak value (θ), and slightly decreased to 6.87% at the residual peak value ($\theta_{residual}$) of the drainage process. Except for mechanical flipping and loosening operations, it is unrealistic and hard to adjust PSD or geometric mean diameter of ore feeds in stacked ore heaps [17], thus it is more feasible with lower cost to interrupt and adjust irrigation conditions to promote the liquid retention of ore packed beds.

3.2. *Residual Liquid Holdup Changes in the Irrigation and Drainage Procedure*

It is no doubt that changing irrigation conditions (such as irrigation rate, mode, etc.) can cause disturbances to the liquid retention condition [42]. To further quantify the liquid holdup and better understand the potential effect of the stepwise irrigation condition on hysteresis behavior, the liquid holdup under controlled irrigation and stacking conditions was carefully measured (Figure 3).

To better explore the effects of stepwise irrigation on ameliorating liquid retention, consistent stacking and irrigation (superficial flow rate) conditions (−19.0 + 13.5 mm, 0.1 mm/s) were used (Figure 3A). The liquid holdup of stepwise irrigation under different geometric particle diameter conditions was also compared (Figure 3B). This illustrated that stepwise irrigation could be divided into two stages distinguished by the u: (1) in the increasing flow rate stage (IFRS, 0-523 min), the u stepwise increased from 0.001 to 0.1 mm/s; (2) in the decreasing flow rate stage (DFRS, 524-800 min), the u stepwise decreased from 0.1 to 0.001 mm/s.

In IFRS, both liquid holdup (θ) and residual liquid holdup ($\theta_{residual}$) of the ore packed bed improved if a higher u was introduced. However, in DFRS, the θ gradually decreased, while the $\theta_{residual}$ roughly kept consistent (4.52%) under the stimulation of a lower u. Therefore, the θ was more sensitive to the u (irrigation rate). The θ tended to decrease, while the $\theta_{residual}$ could not be disturbed if a lower u was introduced in the unsaturated ore heap. By comparing the peak θ and $\theta_{residual}$ of uniform irrigation (Figure 2B) and stepwise irrigation (Figure 3A), the liquid holdup of stepwise irrigation (5.89%) was much higher than uniform irrigation (3.25%), verifying the fact that the ore packed bed could reserve more liquid under stepwise irrigation conditions compared to uniform irrigation. Figure 3A shows that, at 0.005 mm/s, the time to reach a steady state was 50 min of IFRS, which was much longer than 41 min of DFRS. This fact was inferred considering that the historical irrigation could shorten the appearance time of preferential flow paths due to the immobile liquid reserved in heaps.

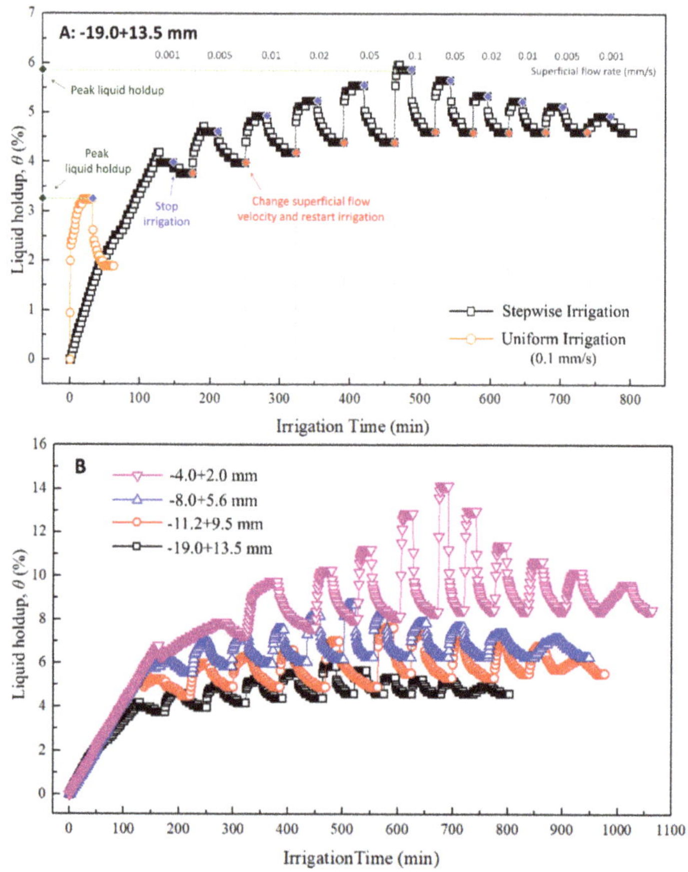

Figure 3. The changes of liquid holdup in a dry ore bed during stepwise irrigation. (**A**): typical changes of liquid holdup in a dry ore bed (−19.0 + 13.5 mm), (**B**): changes of liquid holdup under different particle size conditions.

The liquid holdup of ore heaps is seriously affected by the pore structure, which is tightly related to the geometric mean diameter of packed ore feeds [43–45]. Figure 3B presents the potential effect of geometric mean diameter on liquid holdup under stepwise irrigation conditions. It indicates that the peak residual liquid holdup was negatively related to geometric mean diameter, for instance, the θ of the −4.0 + 2.0 mm packed bed was higher than 14.0%, which was much larger than the θ (5.89%) of the −19.0 + 13.5 mm packed bed. In DFRS, the $\theta_{residual}$ of a different packed bed basically kept consistent. This conveys that the $\theta_{residual}$ was more sensitive to geometric mean diameter, especially at a lower u, while the θ was more sensitive to the u. The liquid hysteresis behavior was clearly observed under stepwise irrigation conditions [41]. The θ of DFRS was slightly higher than the θ of IFRS at the same u. The net difference of liquid holdup indicated that the immobile liquid could not totally flow out and was still retained in intra/inter particle pores if a lower u was introduced [46].

3.3. *Effect of Packed Crushed Ore Diameter on Liquid Retention*

To reveal the potential effect of stepwise irrigation on liquid retention performance, Figure 4 shows the peak weight of liquid retention in the ore heap under different relative porosity conditions. The results showed that the ore heap could reserve more liquid if a higher u was introduced. This phenomenon was more obvious when the finer particles or powders occupied a higher percentage in the ore heap. The liquid holdup at steady state was negatively related to geometric mean diameter, while it was positively related to relative porosity. Thus, the well-developed intra-particle pores benefitted from absorbing more liquid [24]. The irrigation period was much longer in the fine-grained ore heap where there was more liquid retained and the liquid spreading was more sufficient. The time to reach the steady state was positively related to geometric mean diameter. For example, the peak weight of reserved liquid (135 g) was obtained at 676 min in heaps by fine-grained crushed feeds (−4.0+2.0 mm), and the lowest weight of liquid (56 g) was obtained at 471 min in heaps packed by coarse-grained crushed feeds (−19.0 + 13.5 mm).

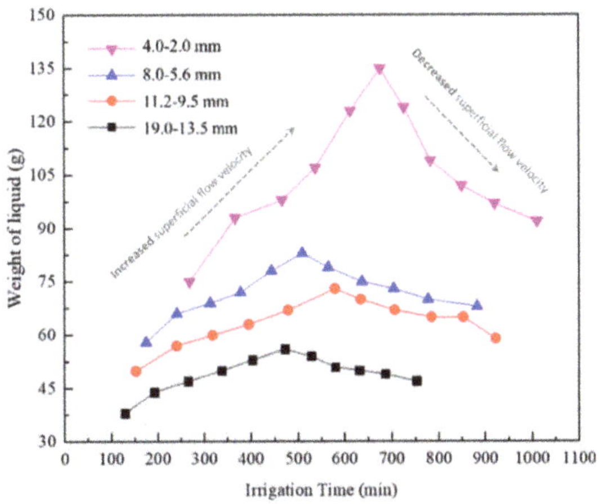

Figure 4. Relationship between irrigation time and weight of liquid retention.

Combined with the stepwise irrigation results, this liquid retention performance under higher v stimulation could be clarified by Figure 5. As Figure 5A shows, large amounts of liquid gathered and transferred along the axis direction below dippers; as a result, the flow paths gradually reached the steady state. This liquid spreading area significantly extended when a higher u was introduced, verifying that the originally existing flow paths rapidly re-formed, and some new flow paths/channels developed at the same time, resulting in desirable liquid retention and a higher steady state liquid holdup (θ).

Different from the heap packed by glass beads [36], the crushed ore heap was observed to have more liquid due to the well-developed intra-particle pores (Figure 5B). When the u decreased from the higher figure to the initial figure, some new paths that formed under the former higher u condition were still reserved [40]. The liquid hysteresis was inferred considering that the response of the liquid drainage process lagged behind the irrigation amelioration [47]. The amount of immobile liquid could not totally flow out of the heap and was still retained in intra-particle pores (Figure 5C). Hence, it could be conjectured that the immobile liquid mainly played three key roles in the liquid retention and the hysteresis behavior: (1) ameliorating the liquid retention of the unsaturated packed bed; (2) accelerating the appearance of potential flow paths/channels via capillary diffusion; (3) pre-wetting ore feeds and promoting new flow paths when a higher u was introduced [48].

3.4. Residual Steady State Liquid Holdup Features under Stepwise Irrigation

To further figure out the residual steady state and better explain the liquid hysteresis behavior that appeared in the drainage process, Figure 6 shows the relationship between the residual steady state liquid holdup value (θ, $\theta_{residual}$) and the superficial flow rate (u) under stepwise irrigation.

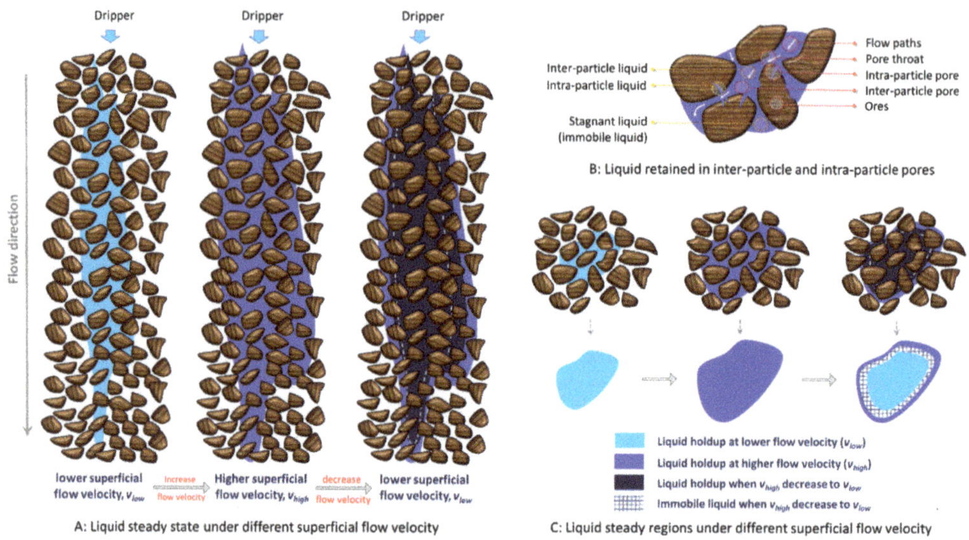

Figure 5. Schematic diagram of liquid holdup and hysteresis behavior in unsaturated packed bed. (**A**): liquid steady state under different superficial flow rate, (**B**): liquid retained in inter-particle and intra-particle pores, (**C**): liquid steady region under different superficial flow rate.

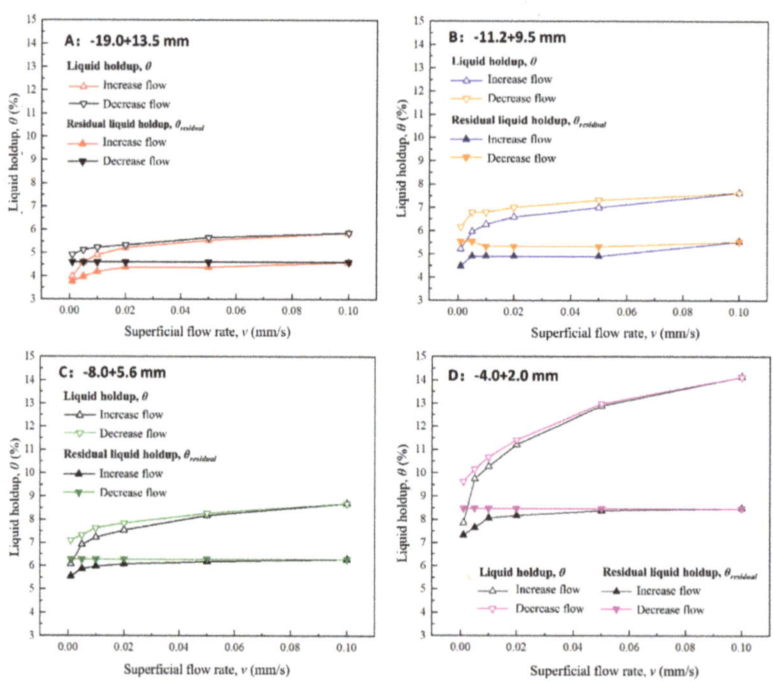

Figure 6. Changes of residual liquid holdup during increase/decrease flow rate arm of stepwise irrigation condition. (**A**): −19.0 + 13.5 mm; (**B**): −11.2 + 9.5 mm; (**C**): −8.0 + 5.6 mm; (**D**): −4.0 + 2.0 mm.

In IFVS, both the θ and the $\theta_{residual}$ increased if a higher u was introduced. For instance, in the −4.0 + 2.0 mm packed bed, θ increased by 6.29% and $\theta_{residual}$ slightly increased by 1.15% when u was improved from 0.001 mm/s to 0.1 mm/s. This means that even θ and $\theta_{residual}$ reached liquid steady state but could still increase if a higher u was introduced. In DFRS, the θ obviously decreased. At a certain u, the θ of DFRS was slightly higher than the previous value of IFVR; the $\theta_{residual}$ kept roughly consistent, especially in the fine-grained heap (Figure 6C,D). In the −4.0 + 2.0 mm packed bed, the θ had a huge decrease of 4.51% when the θ decreased from 0.1 to 0.001 mm/s. At 0.01 mm/s, the θ of DFRS had a smaller net increase (0.42%) due to liquid hysteresis. The maximum net increase of θ and $\theta_{residual}$ was obtained at 0.001 mm/s. The liquid hysteresis easily performed in the fine-grained heap and lower u. This also revealed that θ was positively sensitive to u, but $\theta_{residual}$ was mainly seriously controlled by heap relative porosity.

3.5. *Effect of Historical Irrigation Condition on Liquid Retention*

To better understand the effect of historical irrigation conditions on liquid retention, the relative liquid holdup (θ', defined as the ratio of net liquid holdup and total liquid holdup) was determined. Figure 7 shows the relationship between the u and the θ' of the wetted bed.

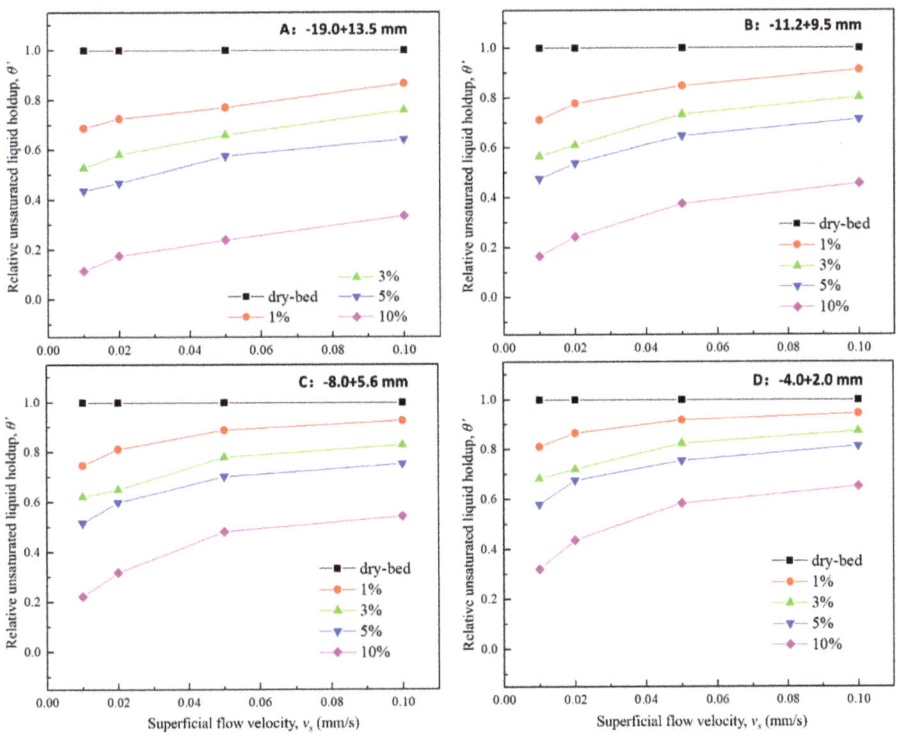

Figure 7. The relationship between superficial flow rate and relative liquid holdup under different geometric mean diameters and initial moisture contents. (**A**): −19.0 + 13.5 mm; (**B**): −11.2 + 9.5 mm; (**C**): −8.0 + 5.6 mm; (**D**): −4.0 + 2.0 mm.

The θ' of the dry bed was consistently 1.0 because the retained liquid was entirely from irrigation. At a certain u, the θ' was lower when the ore heap was gradually pre-wetted. Under a lower u, the differences of θ' were larger in the bed packed by −19.0 + 13.5 mm feeds. Under a higher u, the differences of θ' were smaller, and the θ' value was large in that inter-pore liquid occupied a larger proportion of the total. This also indicates that inter-pore liquid in the dry bed occupied a smaller proportion. Because the intra- and the inter- pores co-existed in industrial heaps [49], the heap with well-developed pores could have good potential to store more liquid. The θ' was positively related to u and negatively related to n^* and initial moisture content. The peak value was obtained in 0.1 mm/s and −4.0 + 2.0 mm conditions. The retained liquid partially filled intra-particle pores via the pre-wetted procedure and directly affected the liquid retention [7].

The peak value of liquid holdup increased when the ore bed was well pre-wetted. This implies that, although the θ gradually increased with the u under stepwise irrigation, the liquid spread along the pre-existing flow paths and created some new, small rivulets under a higher u stimulation. In the pre-wetting process, the liquid spreading was more developed and uniform, massively eliminating the unsaturated liquid regions. The wetted condition potentially affected inter-pore liquid at the steady state: (1) in the poorly wetted bed (0% to 5% initial moisture content), the θ' was much higher than 0.5, meaning that the liquid spreading was undesirable where the inter-pore liquid occupied a dominated percentage of the total pores; (2) however, in the well pre-wetted bed (>10% initial moisture content), the θ' was higher than 0.5 when the u was higher than 0.05 mm/s, which was tightly related to fewer unsaturated regions existing in the bed.

4. Conclusions

To better understand the effects of stepwise irrigation on liquid retention, the responses of liquid holdup and hysteresis behavior in unsaturated heaps to stepwise irrigation were quantitatively discussed via liquid holdup (θ), residual liquid holdup ($\theta_{residual}$), superficial flow rate (u), relative porosity (n^*), and relative liquid holdup (θ') in this paper.

Based on findings of this paper, it revealed that the intra-particle porosity was more developed in ore beds packed by fine-grained feeds. The preferential flow paths/channels easily appeared and developed in well-wetted heaps with a larger n^*, resulting in a desirable θ. Once the liquid retention reached steady state at a consistent u, the fluid flow paths and channels in the ore heap kept stable and were not affected by a lower u but increased when affected by a higher u.

The liquid retention of unsaturated ore packed beds could be ameliorated using stepwise irrigation. These controls of irrigation rate in stepwise irrigation could lead to desirable liquid retention and mass transfer. For instance, the unregular adjustment of the irrigation rate of industrial heaps could disturb preferential flow paths, promote new flow channels growth, and enhance liquid percolation. The drip irrigation with a lower superficial flow rate was suggested to pre-wet ore packed beds and increase the immobile liquid ratio before leaching, resulting in a better solute transport via the liquid film of the leaching interface.

Author Contributions: Conceptualization, L.W.; methodology, L.W.; software, L.W.; validation, L.W. and S.Y.; data curation, L.W., S.Y., and B.D.; writing—original draft preparation, L.W.; writing—review and editing, L.W., S.Y., and B.D.; visualization, L.W.; supervision, S.Y. and L.W.; project administration, L.W. and S.Y.; funding acquisition, S.Y. All authors have read and agreed to the published version of the manuscript.

Funding: This research was funded by the Open Project of Key Laboratory of Green Chemical Engineering Process of Ministry of Education (Grant No. GCP202108), Key Program of National Natural Science Foundation of China (Grant No. 52034001), Funds of key field innovation team of Ministry of Science & Technology of PRC (Grant No. 2018RA4003), the 111 Project (Grant No. B20041), Shandong Provincial Major Science and Technology Innovation Project, China (Grant No. 2019SDZY05), and the Fundamental Research Funds for the Central Universities (Grant No. FRF-TP-18-003C1).

Data Availability Statement: The data are not publicly available due to confidential reasons.

Acknowledgments: This work was carried out in Hydrometallurgy lab at the University of British Columbia (UBC) and mainly funded the scholarship support of the China Scholarship Council (CSC). We sincerely thank Wenying Liu, David Dreisinger, Zihe Ren, Mohsen Hashemzadeh, Kresimir

Ljubetic, Ronny Winarko, and Mohamad Mirazimi from UBC for their valuable advice and technical support throughout the work.

Conflicts of Interest: The authors declare no conflict of interest.

References

1. Petersen, J. Heap leaching as a key technology for recovery of values from low-grade ores—A brief overview. *Hydrometallurgy* **2016**, *165*, 206–212. [CrossRef]
2. Yin, S.; Wang, L.; Kabwe, E.; Chen, X.; Yan, R.; An, K.; Zhang, L.; Wu, A. Copper Bioleaching in China: Review and Prospect. *Minerals* **2018**, *8*, 32. [CrossRef]
3. Wang, L.; Yin, S.; Wu, A.; Chen, W. Synergetic bioleaching of copper sulfides using mixed microorganisms and its community structure succession. *J. Clean. Prod.* **2019**, *245*, 118689. [CrossRef]
4. Watling, H.; Collinson, D.; Li, J.; Mutch, L.; Perrot, F.; Rea, S.; Reith, F.; Watkin, E. Bioleaching of a low-grade copper ore, linking leach chemistry and microbiology. *Miner. Eng.* **2013**, *56*, 35–44. [CrossRef]
5. Yin, S.-H.; Wang, L.-M.; Wu, A.-X.; Chen, X.; Yan, R.-F. Research progress in enhanced bioleaching of copper sulfides under the intervention of microbial communities. *Int. J. Miner. Met. Mater.* **2019**, *26*, 1337–1350. [CrossRef]
6. Van Staden, P.; Petersen, J. Towards fundamentally based heap leaching scale-up. *Miner. Eng.* **2021**, *168*, 106915. [CrossRef]
7. Bouffard, S.C.; West-Sells, P.G. Hydrodynamic behavior of heap leach piles: Influence of testing scale and material properties. *Hydrometallurgy* **2009**, *98*, 136–142. [CrossRef]
8. Brierley, C. Biohydrometallurgical prospects. *Hydrometallurgy* **2010**, *104*, 324–328. [CrossRef]
9. Yin, S.; Wang, L.; Wu, A.; Free, M.L.; Kabwe, E. Enhancement of copper recovery by acid leaching of high-mud copper oxides: A case study at Yangla Copper Mine, China. *J. Clean. Prod.* **2018**, *202*, 321–331. [CrossRef]
10. Wang, L.; Yin, S.; Wu, A. Ore agglomeration behavior and its key controlling factors in heap leaching of low-grade copper minerals. *J. Clean. Prod.* **2020**, *279*, 123705. [CrossRef]
11. Yin, S.; Wang, L.; Wu, A.; Kabwe, E.; Chen, X.; Yan, R. Copper recycle from sulfide tailings using combined leaching of ammonia solution and alkaline bacteria. *J. Clean. Prod.* **2018**, *189*, 746–753. [CrossRef]
12. Van Staden, P.; Petersen, J. The effects of simulated stacking phenomena on the percolation leaching of crushed ore, Part 1: Segregation. *Miner. Eng.* **2018**, *128*, 202–214. [CrossRef]
13. Van Staden, P.; Petersen, J. The effects of simulated stacking phenomena on the percolation leaching of crushed ore, Part 2: Stratification. *Miner. Eng.* **2018**, *131*, 216–229. [CrossRef]
14. Wang, L.; Yin, S.; Wu, A.; Chen, W. Effect of stratified stacks on extraction and surface morphology of copper sulfides. *Hydrometallurgy* **2019**, *191*, 105226. [CrossRef]
15. Wu, A.; Yin, S.; Yang, B.; Wang, J.; Qiu, G. Study on preferential flow in dump leaching of low-grade ores. *Hydrometallurgy* **2007**, *87*, 124–132. [CrossRef]
16. Wu, A.; Yin, S.; Qin, W.; Liu, J.; Qiu, G. The effect of preferential flow on extraction and surface morphology of copper sulphides during heap leaching. *Hydrometallurgy* **2009**, *95*, 76–81. [CrossRef]
17. Lima, L.D.A. Liquid axial dispersion and holdup in column leaching. *Miner. Eng.* **2006**, *19*, 37–47. [CrossRef]
18. Ilankoon, I.; Neethling, S. Transient liquid holdup and drainage variations in gravity dominated non-porous and porous packed beds. *Chem. Eng. Sci.* **2014**, *116*, 398–405. [CrossRef]
19. Solomenko, Z.; Haroun, Y.; Fourati, M.; Larachi, F.; Boyer, C.; Augier, F. Liquid spreading in trickle-bed reactors: Experiments and numerical simulations using Eulerian–Eulerian two-fluid approach. *Chem. Eng. Sci.* **2015**, *126*, 698–710. [CrossRef]
20. McBride, D.; Ilankoon, S.; Neethling, S.; Gebhardt, J.; Cross, M. Preferential flow behaviour in unsaturated packed beds and heaps: Incorporating into a CFD model. *Hydrometallurgy* **2017**, *171*, 402–411. [CrossRef]
21. Ilankoon, I.; Neethling, S. The effect of particle porosity on liquid holdup in heap leaching. *Miner. Eng.* **2013**, *45*, 73–80. [CrossRef]
22. Snyder, V.A. Statistical Hydraulic Conductivity Models and Scaling of Capillary Phenomena in Porous Media. *Soil Science Society of America Journal* **1996**, *60*, 771–774. [CrossRef]
23. Fernando, W.A.M.; Ilankoon, I.; Rabbani, A.; Yellishetty, M. Inter-particle fluid flow visualisation of larger packed beds pertaining to heap leaching using X-ray computed tomography imaging. *Miner. Eng.* **2020**, *151*, 106334. [CrossRef]
24. Teng, J.; Chu, J.-C.; Liu, C.; Xu, T.; Lien, Y.-F.; Cheng, J.-H.; Huang, S.; Jin, S.; Dang, T.; Zhang, C.; et al. Fluid Dynamics in Microchannels. In *Fluid Dynamics, Computational Modeling and Applications*; IntechOpen: London, UK, 2012.
25. Fagan, M.A.; Ngoma, I.E.; Chiume, R.A.; Minnaar, S.; Sederman, A.; Johns, M.L.; Harrison, S.T. MRI and gravimetric studies of hydrology in drip irrigated heaps and its effect on the propagation of bioleaching micro-organisms. *Hydrometallurgy* **2014**, *150*, 210–221. [CrossRef]
26. Velo, A.; Carvalho, D.; Hamada, M. Liquid distribution and holdup in the random packed column. *Flow Meas. Instrum.* **2018**, *62*, 176–185. [CrossRef]
27. Ilankoon, I.; Neethling, S. Inter-Particle liquid spread pertaining to heap leaching using UV fluorescence based image analysis. *Hydrometallurgy* **2018**, *183*, 175–185. [CrossRef]
28. Nimmo, J.R. Preferential flow occurs in unsaturated conditions. *Hydrol. Process.* **2011**, *26*, 786–789. [CrossRef]
29. Cooper, S.; Peraltaa, C. Stepwise pumping approach to improve free phase light hydrocarbon recovery from unconfined aquifers. *J. Contam. Hydrol.* **1995**, *18*, 141–159. [CrossRef]

30. McBride, D.; Gebhardt, J.; Croft, T.; Cross, M. Modeling the hydrodynamics of heap leaching in sub-zero temperatures. *Miner. Eng.* **2016**, *90*, 77–88. [CrossRef]
31. Fernando, W.A.M.; Ilankoon, I.; Chong, M.N.; Syed, T.H. Effects of intermittent liquid addition on heap hydrodynamics. *Miner. Eng.* **2018**, *124*, 108–115. [CrossRef]
32. Fan, Y.; Jacob, K.V.; Freireich, B.; Lueptow, R. Segregation of granular materials in bounded heap flow: A review. *Powder Technol.* **2017**, *312*, 67–88. [CrossRef]
33. Ilankoon, S.K. *Hydrodynamics of Unsaturated Particle Beds Pertaining to Heap Leaching*; Imperial College London: London, UK, 2012.
34. Blackmore, S. *The Role of Hydrology, Geochemistry and Microbiology in Flow and Solute Transport Through Highly Heterogeneous, Unsaturated Waste Rock at Various Test Scales*; The University of British Columbia: Vancouver, BC, Canada, 2015. [CrossRef]
35. Zhang, S.; Liu, W.; Granata, G. Effects of grain size gradation on the porosity of packed heap leach beds. *Hydrometallurgy* **2018**, *179*, 238–244. [CrossRef]
36. Ilankoon, S.; Neethling, S. Hysteresis in unsaturated flow in packed beds and heaps. *Miner. Eng.* **2012**, *35*, 1–8. [CrossRef]
37. Ghorbani, Y.; Becker, M.; Mainza, A.; Franzidis, J.-P.; Petersen, J. Large particle effects in chemical/biochemical heap leach processes—A review. *Miner. Eng.* **2011**, *24*, 1172–1184. [CrossRef]
38. Blackmore, S.; Smith, L.; Mayer, K.U.; Beckie, R.D. Comparison of unsaturated flow and solute transport through waste rock at two experimental scales using temporal moments and numerical modeling. *J. Contam. Hydrol.* **2014**, *171*, 49–65. [CrossRef]
39. Mayergoyz, I. Mathematical models of hysteresis. *IEEE Trans. Magn.* **1986**, *22*, 603–608. [CrossRef]
40. Extrand, C.W. Contact Angles and Their Hysteresis as a Measure of Liquid−Solid Adhesion. *Langmuir* **2004**, *20*, 4017–4021. [CrossRef]
41. Zeidman, B.D.; Lu, N.; Wu, D.T. Hysteresis of liquid adsorption in porous media by coarse-grained Monte Carlo with direct experimental validation. *J. Chem. Phys.* **2016**, *144*, 174709. [CrossRef] [PubMed]
42. Wang, L.-M.; Yin, S.-H.; Wu, A.-X. Visualization of flow behavior in ore-segregated packed beds with fine interlayers. *Int. J. Miner. Met. Mater.* **2020**, *27*, 900–909. [CrossRef]
43. Yin, S.-H.; Wang, L.-M.; Chen, X.; Wu, A.-X. Effect of ore size and heap porosity on capillary process inside leaching heap. *Trans. Nonferr. Met. Soc. China* **2016**, *26*, 835–841. [CrossRef]
44. Nimmo, J.R. Porosity and pore-size distribution. In *Encyclopedia of Soils in the Environment*; Elsevier: Amsterdam, The Netherlands, 2005; pp. 295–303.
45. Ram, R.; Beiza, L.; Becker, M.; Pownceby, M.; Chen, M.; Yang, Y.; Yang, S.; Petersen, J. Study of the leaching and pore evolution in large particles of a sulfide ore. *Hydrometallurgy* **2020**, *192*, 105261. [CrossRef]
46. Robertson, S. Development of an integrated heap leach solution flow and mineral leaching model. *Hydrometallurgy* **2017**, *169*, 79–88. [CrossRef]
47. Chung, T.-W. Predictions of moisture removal efficiencies for packed-bed dehumidification systems. *Gas Sep. Purif.* **1994**, *8*, 265–268. [CrossRef]
48. Wang, J.-P.; Gallo, E.; François, B.; Gabrieli, F.; Lambert, P. Capillary force and rupture of funicular liquid bridges between three spherical bodies. *Powder Technol.* **2017**, *305*, 89–98. [CrossRef]
49. Yin, S.; Wu, A.; Hu, K.; Wang, Y.; Xue, Z. Visualization of flow behavior during bioleaching of waste rock dumps under saturated and unsaturated conditions. *Hydrometallurgy* **2013**, *133*, 1–6. [CrossRef]

minerals

MDPI

Article

The Energy Dissipation, AE Characteristics, and Microcrack Evolution of Rock–Backfill Composite Materials (RBCM)

Jie Wang [1,2,*], Chi Zhang [1,2,*], Weidong Song [1,2] and Yongfang Zhang [1,2]

[1] State Key Laboratory of High-Efficient Mining and Safety of Metal Mines of Ministry of Education, University of Science and Technology Beijing, Beijing 100083, China; songwd@ustb.edu.cn (W.S.); b1903098@ustb.edu.cn (Y.Z.)

[2] School of Civil and Resources Engineering, University of Science and Technology Beijing, Beijing 100083, China

* Correspondence: 18810582761@163.com (J.W.); zhangchi9531@163.com (C.Z.)

Abstract: The backfill in the stope usually forms a composite structure with the surrounding rock in order to bear pressure together to support the goaf and ensure the safe mining of subsequent ores. Based on laboratory tests and theoretical analysis, the energy and damage evolution of the rock–backfill composite materials (RBCM) are studied deeply. The results show that: (1) The σ_p (peak stress), ε_p (peak strain), and E (elasticity modulus) decreased with the increase of the internal backfill diameter. When the diameter of the backfill increases from 10 mm to 40 mm, σ_p decreases from 50.15 MPa to 18.14 MPa, ε_p decreases from 1.246% to 1.017%, and E decreases from 7.51 GPa to 2.33 GPa. The U^T shows an S-shaped distribution, the U^E shows an inverted U-shaped distribution, and the U^D first increases slowly and then increases rapidly. The ${U^T}_p$, ${U^E}_p$, ${U^D}_p$, ${U^E}_p/{U^D}_p$, and ${U^E}_p/{U^T}_p$ decrease by 67.38%, 97.20%, 58.56%, 32.64% and 13.64% respectively, and the ${U^D}_p/{U^T}_p$ increases by 20.93% with the increases of the backfill diameter. (2) A damage constitutive model of the RBCM is established based on the energy consumption characteristics. The damage evolution curve shows an S-shaped distribution, and the damage rate evolution curve shows an inverted U-shaped distribution. (3) The AE correlation fractal dimension decreases with the increase of the strain gradient and damage value, and the AE correlation fractal dimension presents linear and exponential functions with them, respectively. With the increase of stress, microcracks first appear and gather in the internal backfill of the RBCM, and then microcracks appear and gather in the peripheral rock, which together lead to the macro penetration failure of the RBCM.

Keywords: rock–backfill composite materials (RBCM); energy dissipation mechanism; damage constitutive model; microcrack evolution

Citation: Wang, J.; Zhang, C.; Song, W.; Zhang, Y. The Energy Dissipation, AE Characteristics, and Microcrack Evolution of Rock–Backfill Composite Materials (RBCM). *Minerals* **2022**, *12*, 482. https://doi.org/10.3390/min12040482

Academic Editor: Abbas Taheri

Received: 17 March 2022
Accepted: 12 April 2022
Published: 14 April 2022

1. Introduction

The filling mining method involves mixing the tailings solid waste produced by ore mining with cement and other cementing agents, and transporting it to the underground goaf through pipelines. Therefore, its safety is significantly improved, and it has green and environmentally protective characteristics, which are widely used in metal mines at home and abroad, and there is an increasing trend of its use [1–6]. The stope is divided into the one-step room and the two-step pillar. The room is mined first, then the goaf is filled with cemented paste backfill, and then the pillar is mined. In this way, the pillar is mined under the support of the composite structure composed of cemented paste backfill and the surrounding rock. Therefore, the stability of the composite structure of the backfill and surrounding rock is the premise of safe pillar mining. Based on the above reasons, the stability of the pressure-bearing mechanism composed of cemented paste backfill and the surrounding rock must be analyzed in order to ensure safety in the whole pillar mining process.

Fruitful research results have been obtained regarding the stability research of intact rock and backfill [7–13], which not only greatly promoted the development and progress of mine rock mechanics and backfill mechanics but also laid a solid theoretical foundation for future generations to carry out relevant scientific research. Unlike the intact rock and backfill, the RBCM is composed of two different materials, in which the rock is a strong structure and the backfill is a weak structure. Under the load, the two structures interact and depend on each other to jointly resist the external load, and their mechanical properties and damage evolution mechanism will become very complex. However, at present, the research on the bearing mechanism of the rock–backfill composite structure is relatively sparse, but it has gradually attracted the attention of scholars [14–16]. Wang et al. [17] studied the influence of interface roughness and inclination on the strength and failure mode of the rock–backfill composite model. Wu et al. [18] suggested that the failure of the backfill–rock combination was largely caused by the shear failure of the interface. Xiu et al. [19] reported that the dry and wet degree of the interface between the rock and backfill had an important impact on the stability of the interface. Zhao et al. [20] introduced the KCC model, and comprehensively analyzed the influence of backfill inclination on damage initiation and propagation. Yu et al. [21] conducted a triaxial compression test on rock–backfill samples, and CT scanned the samples before and after compression to explore the role of the interface. The lack of basic theoretical research on the rock–backfill composite structure greatly limits its effective application in engineering.

In the process of the compression failure of rock, backfill, and other materials, internal cracks gradually sprout and expand, and release energy in the form of AE. The energy evolution mechanism and the AE characteristic parameters can characterize the damage process to a certain extent. Therefore, there is important significance in the study of the energy consumption and AE characteristics of materials in order to master their damage and instability mechanism [11,22–25]. Hou et al. [26] revealed the influence of the curing age on the energy dissipation characteristics of CPB. Wang et al. [27,28] carried out a triaxial compression test on CPB with different layers, and discussed the influence law of layered numbers and confining pressure on elastic energy and dissipation energy. With the help of numerical technology, Xin et al. [29] studied the influence of waste rock content on the energy evolution characteristics of CPB, and considered that the compressive and tensile energy absorbed by CPB decreased and the interface transition region increased with the increase of waste rock content. Zhao et al. [30] obtained the AE parameter curve by carrying out the uniaxial compression AE test, and deeply studied the change characteristics of the AE correlation fractal dimension during the compression failure of CPB with the help of fractal theory. Zhou et al. [31] deduced the AE energy fractal dimension based on the AE energy counting characteristics, and analyzed the internal relationship between the AE energy fractal dimension and the damage variable. He et al. [32] made rock–backfill combination models and discussed the influence of the cement–tailings ratio on the AE correlation fractal dimension of the combination model.

The failure and instability of CPB and other materials under compression is essentially a process of continuous damage accumulation. In order to understand the damage and instability mechanism of CPB materials, it is very important to establish a reasonable and reliable damage evolution model. Wang et al. [33] obtained the basic mechanical parameters of rock-encased backfill (REB), constructed the damage constitutive model of REB, and deeply discussed the damage evolution mechanism. Ma et al. [34] conducted numerical simulation research on the coal–rock composite structure, established the damage constitutive model of a coal–rock composite structure based on energy consumption theory, and discussed the damage evolution law. Liu et al. [35] studied the damage evolution characteristics of backfill at different temperatures, and considered that the damage evolution process of backfill can be divided into four stages: the nondestructive stage, the initial damage stage, the accelerated damage stage, and the stable damage stage. Cui et al. [36] developed an evolutive elasto-plastic model, and established a damage constitutive model of cemented backfill based on this model; the important role of cemented hydrate in its damage evolution was studied. Sun et al. [12] introduced the disturbance element, connected Burges model, nonlinear viscous damper, and strain trigger, and established the creep disturbance constitutive model of backfill. Scholars have also established many other damage constitutive models of backfill [37–40]. These model construction ideas provide a great reference for the derivation and construction of the damage evolution equation in this paper.

In this paper, a uniaxial compression AE monitoring test of rock–backfill composite materials (RBCM) with different backfill diameters is carried out. Firstly, the stress–strain behavior and mechanical properties of the RBCM are analyzed, and the internal functional relationship between the peak stress, peak strain, elastic modulus and backfill diameter is revealed. Secondly, the energy dissipation law, the AE correlation fractal dimension, the AE events' location characteristics, and the microcrack temporal and spatial evolution of the RBCM are deeply studied. Finally, the damage evolution equation is introduced, and a damage constitutive model of the RBCM is established.

2. Materials and Methods

2.1. Research Background

The room of the filling mining method was first stoped, and then the goaf was filled with tailings cementation. At this time, the cemented paste backfill was surrounded by the surrounding rock, and formed a composite structure (Figure 1) with the surrounding rock to bear pressure together to support the goaf and ensure the safety of the ore mining. Unlike the intact backfill and rock materials, the composite materials were formed by the mutual wrapping of two different media. Their mechanical properties, energy consumption mechanism, and damage evolution need to be further studied.

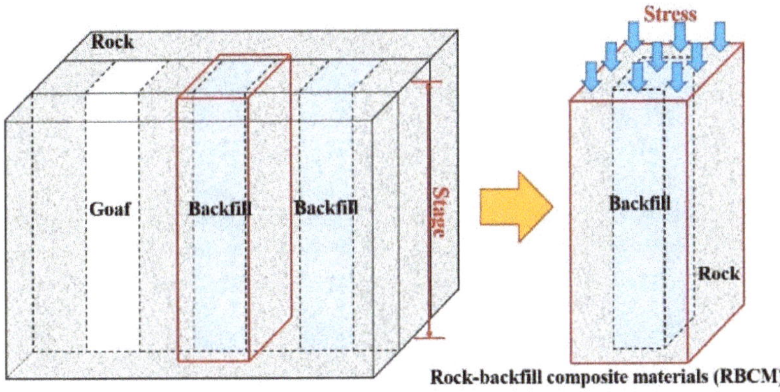

Figure 1. Field model of rock–backfill composite materials.

2.2. Materials

The tailings used for the cemented paste backfill were from a mine in Shandong Province, China, and ordinary Portland cement (OPC) 32.5R was used as a cementing agent. The PSD curve of the tailings was obtained using an LS-POP (9) laser particle size analyzer, as shown in Figure 2. Figure 3 shows the chemical composition of the tailings and OPC obtained by X-ray fluorescence.

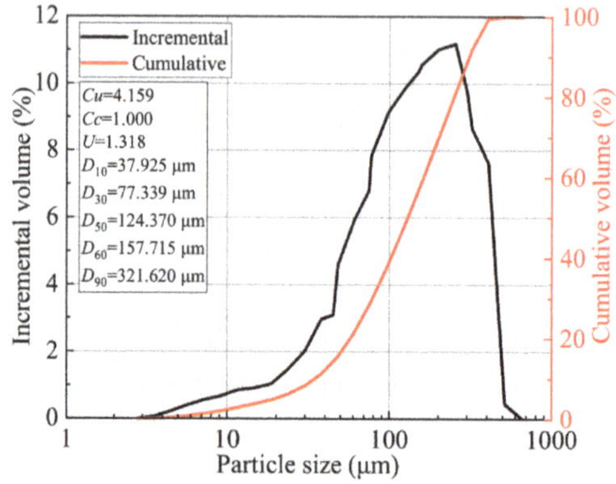

Figure 2. The PSD curves of the tailings.

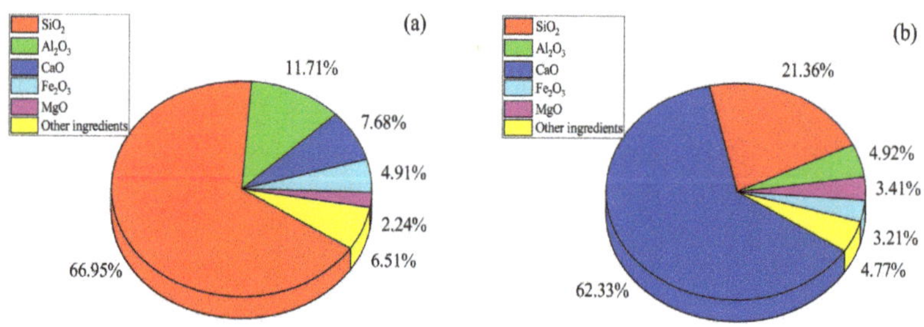

Figure 3. The chemical composition: (**a**) tailings; (**b**) OPC.

2.3. Test Scheme

In this paper, four different types of rock–backfill composite materials (RBCM) were designed and made. The complete rock center was hollowed out by drilling, and the internal backfill was prepared according to the designed ratio to fill the interior of the rock, which was preserved in a standard maintenance box to a specified age. According to the design requirements, the diameter of the peripheral rock was d_r = 50 mm and the height was h_r = 100 mm. The diameters of the internal backfill were d_b = 10 mm, 20 mm, 30 mm, and 40 mm respectively, and the height of the internal backfill was h_b = 100 mm. The test scheme and sample number are shown in Table 1. Each group prepares 3 samples, a total of 12 RBCM samples, and the RBCM samples designed and manufactured are shown in Figure 4.

Table 1. UCS and elastic modulus of different kinds of samples.

Sample ID	Backfill Diameter d_b (mm)	c/t Ratio	Peak Stress (MPa)	Peak Strain (%)	Elastic Modulus (GPa)
RBCM-10	10	1:4	50.15	1.246	7.51
RBCM-20	20		42.47	1.106	5.72
RBCM-30	30		28.73	1.040	3.88
RBCM-40	40		18.14	1.017	2.33

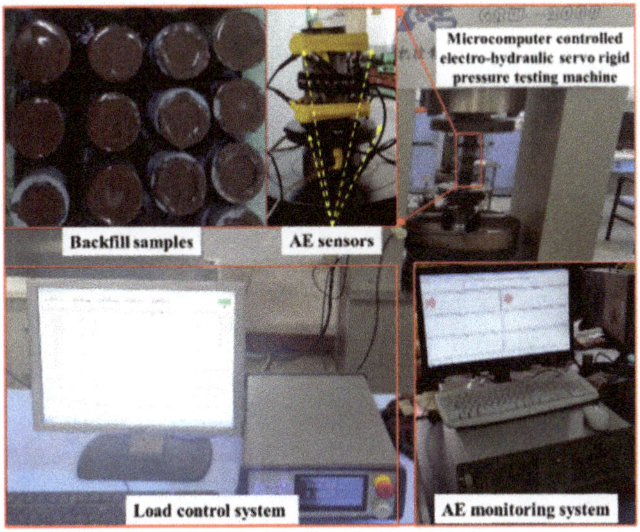

Figure 4. Uniaxial compressive AE test devices.

2.4. Test Devices

The Gaw-2000 system (Figure 4) was used to carry out uniaxial compression AE tests on the RBCM samples. The displacement control mode was adopted, and the loading rate was maintained at 0.5 mm/min. A PCI-2 AE monitoring system was adopted. A total of 6 sensors were used. The sensor frequency was 140 khz and the noise threshold was set to 45 dB. The sensors' layout position is shown in Figure 4. During compression, the stress, strain, and AE data were recorded and exported in Excel format.

3. Results and Analysis

3.1. Mechanical Properties

Figure 5 shows the stress–strain curves of four RBCM samples. The curves can be divided into five different stages. Stage I: The internal pores and cracks of the RBCM are compacted and closed, the sample becomes denser, and the stress–strain curve is concave. Stage II: With the increase of stress, the cracks are almost all closed, the strain increases linearly, the RBCM produces elastic deformation, there is almost no new crack damage, and the stress–strain curve increases linearly. Stage III: The stress gradually exceeds the yield stress of the RBCM, the sample begins to produce plastic deformation, and new cracks and damage are generated; the stress–strain curve is concave, and the RBCM samples' strength begins to approach the peak. Stage IV: When the stress reaches and exceeds the bearing limit of the RBCM, many new cracks and damages sprout on the sample's surface, the macro failure begins to form, and the stress–strain curve reaches the peak and begins to decrease. Stage V: At this stage, the peripheral rocks gradually lose their bearing capacity, but due to the certain bearing capacity of the backfill wrapped in the center, the RBCM has residual stress. It slows down the rate of RBCM deformation and its complete failure, and some samples have double peaks; the stress–strain curve will extend slowly forward.

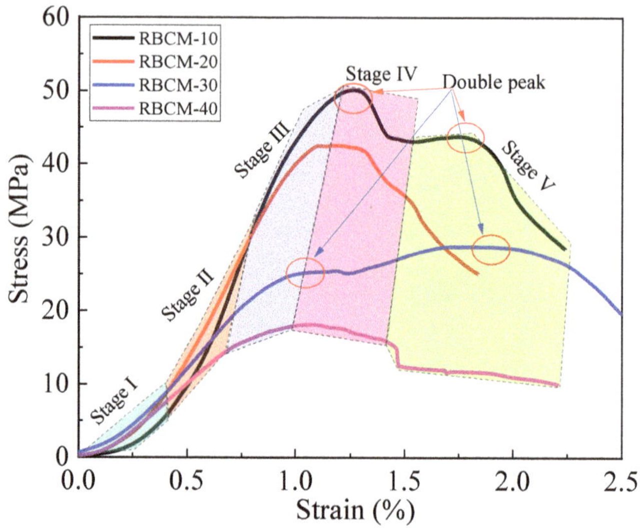

Figure 5. Stress–strain curves of the RBCM samples.

The mechanical properties of the RBCM are closely related to the size of the central backfill. Figure 6 shows the relationship between the peak stress σ_p, peak strain ε_p, elastic modulus E, and the backfill diameter of the RBCM. Figure 6a shows that the σ_p of the RBCM decreases with the increase of the backfill diameter. When the backfill diameter is 10 mm, the σ_p is 50.15 MPa. When the backfill diameter increases to 40 mm, the σ_p of the RBCM decreases to 18.14 MPa. Figure 6b shows that the ε_p of the RBCM also decreases with the increase of the backfill diameter. When the backfill diameter increases from 10 mm to 40 mm, the ε_p of the RBCM decreases from 1.246% to 1.017%. Figure 6c shows that the E of the RBCM also decreases with the increase of the backfill diameter. When the backfill diameter increases from 10 mm to 40 mm, the E decreases from 7.51 GPa to 2.33 GPa.

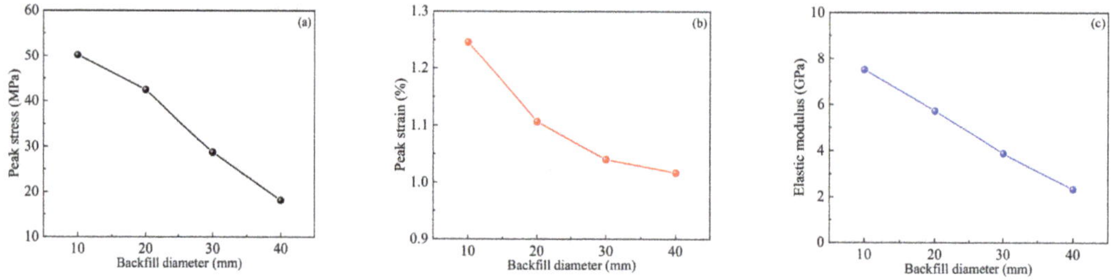

Figure 6. The relationship between the mechanical properties and backfill diameter: (**a**) the σ_p; (**b**) the ε_p; (**c**) the E.

3.2. Energy Dissipation Mechanisms

The essence of the deformation and failure of the RBCM under load is the process of the continuous input of external energy and conversion into elastic energy and dissipative energy. According to energy conservation [41,42]:

$$U^T = U^E + U^D \tag{1}$$

where U^T is the total energy input to the RBCM from the outside, and U^E and U^D are the releasable elastic energy and dissipative energy of the RBCM, respectively.

The total energy U^T absorbed by the RBCM from the outside under uniaxial compression can be expressed by the following formula:

$$U^T = \int_0^{\varepsilon} \sigma d\varepsilon \tag{2}$$

The release elastic energy U^E of the RBCM can be expressed by the following formula:

$$U^E = \frac{\sigma^2}{2\overline{E}} \tag{3}$$

where $\overline{E}$ is unload elastic modulus of the RBCM.

According to Equations (1)–(3) and the stress–strain curves of the RBCM, the energy evolution law of the RBCM with different backfill diameters can be obtained. The results are shown in Figure 7.

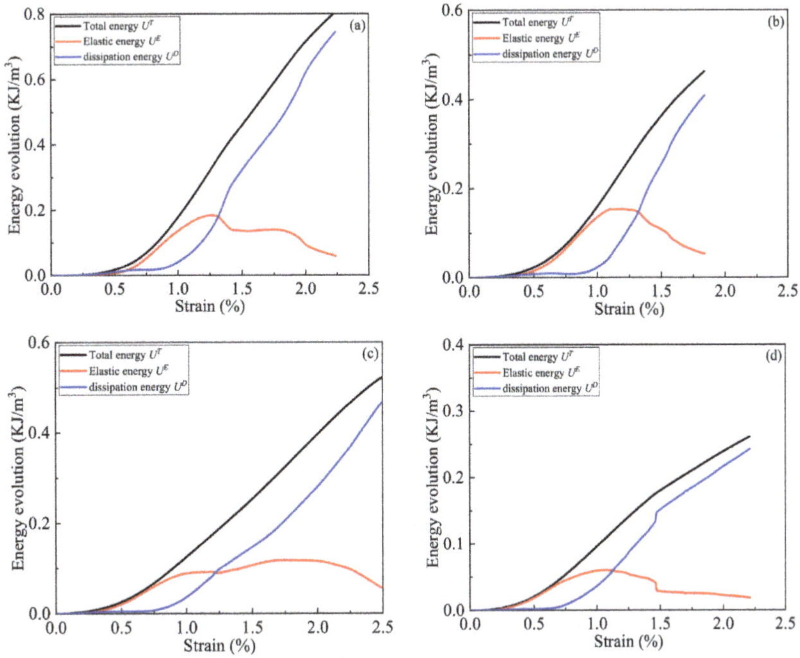

Figure 7. Energy evolution of the different samples: (**a**) RBCM-10; (**b**) RBCM-20; (**c**) RBCM-30; (**d**) RBCM-40.

Figure 7 shows that the total energy input to the RBCM is first converted into releasable elastic energy and stored in the sample, and the releasable elastic energy begins to increase (the red curve begins to rise). At this time, the sample only has elastic deformation, no damage and almost no energy dissipation. Then, the released elastic energy accumulates continuously and gradually exceeds the energy storage limit of the RBCM, the sample begins to produce damage and energy dissipation, and the dissipated energy begins to increase (the blue curve begins to rise). Finally, the released elastic energy decreases, and the dissipated energy increases rapidly until the sample is destroyed.

Figure 8 shows the energy evolution laws of the RBCM. Figure 8a shows that the total energy evolution curve of the input of the RBCM shows an S-shaped distribution. The black line is at the top, indicating that the RBCM-10 sample has the fastest total energy accumulation rate and the largest total energy. The pink line is at the bottom, indicating that the RBCM-40 sample has the slowest total energy accumulation rate and the smallest total energy. Overall, the larger the backfill diameter, the slower the total energy accumulation rate and the smaller the total energy accumulation amount. Figure 8b shows that the releasable elastic energy stored in the RBCM first increases and then decreases, and the releasable elastic energy curve presents an inverted U-shaped distribution. The smaller the backfill diameter, the greater the energy storage limit of the RBCM. It can be seen from Figure 8c that the dissipated energy of the RBCM first increases slowly, and then increases rapidly until the sample is destroyed. The smaller the backfill diameter, the greater the energy dissipation of the RBCM.

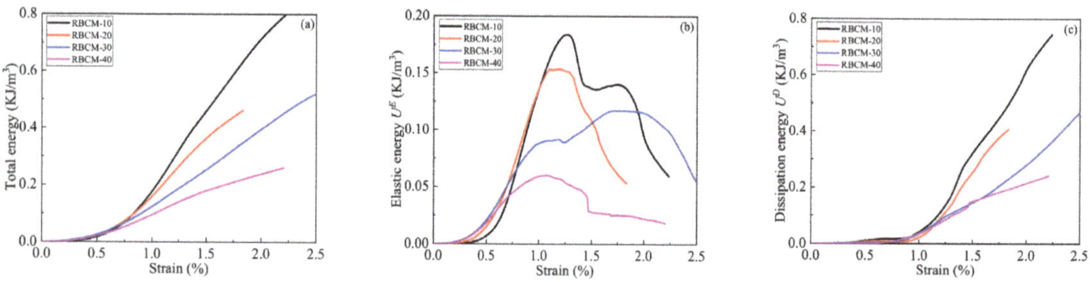

Figure 8. Different types of energy evolution: (**a**) total energy; (**b**) elastic energy; (**c**) dissipation energy.

Figure 9 and Table 2 show the relationship between different types of energy, the energy rate of the RBCM at the peak stress point, and the backfill diameter. Figure 9a shows that the released elastic energy at the peak point is greater than the dissipated energy. In addition, the total energy $U^T{}_p$ input to the RBCM, the releasable elastic energy $U^E{}_p$ stored in the RBCM, and the dissipation energy $U^D{}_p$ all decrease with the increase of the backfill diameter. When the backfill diameter increases from 10 to 40 mm, the $U^T{}_p$, $U^E{}_p$, and $U^D{}_p$ decrease from 0.325 KJ/m^3 to 0.106 KJ/m^3, from 0.214 KJ/m^3 to 0.060 KJ/m^3, and from 0.111 KJ/m^3 to 0.046 KJ/m^3, respectively. As can be seen from Figure 9b, the $U^E{}_p/U^D{}_p$ and $U^E{}_p/U^T{}_p$ decrease with the increase of the backfill diameter. When the backfill diameter increases from 10 to 40 mm, the $U^E{}_p/U^D{}_p$ and $U^E{}_p/U^T{}_p$ decrease from 1.93 to 1.30, and from 0.66 to 0.57, respectively. In addition, the $U^D{}_p/U^T{}_P$ increases with the increase of the backfill diameter. When the backfill diameter increases from 10 to 40 mm, the $U^D{}_p/U^T{}_P$ increases from 0.34 to 0.43.

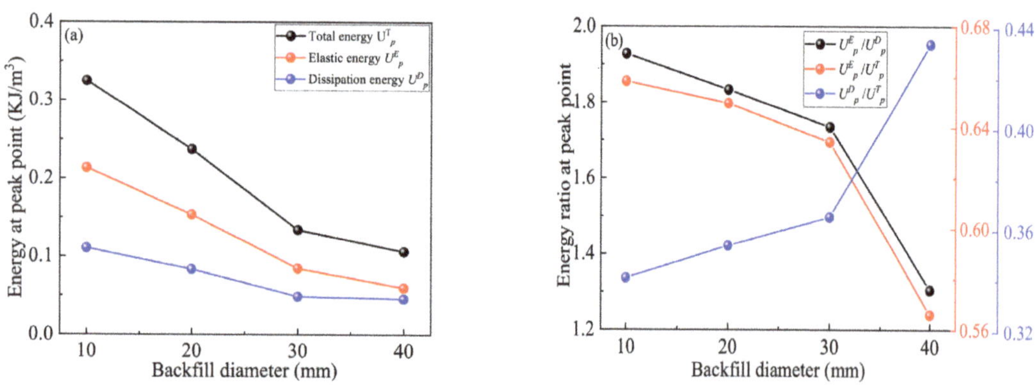

Figure 9. Energy and energy ratio at the peak point: (**a**) energy; (**b**) energy ratio.

Table 2. The $U^T{}_p$, $U^E{}_p$, and $U^D{}_p$ of the RBCM at the peak stress point.

ID	RBCM-10	RBCM-20	RBCM-30	RBCM-40
$U^T{}_p$ (KJ/m^3)	0.325	0.237	0.134	0.106
UEp (KJ/m^3)	0.214	0.154	0.085	0.06
$U^D{}_p$ (KJ/m^3)	0.111	0.084	0.049	0.046
$U^E{}_p/U^D{}_p$	1.93	1.83	1.73	1.30
$U^E{}_p/U^T{}_p$	0.66	0.65	0.63	0.57
$U^D{}_p/U^T{}_p$	0.34	0.35	0.37	0.43

3.3. *Damage Constitutive Model*

3.3.1. Model Constructed

According to the energy dissipation characteristics of the coal–rock combination, Ma et al. [34] considered that all of the energy dissipation generated by materials under load is not used to produce damage, proposed a modified damage variable, and verified the rationality of the damage variable through experimental research:

$$D = (1 - \frac{\sigma_r}{\sigma_p}) \cdot \frac{U^D}{U^D_{\max}} \cdot (1 + \frac{U^D_p}{U^T_p}) \tag{4}$$

where D is the damage variable of the RBCM, σ_r is the residual stress of the RBCM, σ_p is the peak stress of the RBCM, U^D is the dissipation energy, $U^D{}_{max}$ is the cumulative dissipated energy, $U^D{}_p$ is the dissipation energy at the peak point, and $U^T{}_p$ is the total energy input to the RBCM at the peak point.

The damage constitutive model of the RBCM at uniaxial compression is

$$\sigma = (1 - D)E\varepsilon \tag{5}$$

$$\sigma = \left[1 - (1 - \frac{\sigma_r}{\sigma_p}) \cdot \frac{U^D}{U^D_{\max}} \cdot (1 + \frac{U^D_p}{U^T_p})\right] E\varepsilon \tag{6}$$

3.3.2. Model Validation

Figure 10 shows the comparison results between the test curve and the model curve. Figure 10a shows that the coincidence degree of the two curves is poor in the compaction closing stage, but the overall change trend is highly similar. Figure 10b–d show that the model curves are highly consistent with the sample curves.

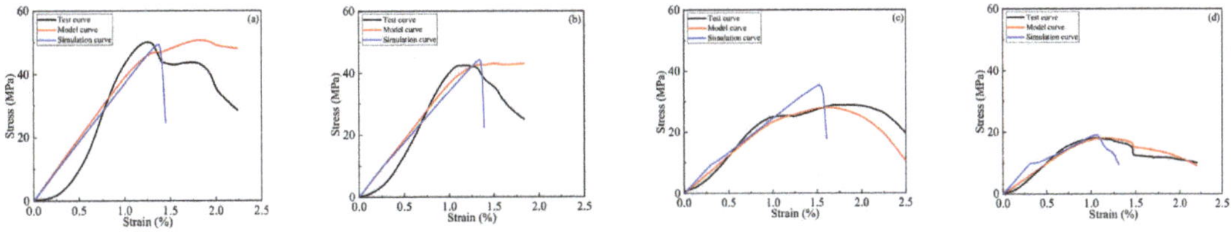

Figure 10. Comparison of the test, model, and simulation curves: (**a**) RBCM-10; (**b**) RBCM-20; (**c**) RBCM-30; (**d**) RBCM-40.

3.3.3. Damage Evolution Analysis

In order to analyze the damage evolution law of the RBCM under load, the damage evolution curves and damage rate evolution curves of the RBCM can be obtained according to Equation (4), and the results are shown in Figure 11. Figure 11a shows that, in the compaction and closing stage, the RBCM has almost no damage, and the damage curve extends forward close to the X-axis. In the linear elastic stage of the RBCM, only a small amount of damage occurs, and the damage curve rises slowly. After exceeding the yield limit of the RBCM, the damage initiates and expands rapidly, and the damage curve rises rapidly. In general, the damage curve first extends horizontally forward, then rises slowly, and finally rises rapidly. Figure 11b shows that the damage rate evolution curves of the RBCM present an inverted U-shaped distribution, which first increases and then decreases. The results indicate that the damage accumulation rate of the RBCM first increases and then decreases.

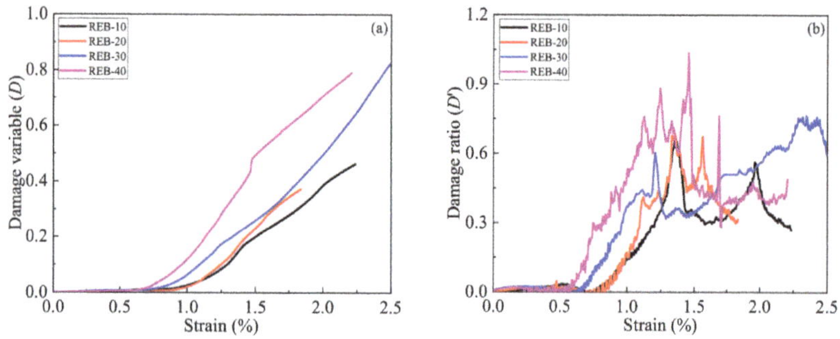

Figure 11. Damage and damage ratio evolution curves: (**a**) damage curve; (**b**) damage ratio curve.

3.4. *AE Characteristics*

3.4.1. AE Fractal Dimension

Fractal dimension is a parameter that quantitatively describes fractal characteristics and represents the complexity of the fractal body [43,44]. The correlation fractal dimension is used to describe the AE fractal characteristics of the RBCM in the failure process under uniaxial compression. The common calculation method of the correlation fractal dimension is the G-P algorithm. The G-P algorithm takes the AE parameter sequence as its research object, and each AE parameter sequence corresponds to a sequence set X_n with a capacity of n:

$$X_n = \{x_1, x_2 \cdots x_n\} \tag{7}$$

where x_i is the AE parameter, $i = 1, 2, \cdots n$. Take the first m points in Equation (8) to form the first m-dimensional vector space, which is recorded as

$$X_1 = \{x_1, x_2 \cdots x_m\} \tag{8}$$

Then move one point backward and take m points to form the second m-dimensional vector space. By analogy, Equation (7) can form a total of $N = n - m + 1$ m-dimensional vector spaces:

$$X = \begin{bmatrix} x_1 & x_2 & \cdots & x_m \\ x_2 & x_3 & \cdots & x_{m+1} \\ \vdots & \vdots & \cdots & \vdots \\ x_{n-m+1} & x_{n-m+2} & \cdots & x_n \end{bmatrix} \tag{9}$$

The correlation function $C(r)$ is

$$C(r) = \frac{1}{N^2} \sum_{i=1}^{N} \sum_{j=1}^{N} H(r - \|X_i - X_j\|) \tag{10}$$

where H is the Heaviside function, X_i and X_j are the ith and jth vector spaces respectively, and r is the given scale:

$$H(x) = \begin{cases} 1, x > 0 \\ 0, x \le 0 \end{cases} \tag{11}$$

$$r = k\frac{1}{N^2}\sum_{i=1}^{N}\sum_{j=1}^{N}\|X_i - X_j\| \tag{12}$$

where k is coefficient. For a given r, there is always a corresponding correlation function $C(r)$. By substituting the $C(r)$ and r into the double logarithmic coordinate system, the coordinate points ($\lg C(r)$, $\lg r$) can be obtained. Taking different k values, respectively, multiple coordinate points can be obtained in the double logarithmic coordinate system. The obtained coordinate points are subjected to univariate linear regression, and the slope of the regression function is the AE correlation fractal dimension D_f.

Figure 12 shows the relationship curve between the AE ringing count correlation fractal dimension and the damage variable at different strain stages. With the increase of the strain, the AE correlation fractal dimension decreases, while the damage variable continues to increase. At the initial stage of loading, the internal microcracks of the RBCM are gradually compacted and closed, almost no damage is generated, the AE ringing counts evolve irregularly, and the correlation fractal dimension of the AE parameters is large. In the middle of loading, the internal crack began to evolve and expand gradually, the internal damage began to accumulate slowly, the AE ringing counts were released to the outside world according to a certain law, and the AE correlation fractal dimension gradually decreased. In the later stage of loading, the internal meso cracks evolve rapidly and become macro through failure, the internal damage accumulates rapidly, the AE ringing counts are released to the outside in an orderly manner, and the AE correlation fractal dimension is further reduced. This result is similar to the research results of Zhao [30] and Guo [43].

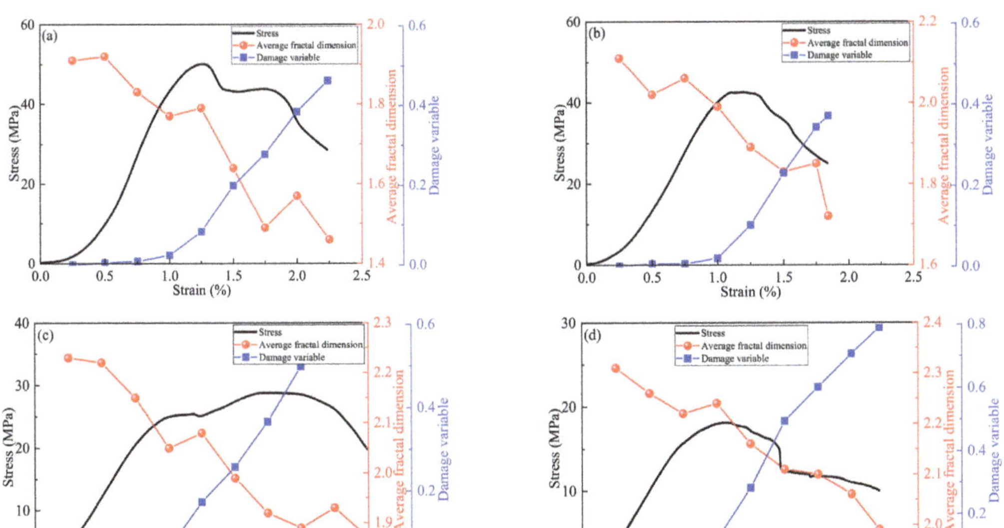

Figure 12. AE average fractal dimension: (**a**) RBCM-10; (**b**) RBCM-20; (**c**) RBCM-30; (**d**) RBCM-40.

3.4.2. AE Event Location

AE location technology can intuitively show the internal failure process of the RBCM. Figure 13 shows the AE event location of the RBCM-10 and RBCM-30. At the initial stage of loading (0%–30% ε_f), the internal cracks of the RBCM were closed by compression, and only a few AE events occurred in the internal backfill. With continuous loading (30%–50%ε_f), the internal cracks of the RBCM began to evolve gradually, the AE events began to increase gradually, and they mainly gathered in the internal backfill. When loading to 50%–80% ε_f, the cracks further increase, the AE events also increase, and they gradually appear in the external rock. When loading to 80%–100% ε_f, AE events begin to gather in the external rock, and macro penetration failure also occurs. At the later stage of loading (100%–120% ε_f), AE events do not increase significantly. Because the RBCM has not completely lost its bearing capacity, the macro crack will continue to expand, and a small number of AE events will appear in the external rock.

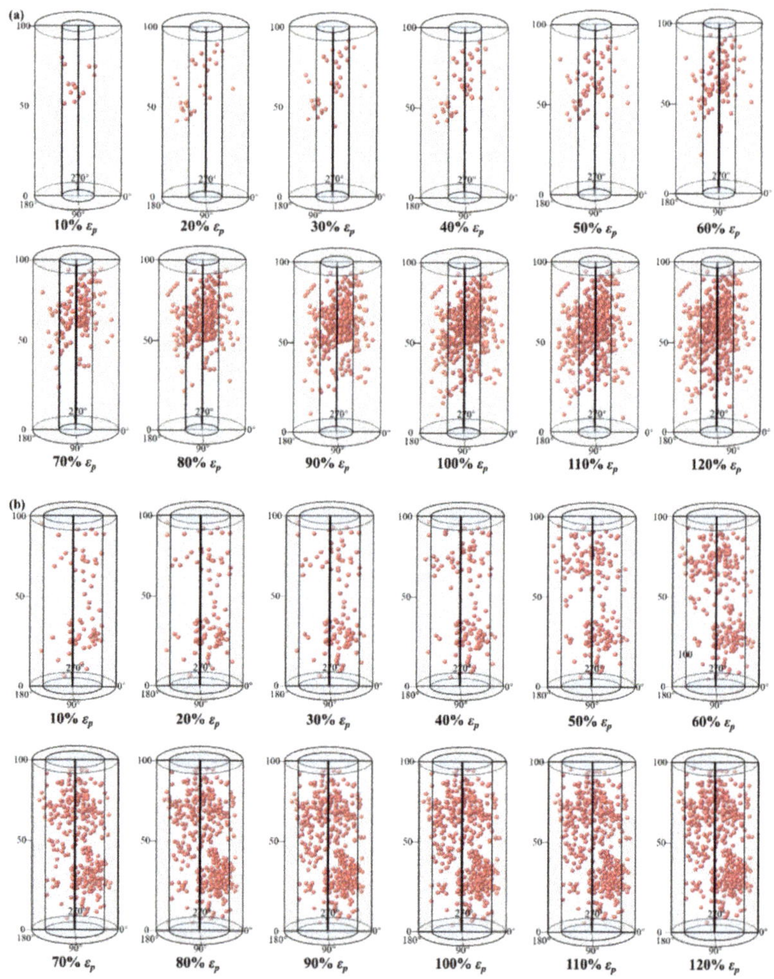

Figure 13. Temporal and spatial evolution of the AE events: (**a**) RBCM-10; (**b**) RBCM-30.

3.5. Temporal and Spatial Evolution of Microcracks

The laboratory test is the most direct means to study the internal damage evolution of the RBCM, but experimental research often cannot intuitively show the spatio-temporal evolution characteristics of internal cracks in the RBCM. At present, scholars at home and abroad often use Particle Flow Code (PFC) software to reveal the internal damage mechanism of materials. By establishing a numerical model consistent with the test, and then giving reasonable parameters, the results can also be very accurate. In addition, by setting a specific program, the PFC can intuitively show when and where the sample will be damaged, and can simulate the damage crack shape through a specific module [45,46].

Unlike other continuous models, PFC software can accurately reveal the basic mechanical behavior of rock-like materials under uniaxial compression, and can intuitively show the fracture evolution process in the sample. When using PFC-3D software to simulate rock-like materials, the contact model between particles usually adopts a parallel bonding model (Figure 14a) [46]. Under axial load, if the local contact force between particles exceeds the shear strength of the contact model, a shear crack will occur between the two particles. In the model, the shear crack is represented by a green circle (Figure 14b). If the local contact force between particles exceeds the tensile strength of the contact model, a tensile crack will occur between the two particles. In the model, the tensile crack is represented by a red circle (Figure 14b).

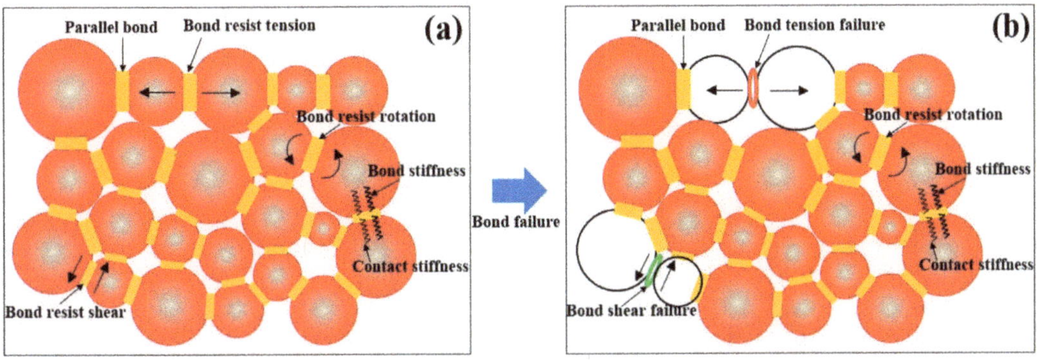

Figure 14. Parallel bonding model: (**a**) The stress pattern; (**b**) Failure mode.

The random function in PFC software is used to generate particles with different diameters in the calculation range, and the balance calculation is carried out to make the particles fill the whole columnar calculation range. After the initial equilibrium, a parallel bond model was set for the peripheral rocks and the internal backfill, respectively, and the loading rate of the model was determined. A monitoring system was arranged to monitor the number of cracks, stress and strain of the model in real time. The final stress–strain curve is shown in Figure 9 after multiple corrections of the model's mesoscopic parameters, and the evolution process of internal cracks in the RBCM sample is finally obtained. The results are shown in Figure 15 (RBCM-10 and RBCM-30). At the initial stage of loading, the local contact force between the RBCM particles is less than its bonding strength, and the stress and strain of the RBCM increase, but there are almost no cracks. With the continued loading of the RBCM, the local contact force between the particles is gradually greater than the bond strength, the bond between some particles is damaged, and cracks are generated. Because the bond strength between the backfill particles is less than that between the rock particles, the cracks first appear between the internal backfill particles. In the middle stage of loading, the bond between most of the backfill particles was destroyed, and the cracks spread all over the backfill. At the same time, the local contact force between rock particles begins to exceed its bond strength, and the cracks begin to appear in the surrounding rocks. In the later stage of loading, the contact force between rock particles is generally greater than its bond strength. The crack evolves on a large scale in the surrounding rock, and gradually forms a macro failure through the crack, which eventually leads to the overall instability and failure of the RBCM.

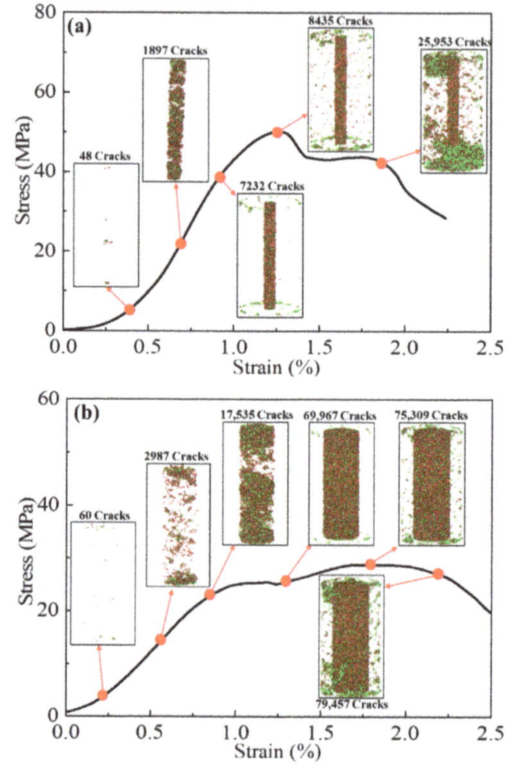

Figure 15. Temporal and spatial evolution of a crack: (**a**) RBCM-10; (**b**) RBCM-30.

In order to study the spatial distribution of the internal cracks of the RBCM after compression failure, 10 sections were selected for the numerical model after RBCM model failure, and the results are shown in Figure 16.

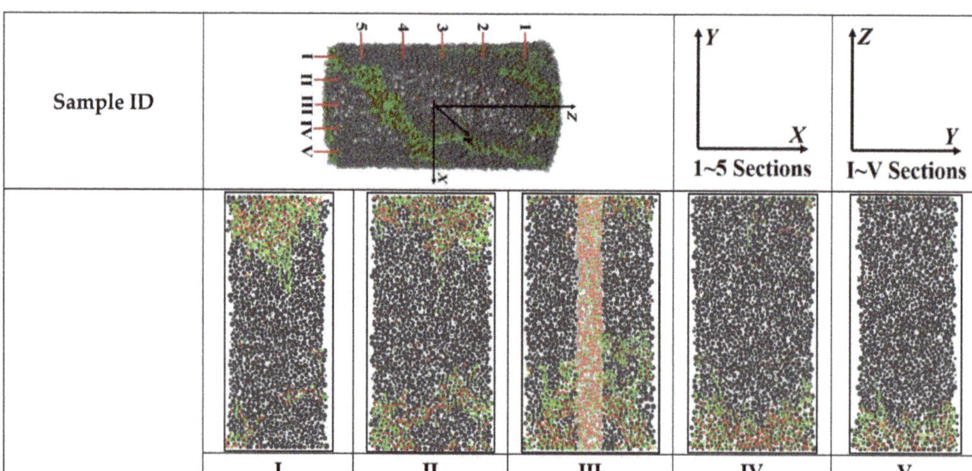

Figure 16. *Cont.*

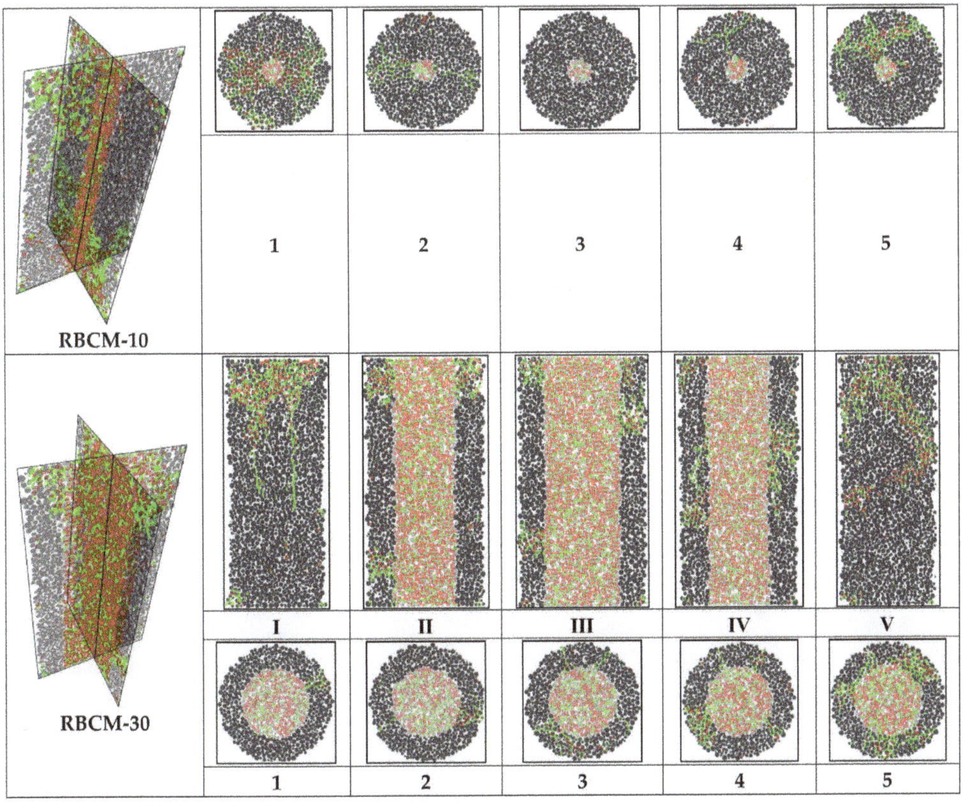

Figure 16. Spatial distribution of the cracks after the samples' failure.

For the RBCM-10, the cracks of section I are mainly distributed at the top of the sample. The cracks of section II are mainly distributed at the upper and lower ends of the sample. Section III contains two parts of rock and backfill; the cracks are mainly distributed in the internal backfill, and some cracks are also distributed at the bottom of the peripheral rock. The cracks of sections IV and V are mainly distributed at the bottom of the sample. In addition, the cracks of sections 1 and 2 are mainly distributed in the internal backfill and the left and right end of the peripheral rock. The cracks of section 3 are mainly distributed in the internal backfill, and there are almost no cracks in the peripheral rock. The cracks of sections 4 and 5 are mainly distributed in the internal backfill and at the top of the peripheral rock. In general, the cracks of RBCM-10 are mainly distributed in the internal backfill and the upper and lower ends of the peripheral rock.

For the RBCM-30, the cracks of sections of I and V are mainly distributed at the top of the rock. The cracks of sections II–IV are mainly distributed in the internal backfill and the top of the peripheral rock. The cracks of sections 1–3 are mainly distributed in the internal backfill, and there are few cracks in the peripheral rock. The cracks of section 4 are mainly distributed in the internal backfill and the left end of the peripheral rock. The cracks of section 5 are mainly distributed in the internal backfill and the upper and lower ends of the peripheral rock. In general, the cracks of the RBCM-30 are mainly distributed in the internal backfill and the top and bottom ends of the peripheral rocks.

Considering the spatial distribution law of cracks in RBCM-10 and RBCM-30, most of the cracks are distributed in the internal backfill, and a few cracks are distributed at the upper and lower ends of the peripheral rock. The evolution of meso tension cracks and shear cracks in the model leads to macro penetration failure.

4. Discussion

In Section 3.3, the damage evolution equation of the RBCM was established, and the damage evolution characteristics of the RBCM were studied. In Section 3.4, the AE correlation fractal dimension of the RBCM was calculated, and the qualitative relationship between AE correlation fractal dimension, damage evolution characteristics, and strain gradient were analyzed. The analysis results show that the AE correlation fractal dimension is inversely proportional to the strain gradient, while the damage evolution characteristics are directly proportional to the strain gradient. The damage directly represents the evolution process of cracks in the RBCM, and the AE correlation fractal dimension is a summary of the AE parameters released by the RBCM during loading. Therefore, the AE correlation fractal dimension can indirectly represent the damage evolution process of the RBCM to a certain extent. The AE correlation fractal dimension must have some internal relationship with the damage evolution characteristics.

In order to study the internal functional relationship between the AE correlation fractal dimension and damage value, the relationship curve between the AE correlation fractal dimension and damage evolution characteristics was established. The results are shown in Figure 17. The AE correlation fractal dimension is inversely proportional to the damage evolution. This is consistent with the research conclusion of reference [31]. Reference [31] shows that with the increase of stress percentage, the fractal dimension of the CPB decreases under uniaxial stress, while the damage variable of the CPB increases gradually. In addition, the damage variable increases more rapidly. When the fractal dimension decreases to a certain level, the fractal dimension no longer decreases significantly, and the CPB damage variable increases rapidly.

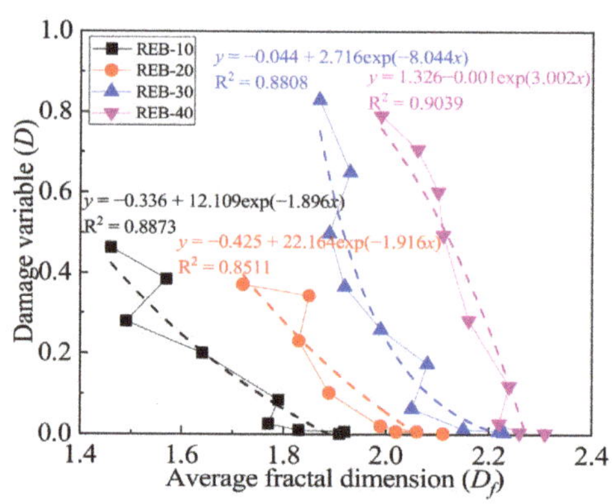

Figure 17. The relationship between the damage variable and the average fractal dimension.

Take the RBCM-20 as an example; when the AE correlation fractal dimension is 1.72, the damage value is 0.37. When the AE correlation fractal dimension increases to 1.85, 1.89, and 2.06, respectively, the damage values decrease to 0.34, 0.10, and 0.01, respectively. A variety of functions are used to fit the relationship between the AE correlation fractal dimension and damage value. The results show that the exponential function can better characterize the relationship between them:

$$D = a_4 + b_4 \cdot \exp(c_4 \cdot D_f) \tag{13}$$

where D is the damage variable of the RBCM, D_f is the AE correlation fractal dimension of the RBCM, and a_4, b_4, and c_4 are relevant parameters.

Under the action of a load, the internal stress of the RBCM increases continuously and gradually exceeds the bond strength between the particles. Cracks are generated between the particles, resulting in damage accumulation, and releasing energy to the outside in the form of acoustic emission. Therefore, during compression, stress behavior, damage characteristics, AE events, and microcracks must be interrelated. Figure 18 shows the composite diagram of the stress–strain, damage value, AE event, and cracks of the RBCM-30. When the internal stress of RBCM-30 increases to 4.57 MPa, its damage value evolves to 0.01, the number of microcracks increases to 60, and the AE events are mainly concentrated in the internal backfill. When the internal stress increases to 12.42 MPa, the damage value evolves to 0.02, the number of microcracks increases to 2987, and the AE events accumulate rapidly in the internal backfill. When the internal stress increases to 21.92 MPa, the damage value evolves to 0.06, the number of microcracks increases to 17,535, and the AE event begins to expand from the internal backfill to the surrounding rock. When the internal stress increases to 28.63 MPa, the damage value evolves to 0.34, the number of microcracks increases to 75,309, and the increment of the microcracks is the largest at this stage; the AE events accumulate rapidly in the surrounding rocks. When the internal stress reaches the peak and decreases to 26.83 MPa, the damage value evolves to 0.65, the number of microcracks increases slightly to 79,457, and the AE events increase slightly in the whole sample.

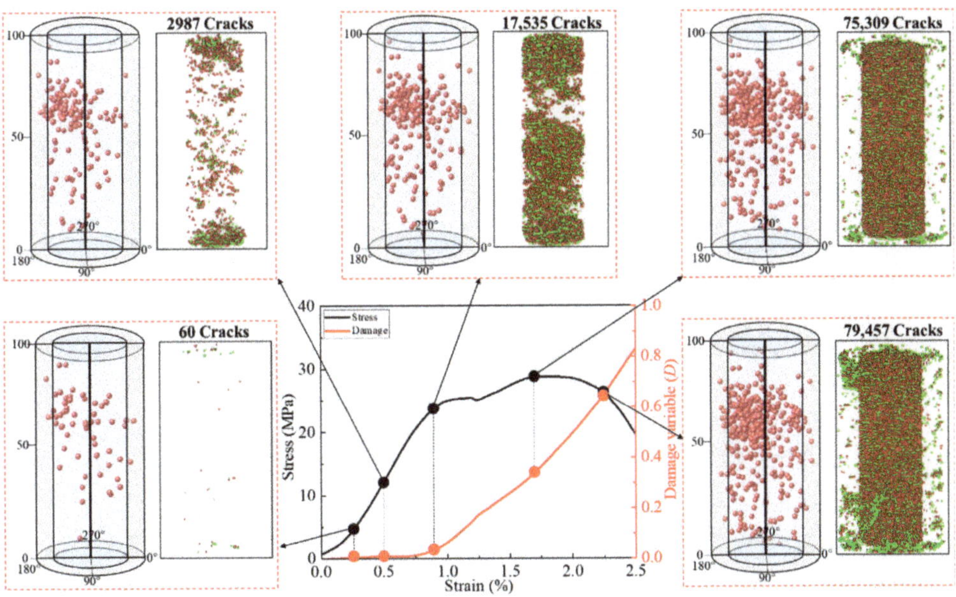

Figure 18. The composite diagram of the stress, damage, AE events, and cracks of the RBCM-30.

5. Conclusions

In this paper, firstly, the macroscopic mechanical behavior of the RBCM was analyzed, and the damage constitutive model of the RBCM was established. Secondly, the AE ring counting correlation fractal dimension was deduced. Finally, with the help of PFC software, the microcrack distribution was dynamically displayed. The main conclusions are as follows:

(1) The stress–strain curve of the RBCM can be divided into five stages: the pore compaction closure stage, the linear elasticity stage, the yield stage, the peak strength stage, and the residual stress stage; some stress–strain curves show double peaks. The σ^p, ε^p, and E decreased with the increase of the internal backfill diameter.

(2) The U^T shows an S-shaped distribution. The U^E shows an inverted U-shaped distribution. The U^D first increases slowly and then increases rapidly. The $U^T{}_p$, $U^E{}_p$, and $U^D{}_p$ all decrease with the increase of the backfill diameter. The $U^E{}_p/U^D{}_p$ and $U^E{}_p/U^T{}_p$ decrease and the $U^D{}_p/U^T{}_p$ increases with the increases of the backfill diameter.
(3) A damage constitutive model of the RBCM was established. The damage evolution curves and damage rate evolution curves of the RBCM show S-shaped and U-shaped distributions, respectively.
(4) The AE correlation fractal dimension of the RBCM was deduced and calculated. The AE correlation fractal dimension decreases with the increase of the strain gradient and damage value, and the AE correlation fractal dimension presents a linear and an exponential function with them, respectively. The AE events' localization results show that the damage first accumulates in the internal backfill. Then, the damage extends into the surrounding rock. Finally, the damage accumulates rapidly in the surrounding rock.
(5) The internal microcracks of the RBCM are caused by the local tensile stress or shear stress between particles exceeding the bond strength between particles. With the increase of stress, microcracks first appear and gather in the internal backfill of the RBCM, and then microcracks also appear and gather in the peripheral rock, which together leads to the macro penetration failure of the RBCM.

Author Contributions: Conceptualization, J.W. and W.S.; methodology, C.Z.; software, J.W.; validation, Y.Z.; formal analysis, Y.Z.; investigation, C.Z.; resources, W.S.; data curation, J.W.; writing-original draft preparation, C.Z.; writing-review and editing, J.W.; visualization, J.W.; supervision, Y.Z.; project administration, W.S; funding acquisition, W.S. All authors have read and agreed to the published version of the manuscript.

Funding: This experimental work was financially supported by the National Natural Science Foundation of China (Grant No. 51974012), the National Natural Science Foundation of China Youth Fund (51804016), and the China Postdoctoral Science Foundation (2021M690361).

Acknowledgments: The experimental works described in this study were conducted at the Key Laboratory of High-Efficient Mining and Safety of Metal Mines of the Ministry of Education in the University of Science and Technology Beijing. The authors gratefully acknowledge the staff and students at the laboratory for their technical help during the testing.

Conflicts of Interest: The authors declare no conflict of interest.

References

1. Wang, J.; Fu, J.; Song, W. Mechanical properties and microstructure of layered cemented paste backfill under triaxial cyclic loading and unloading. *Constr. Build. Mater.* **2020**, *257*, 119540. [CrossRef]
2. Wang, J.; Fu, J.; Song, W.; Zhang, Y.; Wu, S. Acoustic emission characteristics and damage evolution process of layered cemented tailings backfill under uniaxial compression. *Constr. Build. Mater.* **2021**, *295*, 123663. [CrossRef]
3. Xu, W.; Cao, Y.; Liu, B. Strength efficiency evaluation of cemented tailings backfill with different stratified structures. *Eng. Struct.* **2019**, *180*, 18–28. [CrossRef]
4. Xu, W.-B.; Liu, B.; Wu, W.-L. Strength and deformation behaviors of cemented tailings backfill under triaxial compression. *J. Central South Univ.* **2020**, *27*, 3531–3543. [CrossRef]
5. Wang, J.; Fu, J.; Song, W.; Zhang, Y. Viscosity and Strength Properties of Cemented Tailings Backfill with Fly Ash and Its Strength Predicted. *Minerals* **2021**, *11*, 78. [CrossRef]
6. Ferdi, C.; Yunus, A.; Marco, C. Effect of Desliming of Tailings on the Fresh and Hardened Properties of Paste Backfill Made from Alkali-Activated Slag. *Adv. Mater. Sci. Eng.* **2020**, *2020*, 4536257.
7. Hou, J.; Guo, Z.; Li, J.; Zhao, L. Study on triaxial creep test and theoretical model of cemented gangue-fly ash backfill under seepage-stress coupling. *Constr. Build. Mater.* **2021**, *273*, 121722. [CrossRef]
8. Xiu, Z.; Wang, S.; Ji, Y.; Wang, F.; Ren, F.; Nguyen, V.-T. Loading rate effect on the uniaxial compressive strength (UCS) behavior of cemented paste backfill (CPB). *Constr. Build. Mater.* **2021**, *271*, 121526. [CrossRef]
9. Zhao, K.; Yu, X.; Zhu, S.T.; Zhou, Y.; Wang, Q.; Wang, J.Q. Acoustic emission investigation of cemented paste backfill pre-pared with tantalum-niobium tailings. *Constr. Build. Mater.* **2020**, *237*, 117523. [CrossRef]
10. Doherty, J.P.; Wood, D.M. A bonding and damage constitutive model for lightly cemented granular material. *Comput. Geotech.* **2020**, *127*, 103732. [CrossRef]

11. Dzaye, E.D.; De Schutter, G.; Aggelis, D.G. Monitoring early-age acoustic emission of cement paste and fly ash paste. *Cem. Concr. Res.* **2020**, *129*, 105964. [CrossRef]
12. Sun, Q.; Li, B.; Tian, S.; Cai, C.; Xia, Y. Creep properties of geopolymer cemented coal gangue-fly ash backfill under dynamic disturbance. *Constr. Build. Mater.* **2018**, *191*, 644–654. [CrossRef]
13. Li, H.; Wu, A.; Wang, H. Evaluation of short-term strength development of cemented backfill with varying sulphide contents and the use of additives. *J. Environ. Manag.* **2019**, *239*, 279–286. [CrossRef]
14. Fang, K.; Fall, M. Insight into the mode I and mode II fracture toughness of the cemented backfill-rock interface: Effect of time, temperature and sulphate. *Constr. Build. Mater.* **2020**, *262*, 120860. [CrossRef]
15. Fang, K.; Cui, L.; Fall, M. A coupled chemo-elastic cohesive zone model for backfill-rock interface. *Comput. Geotech.* **2020**, *125*, 103666. [CrossRef]
16. Koupouli, N.J.; Belem, T.; Rivard, P.; Effenguet, H. Direct shear tests on cemented paste backfill–rock wall and cemented paste backfill–backfill interfaces. *J. Rock Mech. Geotech. Eng.* **2016**, *8*, 472–479. [CrossRef]
17. Wang, Y.R.; Lu, H.J.; Wu, J. Experimental investigation on strength and failure characteristics of cemented paste back-fill-rock composite under uniaxial compression. *Constr. Build. Mater.* **2021**, *304*, 124629. [CrossRef]
18. Wu, W.L.; Xu, W.B.; Zuo, J.P. Effect of inclined interface angle on shear strength and deformation response of cemented paste backfill-rock under triaxial compression. *Constr. Build. Mater.* **2021**, *279*, 122478.
19. Xiu, Z.; Wang, S.; Ji, Y.; Wang, F.; Ren, F.; Nguyen, V.-T. The effects of dry and wet rock surfaces on shear behavior of the interface between rock and cemented paste backfill. *Powder Technol.* **2021**, *381*, 324–337. [CrossRef]
20. Zhao, L. Numerical investigation on the mechanical behaviour of combined backfill-rock structure with KCC model. *Constr. Build. Mater.* **2021**, *283*, 122782. [CrossRef]
21. Yu, X.; Kemeny, J.; Tan, Y.; Song, W.; Huang, K. Mechanical properties and fracturing of rock-backfill composite specimens under triaxial compression. *Constr. Build. Mater.* **2021**, *304*, 124577. [CrossRef]
22. Wang, J.; Fu, J.X.; Song, W.D.; Zhang, Y.F.; Wang, Y. Mechanical behavior, acoustic emission properties and damage evo-lution of cemented paste backfill considering structural feature. *Constr. Build. Mater.* **2010**, *261*, 119958. [CrossRef]
23. Shah, S.G.; Kishen, J.M.C. Fracture behavior of concrete-concrete interface using acoustic emission technique. *Eng. Fract. Mech.* **2010**, *77*, 908–924. [CrossRef]
24. Liu, S.; Li, X.; Li, Z.; Chen, P.; Yang, X.; Liu, Y. Energy distribution and fractal characterization of acoustic emission (AE) during coal deformation and fracturing. *Measurement* **2019**, *136*, 122–131. [CrossRef]
25. Anay, R.; Soltangharaei, V.; Assi, L.; DeVol, T.; Ziehl, P. Identification of damage mechanisms in cement paste based on acoustic emission. *Constr. Build. Mater.* **2018**, *164*, 286–296. [CrossRef]
26. Hou, Y.; Yin, S.; Chen, X.; Zhang, M.; Yang, S. Study on characteristic stress and energy damage evolution mechanism of cemented tailings backfill under uniaxial compression. *Constr. Build. Mater.* **2021**, *301*, 124333. [CrossRef]
27. Wang, J.; Song, W.; Cao, S.; Tan, Y. Mechanical properties and failure modes of stratified backfill under triaxial cyclic loading and unloading. *Int. J. Min. Sci. Technol.* **2019**, *29*, 809–814. [CrossRef]
28. Song, W.D.; Wang, J.; Tan, Y.Y.; Cao, S. Energy consumption and damage characteristics of stratified backfill under triaxial loading and unloading. *J. China Univ. Min. Technol.* **2017**, *45*, 1050–1057.
29. Xin, L. Meso-scale modeling of the influence of waste rock content on mechanical behavior of cemented tailings backfill. *Constr. Build. Mater.* **2021**, *307*, 124473. [CrossRef]
30. Zhao, K.; Yu, X.; Zhu, S.; Yan, Y.; Zhou, Y.; He, Z.; Song, Y.; Huang, M. Acoustic emission fractal characteristics and mechanical damage mechanism of cemented paste backfill prepared with tantalum niobium mine tailings. *Constr. Build. Mater.* **2020**, *258*, 119720. [CrossRef]
31. Zhou, Y.; Yan, Y.; Zhao, K.; Yu, X.; Song, Y.; Wang, J.; Suo, T. Study of the effect of loading modes on the acoustic emission fractal and damage characteristics of cemented paste backfill. *Constr. Build. Mater.* **2021**, *277*, 122311. [CrossRef]
32. He, Z.; Zhao, K.; Yan, Y.; Ning, F.; Zhou, Y.; Song, Y. Mechanical response and acoustic emission characteristics of cement paste backfill and rock combination. *Constr. Build. Mater.* **2021**, *288*, 123119. [CrossRef]
33. Wang, J.; Fu, J.; Song, W.; Zhang, Y. Mechanical properties, damage evolution, and constitutive model of rock-encased backfill under uniaxial compression. *Constr. Build. Mater.* **2021**, *285*, 122898. [CrossRef]
34. Ma, Q.; Tan, Y.; Liu, X.; Gu, Q.; Li, X. Effect of coal thicknesses on energy evolution characteristics of roof rock-coal-floor rock sandwich composite structure and its damage constitutive model. *Compos. Part B Eng.* **2020**, *198*, 108086. [CrossRef]
35. Liu, W.Z.; Chen, J.T.; Guo, Z.P.; Yang, H.Z.; Xie, W.W.; Zhang, Y.D. Mechanical properties and damage evolution of ce-mented coal gangue-fly ash backfill under uniaxial compression: Effects of different curing temperatures. *Constr. Build. Mater.* **2021**, *305*, 124820. [CrossRef]
36. Cui, L.; Fall, M. An evolutive elasto-plastic model for cemented paste backfill. *Comput. Geotech.* **2016**, *71*, 19–29. [CrossRef]
37. Zhou, Y.; Yu, X.; Guo, Z.; Yan, Y.; Zhao, K.; Wang, J.; Zhu, S. On acoustic emission characteristics, initiation crack intensity, and damage evolution of cement-paste backfill under uniaxial compression. *Constr. Build. Mater.* **2021**, *269*, 121261. [CrossRef]
38. Qiu, J.; Zhou, Y.; Vatin, N.I.; Guan, X.; Sultanov, S.; Khemarak, K. Damage constitutive model of coal gangue concrete under freeze-thaw cycles. *Constr. Build. Mater.* **2020**, *264*, 120720. [CrossRef]
39. Liu, X.S.; Tan, Y.L.; Ning, J.G.; Lu, Y.W.; Gu, Q.H. Mechanical properties and damage constitutive model of coal in coal-rock combined body. *Int. J. Rock Mech. Min. Sci.* **2018**, *110*, 140–150. [CrossRef]

40. Fu, J.; Wang, J.; Song, W. Damage constitutive model and strength criterion of cemented paste backfill based on layered effect considerations. *J. Mater. Res. Technol.* **2020**, *9*, 6073–6084. [CrossRef]
41. Xie, H.P.; Ju, Y.; Li, L.Y. Criteria for strength and structural failure of rocks based on energy dissipation and energy release principles. *Chin. J. Rock Mech. Eng.* **2005**, *24*, 3003–3010.
42. Jia, Z.Q.; Li, C.B.; Zhang, R.; Wang, M.; Gao, M.Z.; Zhang, Z.T.; Zhang, Z.P.; Ren, L.; Xie, J. Energy Evolution of Coal at Dif-ferent Depths Under Unloading Conditions. *Rock Mech. Rock Eng.* **2019**, *52*, 4637–4649. [CrossRef]
43. Guo, H.F.; Song, D.Z.; He, X.Q.; Lou, Q.; Qiu, L.M. Fractal characteristics of acoustic emission in different damage degrees of impact coal. *Coal Sci. Technol.* **2021**, *49*, 38–46.
44. Gong, C.; Zhao, K.; Bao, H.; Zhao, K.; Zeng, P.; Wang, W.J. Acoustic emission source evolution characteristics and fractal features during creep failure of red sandstone. *Rock Soil Mech.* **2021**, *42*, 1–13.
45. Wang, J.; Fu, J.X.; Song, W.D.; Zhang, Y.F.; Wang, Y. Particle flow simulation of mechanical properties and microcrack evolution characteristics of rock-backfill combined model. *J. China Univ. Min. Technol.* **2020**, *49*, 453–462.
46. Cheng, A.; Shu, P.; Deng, D.; Zhou, C.; Huang, S.; Ye, Z. Microscopic acoustic emission simulation and fracture mechanism of cemented tailings backfill based on moment tensor theory. *Constr. Build. Mater.* **2021**, *308*, 125069. [CrossRef]

minerals

MDPI

Article

The Energy Dissipation Mechanism and Damage Constitutive Model of Roof–CPB–Floor (RCF) Layered Composite Materials

Jie Wang [1,2,*], Chi Zhang [1,2,*], Jianxin Fu [1,2], Weidong Song [1,2] and Yongfang Zhang [1,2]

1 School of Civil and Resources Engineering, University of Science and Technology Beijing, Beijing 100083, China; fjx@ustb.edu.cn (J.F.); songwd@ustb.edu.cn (W.S.); b1903098@ustb.edu.cn (Y.Z.)
2 State Key Laboratory of High-Efficient Mining and Safety of Metal Mines of Ministry of Education, University of Science and Technology Beijing, Beijing 100083, China
* Correspondence: wangjie123@ustb.edu.cn (J.W.); d202110040@xs.ustb.edu.cn (C.Z.)

Abstract: The stability of composite material that is composed of roof rock, cemented paste backfill (CPB), and floor rock has an important impact on safe mining within metal mines. In order to explore the mechanical properties, acoustic emission (AE), energy dissipation, and damage evolution of roof–CPB–floor (RCF) layered composite materials, uniaxial compression (loading rate 0.02 mm/min) AE tests on RCF materials with different CPB height ratios were performed. The test results show that: (1) the uniaxial compressive strength (*UCS*) and elastic modulus (E_R) of the RCF material were lower than those of the roof or floor rock and higher than that of the CPB. With the increase in the CPB's height ratio from 0.2 to 0.7, the (*UCS* and the E_R decreased from 18.42 MPa to 10.08 MPa and 3.15 GPa to 1.79 GPa, respectively, and the peak strain first decreased from 0.695 to 0.510 and then increased from 0.510 to 0.595. The (*UCS* increased as a polynomial function with the increase in the E_R. (2) The AE ring count first increased slowly, then increased rapidly, and finally maintained a high-speed increase. The AE cumulative ring count at the peak point decreased with the increase in the CPB height ratio. The energy dissipation showed that the elastic energy U_E accumulated slowly at first, then the dissipated energy U_D increased, and finally the U_E decreased and the U_D increased almost linearly. The U_T, U_E, U_D, U_E–U_T ratio and U_D–U_T ratio showed a decreasing trend, and the U_E–U_D ratio showed an increasing trend at the peak point with the increase in the CPB height ratio. (3) Two damage constitutive models were established based on the AE ring count and energy principle. The damage evolution process of RCF materials can be divided into three stages: the slow damage accumulation stage, stable damage growth stage, and rapid damage accumulation stage.

Keywords: RCF layered composite material (RCF); mechanical properties; acoustic emission (AE); energy dissipation; damage constitutive model

Citation: Wang, J.; Zhang, C.; Fu, J.; Song, W.; Zhang, Y. The Energy Dissipation Mechanism and Damage Constitutive Model of Roof–CPB–Floor (RCF) Layered Composite Materials. *Minerals* **2022**, *12*, 419. https://doi.org/10.3390/min12040419

Academic Editor: Abbas Taheri

Received: 15 February 2022
Accepted: 23 March 2022
Published: 29 March 2022

1. Introduction

Affected by the causes of its formation, the spatial shape of a gold ore body is usually complex, the dip angle of the ore body is usually small and approximately horizontal in its distribution, the thickness of the ore body is small, the thickness distribution is very uneven, and the surrounding rock is usually very broken [1–3]. Therefore, gold ore bodies are usually mined by the upward horizontal drift filling method. The rooms and pillars are arranged at intervals. First, the room is mined, then the goaf is filled with the tailings, and then the adjacent pillars are mined. In the process of pillar mining, the adjacent sides of the pillar are filled with cemented paste backfill (CPB), which forms a layered composite material with the roof and floor rock in order to bear pressure together to support the goaf and prevent the goaf from collapsing, so as to ensure the safety of the pillar mining [4–6]. In this situation, the overall stability of the layered composite material that is composed of roof rock, CPB, and floor rock is related to the safe mining and sustainable development of the mine. In addition, in the mining process, the thickness of the ore body will continue to change, so the CPB height of the layered composite material will also vary with the different mining areas. Therefore, it is of great theoretical and practical significance to study the influence of the CPB height on the overall stability of layered composite materials.

Recently, domestic and foreign scholars have conducted in-depth research on the common pressure bearing mechanism of backfill and rock composite structures by means of an indoor test, theoretical research, and numerical simulation and they have achieved fruitful research results [7–12], which provide a lot of reference basis for the research of this paper. Wang et al. [13] studied the effects of roughness and the inclination of a nonlinear interface on the strength and failure of the backfill–rock combination under uniaxial compression. He et al. [14] carried out a uniaxial compression test on a CPB–rock combination, discussed the influence of various CPB cement tailings ratios on the mechanical properties of the CPB–rock combination, and analyzed the crack evolution law with the help of fractal theory. Wu et al. [15] analyzed the influence of the interface angle on the shear strength and deformation behavior of a backfill–rock combination under triaxial compression. Fang et al. [16,17] studied the shear characteristics of the rock–backfill interface with the help of a chemo–elastic coupling cohesive zone model. Yu et al. [18] investigated the effects of the rock–backfill volume fraction and confining pressure on the strength characteristics and failure modes of the rock–backfill combination model with the help of a triaxial compression test and CT scanning. Zhao et al. [19] used the Karagozian–Case concrete (KCC) model to analyze the uniaxial compressive mechanical behavior of the backfill–rock composite structure. However, through the above reference analysis, almost all of the studies on the backfill–rock combination structure focus on the two-body model. In the actual filling mining process, the CPB usually forms a three-body composite structure with the roof rock and floor rock in order for the three elements to bear the pressure together. There are great differences in the mechanical properties and deformation behaviors between the three-body composite structure, the single structure, and the two-body composite structure. Only studying the single structure or the two-body composite structure cannot fully characterize the actual situation. Therefore, it is necessary to study the mechanical properties of a roof rock–CPB–floor rock (RCF) layered composite material.

Furthermore, with the popularization and application of acoustic emission (AE) technology, more and more scholars use AE technology to study the evolution law of the AE parameters of rock, backfill, and other materials in the compression process, so as to analyze the evolution process of internal damage in materials [20–24]. Wang et al. [25] carried out a uniaxial compression AE test on three-layered cemented paste backfill, discussed the evolution law of the AE ring count, energy count, and amplitude during loading, and analyzed the temporal and spatial evolution law of cracks during loading with the help of AE positioning technology. Cheng et al. [26] utilized laboratory testing and PFC numerical simulation and analyzed the temporal and spatial evolution law of AE events under the uniaxial compression of cemented tailings backfill (CTB). Zhou et al. [27] carried out uniaxial compression and Brazilian splitting AE tests on CPB and studied the relationships between the AE fractal dimension, energy dissipation, and damage variables. Zhao et al. [28] investigated the correlation between the AE fractal dimension and mechanical damage of cemented tailings backfill with different cement–sand ratios. Wang et al. [29] studied the influence of joint inclination on the dynamic evolution law of rock cracks with the help of AE acoustic emission positioning technology. In summary, these scholars have derived fruitful research results regarding the AE characteristics of backfill and rock; however, there are relatively few studies on the AE characteristics of the rock and backfill combined structure, which has a wide research prospect.

In addition, whether it is filling material or rock material, the process of internal crack initiation, propagation, and penetration under a load is the process of the continuous conversion of internal energy. The energy absorption and storage characteristics of the filling materials determine their strength characteristics and failure modes [30–34]. Hou et al. [35,36] revealed the relationship between the internal energy evolution and stress development of the CPB during the whole loading process. Qiu et al. [37] elaborated the AE energy dissipation characteristics of the CPB under uniaxial compression with the help of AE monitoring means. Xin et al. [38] studied the internal relationship between the internal compression energy, tensile energy, and waste rock content.

As many studies have analyzed the AE and internal energy evolution mechanism of filling materials by different means and from different angles, they have revealed the internal relationship between the damage, failure, and energy dissipation of filling materials. However, the above research objects were single medium materials or dual medium materials, which are characterized by a relatively uniform internal structure. Therefore, under the action of a load, their AE and internal energy distribution are also relatively uniform, and the relationship between their AE and energy dissipation is relatively simple [39]. The AE and energy consumption mechanism of rock–CPB–rock composite material are completely different from those of single-medium materials and dual-medium materials [40]. Firstly, there are more factors affecting the energy evolution of composite materials (rock properties, CPB properties, structural plane properties, etc.) [41,42]. Secondly, the internal structure of the composite materials is asymmetric and non-uniform and the stress distribution and deformation characteristics are more complex, resulting in very complex AE and energy evolution [10]. Finally, the composite materials have non-uniform structural planes and the energy transfer relationship between them is more complex and changeable [43]. Therefore, the AE and energy evolution of rock–CPB–rock composite materials need to be deeply studied.

The process of the gradual instability and failure of rock, backfill, and other materials under load is essentially a process of the continuous accumulation of internal damage [44]. Mastering the damage evolution process of rock and backfill is very important for its overall stability control. Wang et al. [45] constructed the damage constitutive model of rock-encased backfill with the help of the energy principle. Fu et al. [46] constructed the damage constitutive model of layered cemented tailings backfill considering its layered structure and analyzed the effects of layer numbers and the confining pressure on the damage evolution law. Liu et al. [47] constructed the damage constitutive model of cemented coal gangue–fly ash backfill and discussed the influence of curing temperature on damage evolution. Hou et al. [48] established a triaxial creep damage constitutive model of cemented gangue–fly ash backfill and analyzed its creep damage characteristics. Lin et al. [49] deduced the fatigue constitutive model of yellow sandstone based on the macro–micro coupled damage.

Consequently, the present AE detection and uniaxial compression experiments were carried out on RCF samples with the CPB height ratio of 0.2, 0.3, 0.4, 0.5, 0.6, and 0.7, respectively. Firstly, the (*UCS*, elastic modulus, and peak strain of each of the RCF samples were studied. The relationships between the (*UCS*, elastic modulus, and CPB height ratio, and the (*UCS* and elastic modulus were characterized. Secondly, the AE ring count characteristic and energy dissipation mechanism of the RCF samples with different CPB height ratios are analyzed. Finally, two damage constitutive models of the RCF materials have been established based on the AE and energy principle, and its damage evolution law is deeply discussed.

2. Materials and Methods

2.1. Research Background

The background of this present research is the Xinhui gold mine in Shandong Province, China. The length of the ore body is about 270 m, the thickness of the ore body in the gold mine in Shandong is 1–2.5 m, the dip angle of the ore body is about 20°, and the grade is about 3.15 g/t. The ore body is moderately stable and the stability of the hanging wall rock is poor. An upward horizontal drift filling mining method was adopted and the ore rooms and ore pillars were mined step by step. In the first step, the ore room is mined and the ore room goaf is filled with cemented paste backfill (CPB). In the second step, the ore pillar is mined. In the process of pillar mining, the CPB, roof rock, and floor rock form a layered composite material in order to bear the pressure together, so as to ensure the safety of the pillar mining. Due to the uneven thickness of the ore body, the height ratio of the layered composite material is in dynamic change. The actual model and laboratory model of the roof rock–CPB–floor rock (RCF) layered composite materials are shown in Figure 1. The stability of the RCF layered composite material has an important impact on two-step pillar mining, so it is of great theoretical and practical significance to study its mechanical properties and damage evolution.

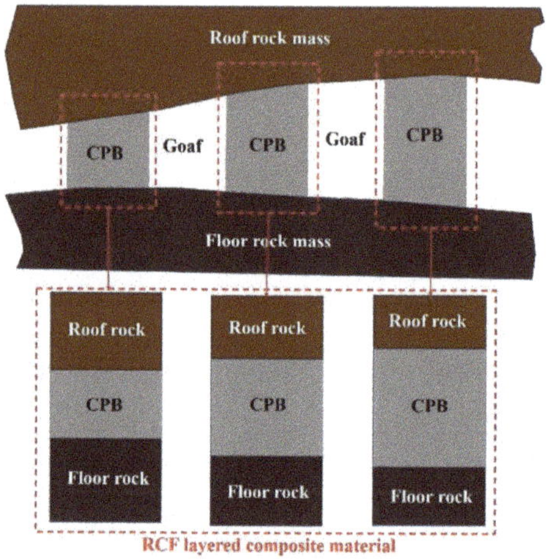

Figure 1. Layered composite material in mining field and laboratory of roof rock–CPB–floor rock (RCF).

2.2. Experimental Materials and Characteristics

The roof rock (Phyllite) that was used in the test came from the roof rock mass of the gold mine, as shown in Figure 2a. The floor rock (Hematite) that was used in the test came from the floor rock mass of the gold mine, as shown in Figure 2b. The tailings that were used in the test were from the full tailings of the mine. The cement that was used in the test was ordinary Portland cement (OPC) 42.5R. The water that was used in the test was laboratory tap water. The particle size distribution (PSD) curves of the full tailings and OPC are shown in Figure 3. The chemical compositions of the full tailings and OPC are shown in Table 1.

Figure 2. Test samples: (**a**) Roof rock sample; (**b**) Floor rock sample; (**c**) CPB sample; (**d**) RCF sample.

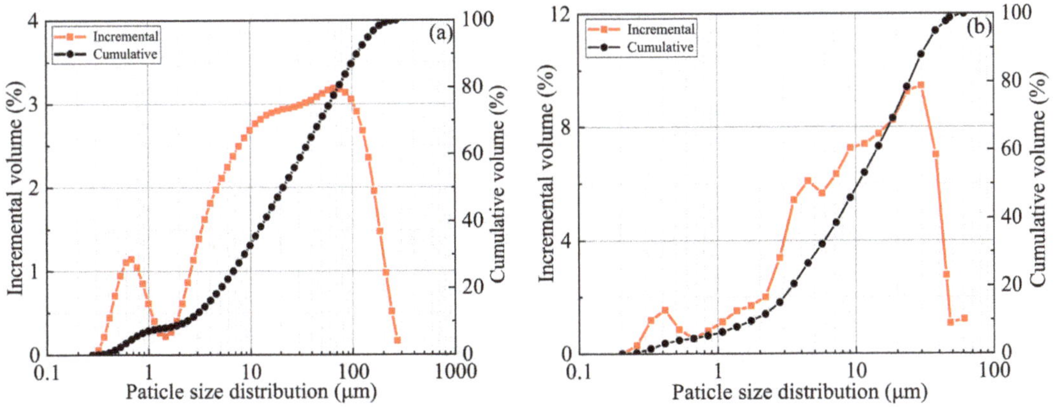

Figure 3. The PSD curves: (**a**) full tailings; (**b**) OPC.

Table 1. Chemical composition of the raw materials.

Sample	SiO_2	K_2O	CaO	P_2O_5	MgO	Al_2O_3	Fe_2O_3	SO_3
OPC (%)	20.10	0.37	61.8	/	1.57	5.11	2.91	1.98
Full tailing (%)	66.95	4.40	7.68	0.15	2.24	11.71	4.91	0.20

2.3. Sample Preparation

Four types of samples were prepared: (1) the intact roof rock sample (Figure 2a); (2) the intact floor rock sample (Figure 2b); (3) the intact CPB sample (Figure 2c) (the cement-to-tailings ratio of which was 1:4, the slurry concentration of which was 75%, and the curing age of which was 28 days); (4) the RCF layered composite material sample (Figure 2d). The overall size of the four types of samples was the same, with a diameter of Φ = 50 mm and a height of H = 100 mm. The height of the CPB in the middle of the RCF sample was set at 20, 30, 40, 50, 60, and 70 mm and the heights of the roof rock and floor rock were the same, which were set at 40, 35, 30, 25, 20, and 15 mm, respectively. The manufacturing process of the RCF sample was executed as follows: First, the prepared intact CPB sample and rock sample were cut according to the specified height. Then, the tailing slurry with the cement-to-tailings ratio of 1:4 and slurry concentration of 75% was prepared. Finally, the prepared slurry was used to paste the cut roof rock, CPB, and floor rock together to form an RCF sample with a total height of 100 mm. The sample design scheme and mechanical parameters are shown in Tables 2 and 3.

Table 2. Mechanical parameters and number of intact samples.

Intact Sample	Sample No.	(*UCS* (MPa)	Elastic Modulus (GPa)	Peak Strain (%)
Roof rock	RR	23.91	4.13	0.71
Cemented paste backfill	CPB	7.25	1.65	0.57
Floor rock	FR	41.00	5.64	0.92

2.4. Uniaxial Compression AE Test

A GAW-2000rock mechanics testing machine made by Chaoyang Test Instrument Co., Ltd (Changchun, China) was used to carry out a uniaxial compression test. The loading mode was set to displacement control and the loading rate was 0.5 mm/min. During the loading process, the computer system recorded the displacement and load of the testing machine in real-time. After the loading, the size parameters of the sample were input into the system and the system automatically calculated the strain and strain value of the sample throughout the whole loading process. At the same time, the peak stress, peak strain, and other parameters of the sample were also recorded. After the test, the data file was exported and could then be used in the various required formats. For the whole process of uniaxial compression, PCI-2 AE equipment made by Physical Acoustics Co., Ltd (Princeton, NJ, USA) was used for the AE signal acquisition. The frequency of the AE sensor was 140 kHz and the noise threshold was set to 45 dB.

Table 3. Mechanical parameters and number of RCF samples.

CPB Height (*h*/*H*) Ratio	**0.2**	**0.3**	**0.4**	**0.5**	**0.6**	**0.7**
Sample No.	RCF2	RCF3	RCF4	RCF5	RCF6	RCF7
RCF samples models		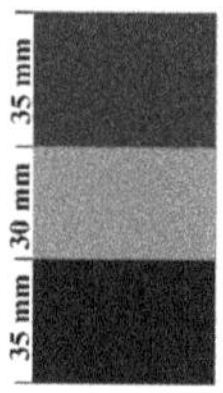	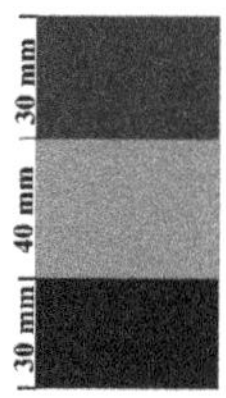		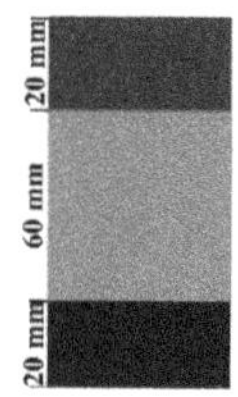	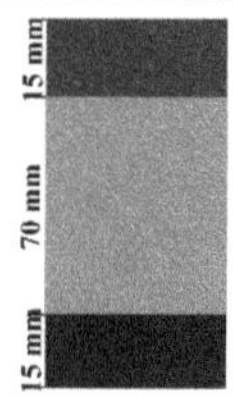
(*UCS* (MPa)	18.42	13.54	10.69	10.35	10.28	10.18
EM (GPa)	3.15	2.74	2.44	2.16	1.96	1.79
Peak strain (%)	0.695	0.567	0.514	0.510	0.542	0.595

3. Results and Analysis

3.1. Mechanical Properties of RCF Materials

The mechanical properties of the RCF layered composite materials are related to the safety and stability of the goaf roof. As can be seen from Table 1, the uniaxial compressive strength ((*UCS*) and elastic modulus (E_R) of the floor rock were the largest, followed by those of the roof rock, and then by those of the (*UCS* and E_R of the cemented paste backfill (CPB), which were the smallest. As can be seen from Table 2, the (*UCS* and E_R of the RCF material decreased with the increase in the height (h/H) ratio of the middle CPB and the (*UCS* and E_R of the RCF material were less than those of the roof rock and floor rock, but greater than those of the CPB.

3.1.1. Uniaxial Compressive Strength

Figure 4 shows the relationships between the (*UCS*′ (The normalized value of uniaxial compressive strength), (*UCS* ratio, and CPB height (h/H) ratio. As can be seen from Figure 4a, the (*UCS*′ of the RCF material decreased with the increase in the CPB height ratio. When the CPB height ratio increased from 0.2 to 0.7, the (*UCS*′ of the RCF material decreased from 1 to 0.41, 0.06, 0.02, 0.01, and 0. With the increase in the CPB height ratio of the RCF material, the proportion of the CPB's weak sandwich increased. Under the action of the load, the overall bearing capacity decreased and the (*UCS*′ decreased. An exponential function can be used to characterize the relationship between the (*UCS*′ of the RCF material and the CPB height ratio:

$$(UCS' = -0.02 + 7.29 \times \exp[-9.77 \times (h/H)] \tag{1}$$

where $(UCS'$ is the normalized value of uniaxial compressive strength of the RCF material and h/H is the CPB height ratio.

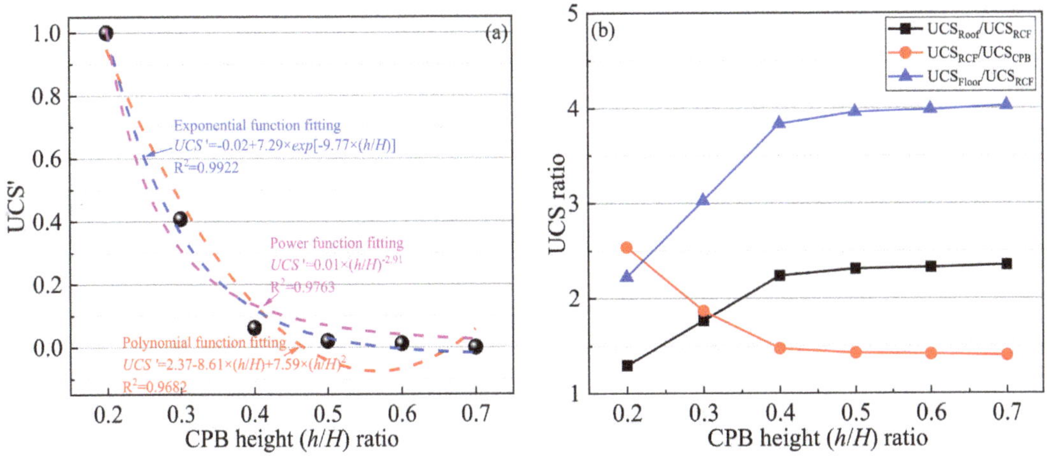

Figure 4. (*UCS* of RCF material: (**a**) (*UCS* and h/H ratio of RCF material; (**b**) (*UCS* ratio and h/H ratio.

The (*UCS* ratios of the roof rock (RR) and floor rock (FR) to the RCF material represents the (*UCS* deterioration degree of the RCF material on the RR and FR, respectively. The (*UCS* ratio of the RCF material to CPB represents the (*UCS* strengthening degree of the RCF material on the CPB. Figure 4b shows that the (UCS_{Roof}–(UCS_{RCF} ratio and the (UCS_{Floor}–(UCS_{RCF} ratio increased with the increase in the CPB height ratio. When the CPB height ratio increased from 0.2 to 0.7, the (UCS_{Roof}–(UCS_{RCF} ratio increased from 1.298 to 2.349 and the (UCS_{Floor}–(UCS_{RCF} ratio increased from 2.226 to 4.028. In addition, the (UCS_{RCF}–(UCS_{CPB} ratio decreased with the increase in the CPB height ratio. When the CPB height ratio increased from 0.2 to 0.7, the (UCS_{RCF}–(UCS_{CPB} ratio decreased from 2.541 to 1.404. When the CPB height ratio was greater than 0.4, the strength of the RCF material approached the strength of the CPB, the change rate of the RCF material's strength decreased, and the influence of the CPB height ratio on the RCF material's strength decreased significantly. This is because the height of the CPB layer began to be greater than that of the roof rock or floor rock layers when the CPB height ratio was greater than 0.4. the CPB, as the weak structure, began to be the main part of the RCF material and the failure of the CPB layer can be seen to be the decisive factor for the failure of the RCF structure. Therefore, the change of the RCF material's strength decreased significantly and was close to zero when the CPB height ratio was greater than 0.4.

3.1.2. Elastic Modulus

Figure 5 shows the relationship between the E_R'(The normalized value of elastic modulus), E_R ratio of the RCF material, and the CPB height ratio. As can be seen from Figure 5a, the E_R' of the RCF material decreased with the increase in the CPB height ratio. When the CPB height ratio increased from 0.2 to 0.7, the E_R' of the RCF material decreased from 1 to 0. With the increase in the CPB height ratio, the proportion of the CPB's weak sandwich increased, resulting in a decrease in the overall stiffness of the RCF samples. In addition, a polynomial function can be used to characterize the relationship between the E_R' of the RCF material and the CPB height ratio:

$$E_R' = 1.68 - 3.87 \times (h/H) + 2.10 \times (h/H)^2 \quad (2)$$

where E_R' is the normalized value of elastic modulus of the RCF material.

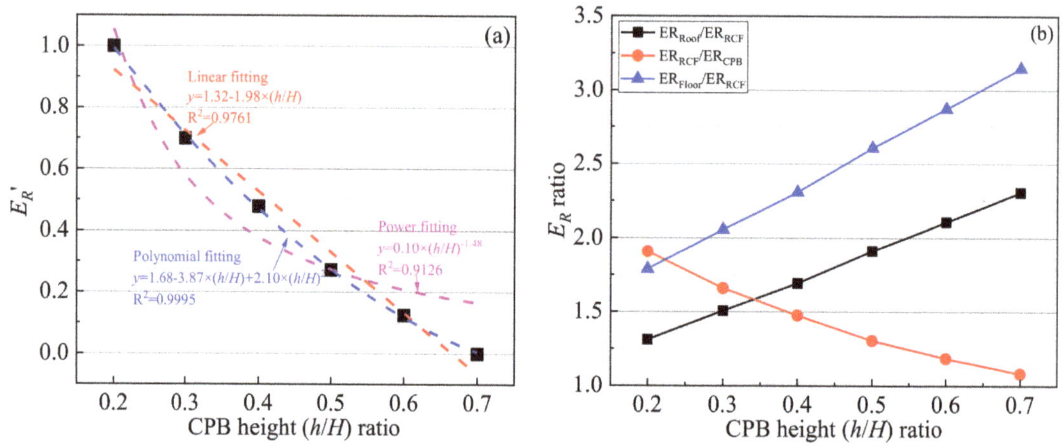

Figure 5. Elastic modulus of RCF material: (**a**) E_R and h/H ratio; (**b**) E_R ratio and h/H ratio.

Similarly, the E_R ratio of the RR and FR to the RCF material represents the E_R deterioration degree of the RCF material on the RR and FR, respectively. The E_R ratio of the RCF material to the CPB represents the E_R strengthening degree of the RCF material on the CPB. Figure 5b shows the E_R ratio. The ER_{Roof}–ER_{RCF} ratio and the ER_{Floor}–ER_{RCF} ratio increased with the increase in the CPB height ratio and the ER_{RCF}–ER_{CPB} ratio decreased with the increase in the CPB height ratio. When the CPB height ratio increased from 0.2 to 0.7, the ER_{Roof}–ER_{RCF} ratio and the ER_{Floor}–ER_{RCF} ratio increased from 1.311 to 2.307 and 1.790 to 3.151, and the ER_{RCF}–ER_{CPB} ratio decreased from 1.909 to 1.085.

3.1.3. Peak Strain

Figure 6 shows the relationship between the peak strain of the RCF material (the strain at its peak stress point) and the CPB height ratio. The peak strain of the RCF material first decreased and then increased with the increase in the CPB height ratio. When the CPB height ratio increased from 0.2 to 0.5, the peak strain of the RCF material decreased from 0.695% to 0.510%. When the CPB height ratio increased from 0.5 to 0.7, the peak strain of the RCF material began to increase from 0.510% to 0.577%. The peak strain of the RCF material reached its maximum value of 0.695% when the CPB height ratio was 0.2 and its minimum value of 0.510% when the CPB height ratio was 0.5. When the CPB height ratio was in the range of 0.2–0.5, with the increase in the CPB height ratio, the proportion of the rock layer (i.e., the roof rock and floor rock layers) decreased, the proportion of the CPB layer increased, and the decrease in the strength of the RCF material was greater than that of the elastic modulus. Therefore, the peak strain of the RCF material decreased gradually at this stage. When the CPB height ratio was greater than 0.5, the CPB layer became the main part of the RCF material and the influence of the rock layer on the RCF material was much less than that of the CPB layer. At this time, the decrease in the elastic modulus was the main change in the RCF material and the change in its strength was very small (Section 3.1.1). Therefore, with the increase in the proportion of the CPB layer, the peak strain of the RCF material began to increase.

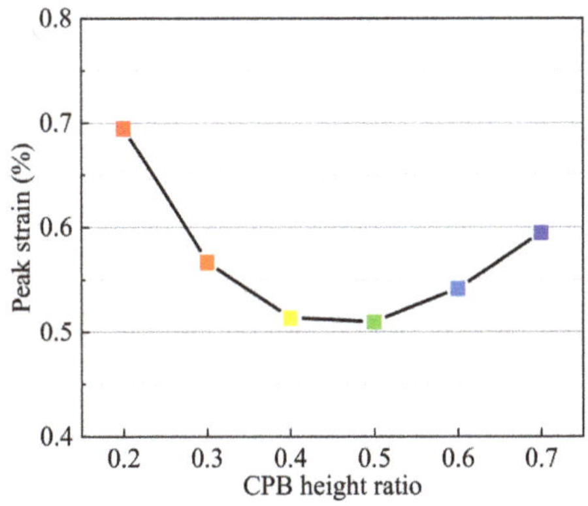

Figure 6. Peak strain and CPB height ratio of RCF material.

3.2. AE Ring Count Characteristic of RCF Materials

The stress-AE ring count characteristics of RCF materials with different CPB height ratios are shown in Figure 7. As can be seen from Figure 7, the change law of the AE ring counts of the different RCF samples is basically similar. At the initial stage of loading, the stress was small, the primary cracks in the RCF sample were closed, no new cracks were generated, and the AE ring count was small. With the increase in the load, the internal stress of the RCF sample increased, the primary crack began to expand, new cracks began to sprout, and the AE ring count increased slowly. With the continued increase in the load, the internal stress of the RCF sample gradually exceeded its yield limit, the sample showed yield failure, the cracks began to expand rapidly, the AE ring count increased rapidly, and the AE accumulation curve rose rapidly. When the load exceeded the ultimate load of the RCF sample, macro cracks began to appear, the number of the AE ring count continued to increase, and the AE ring count accumulation curve increased almost linearly. With the further application of the load, the RCF sample became unstable and damaged, but there was still a small AE ring count due to the cohesion and friction in the sample.

Furthermore, Figure 7a,b show that there were multiple peaks in the AE ring count curves of the RCF material, indicating that there were stress concentration phenomena in the RCF sample, resulting in a sudden increase in the cracks in the sample, and the stress and AE ring count accumulation curves show a steep upward trend. It can be seen from Figure 7c–f that the AE ring count curve of the RCF material continued to rise without multiple peaks and the stress and AE ring count accumulation curves showed a smooth upward trend.

Figure 8 shows the relationship between the AE cumulative ring count of the RCF material at its peak stress point and the CPB height ratio. The AE cumulative ring count decreased with the increase in the CPB height ratio. When the CPB height ratio increased from 0.2 to 0.7, the AE cumulative ring count of the RCF material at the peak stress point decreased from 925 to 735. When the CPB height ratio increased, the overall strength of the RCF material decreased, the internal released energy decreased, and the AE cumulative ring count decreased.

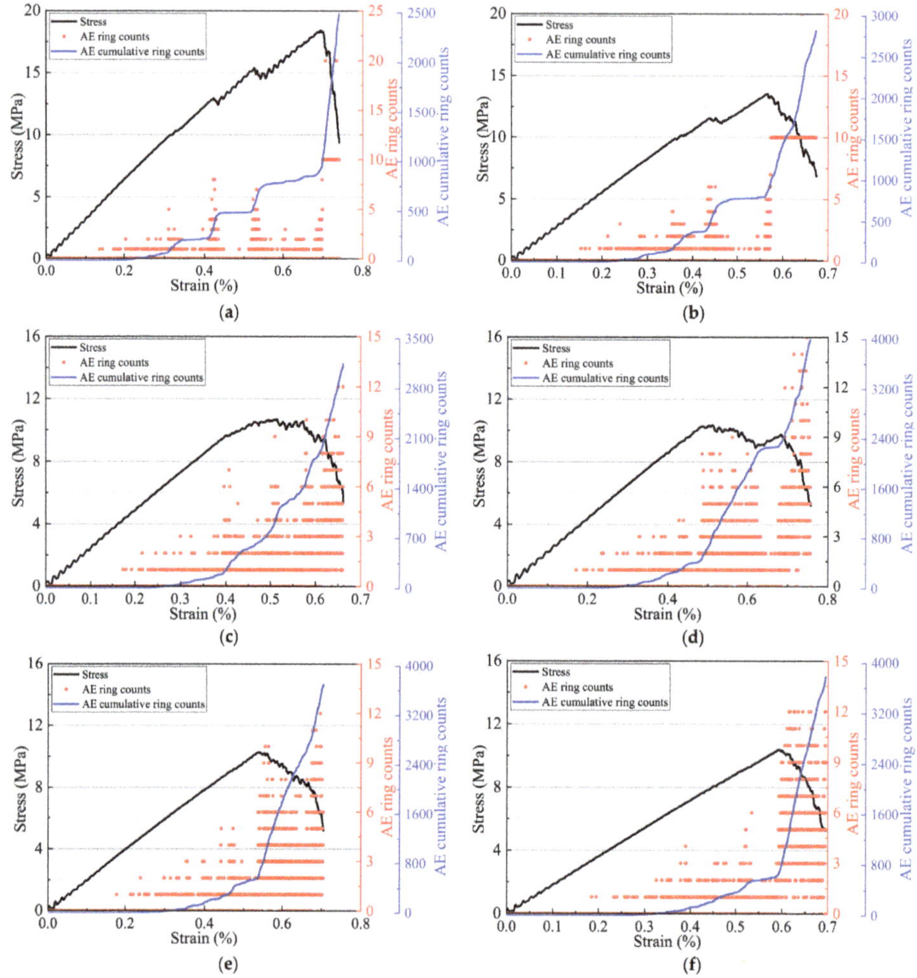

Figure 7. Stress-AE ring count–AE cumulative ring count curves: (**a–f**): RCF2–RCF7.

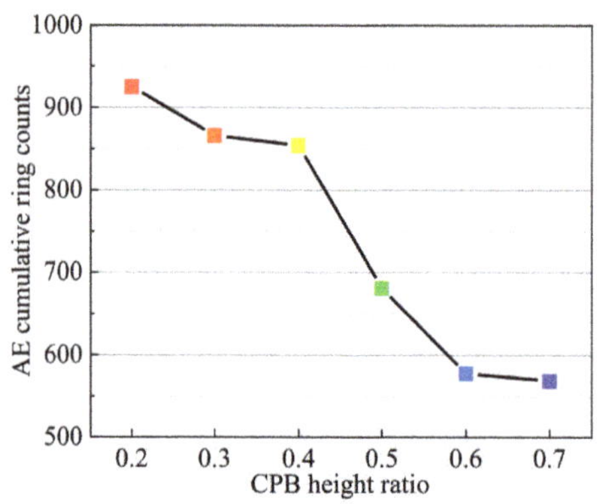

Figure 8. AE cumulative ring count of RCF material at peak point.

3.3. Energy Dissipation Mechanism of RCF Materials

The deformation and failure of the RCF samples is a process of continuous energy exchange with the external environment. During the loading process, the RCF sample continuously absorbs energy and converts that energy into releasable elastic energy and dissipated energy. When the energy that is absorbed by the RCF sample exceeds its energy storage limit, part of the releasable elastic energy will be released, resulting in the overall failure of the RCF sample. According to energy conservation [35,50]:

$$U_T = U_E + U_D \tag{3}$$

where U_T is the total energy that is absorbed by the sample, U_E is the releasable elastic properties that are stored in the sample, and U_D is the dissipated energy of the sample.

Figure 9 shows the internal energy evolution process of the RCF sample. In the initial stage, the RCF sample absorbs the energy and converts it into releasable elastic energy, which is stored in the sample. In the middle stage, the releasable elastic energy that is stored in the RCF sample exceeds its elastic limit and part of that absorbed energy is converted into dissipated energy and released to the outside world. In the later stage, the energy that is stored inside the RCF sample exceeds its energy storage limit, resulting in the failure of the sample and the energy that is stored in the sample is quickly released to the outside. The energy balance equation of the RCF sample under uniaxial compression is [50–52]:

$$U_T = \int_0^\varepsilon \sigma d\varepsilon \tag{4}$$

$$U_E = \frac{\sigma^2}{2E_R} \tag{5}$$

where E_R is the elastic modulus of the RCF sample.

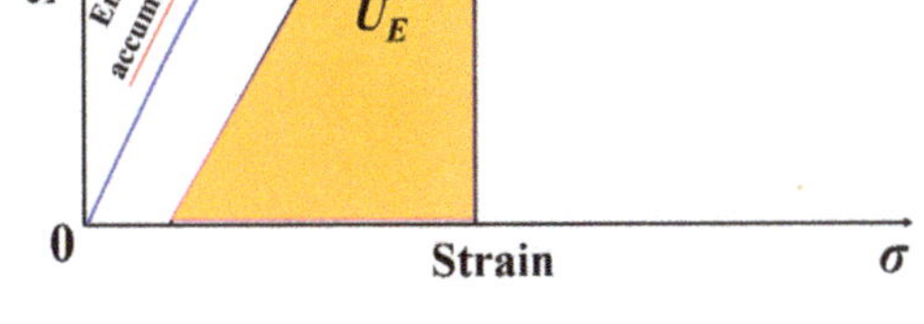

Figure 9. RCF sample energy evolution process.

According to Equations (4) and (5), the evolution curves of the U_T, U_E, and U_D can be calculated combined with the stress-strain curve of the RCF sample and the results of these calculations are shown in Figure 10. In the initial stage, all of the energy that was absorbed by the sample was converted into releasable elastic energy, all of the deformation of the sample was elastic deformation, and there was no damage inside the sample, so the dissipated energy at this stage was almost zero. In the middle stage, the load gradually exceeded the yield limit of the sample, the damage began to occur in the sample, and the dissipated energy began to increase. In the later stage, the damage to the sample accumulated rapidly, the releasable elastic energy began to decrease and the dissipated energy increased rapidly.

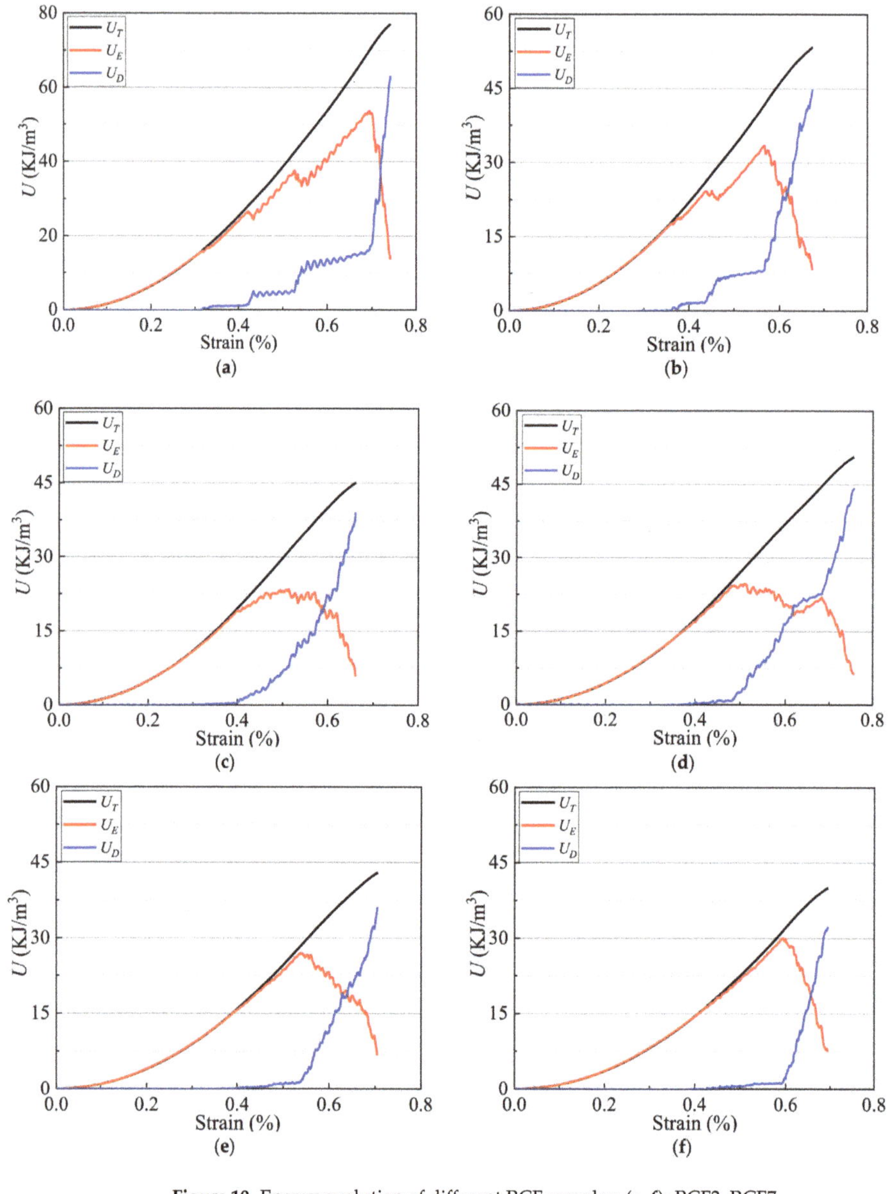

Figure 10. Energy evolution of different RCF samples: (**a**–**f**): RCF2–RCF7.

Figure 11 shows the three energies evolution curves with strain. Figure 11a shows that the evolution of the total energy U_T that was absorbed by the different RCF samples was basically similar, increasing slowly at first and then rapidly. In addition, the greater the CPB height ratio, the slower the cumulative speed of the U_T, and the smaller the U_T. Figure 11b shows that the storage process of the U_E of the different RCF samples was also basically similar. In the initial stage, the sample was compacted and the U_E accumulated slowly. In the middle stage, elastic deformation occurred, the U_E accumulated rapidly, and the elastic energy curve rose almost linearly. In the later stage, the load exceeded the bearing limit of the sample, resulting in the failure of the sample, the sample's release of energy to the outside, and a rapid decrease in the elastic energy. Figure 11c shows that, in the initial stage, there was no damage inside the sample and the U_D was almost zero. In the middle stage, the damage to the sample gradually accumulated and the U_D began to increase slowly. In the later stage, the damage accumulated rapidly and the U_D increased rapidly. The dissipated energy curves of the different RCF samples are not very distinguishable.

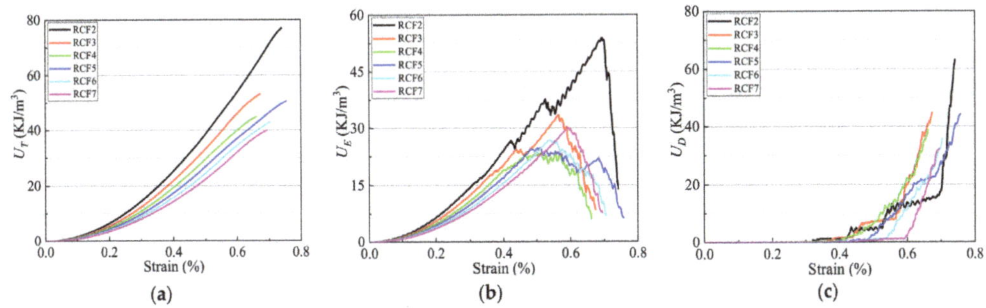

Figure 11. Different types energy evolution of RCF samples: (**a**) U_T; (**b**) U_E; (**c**) U_D.

Figure 12 shows the variation of the three energies at the peak point of the RCF samples with their CPB height ratio. Figure 12a shows that the U_T, U_E, and U_D at the peak point decreased with the increase in the CPB height ratio. When the CPB height ratio was 0.2, the U_T, U_E, and U_D were 70.07 KJ/m^3, 53.84 KJ/m^3, and 16.23 KJ/m^3, respectively. When the CPB height ratio increased to 0.3, the three energies decreased to 41.80 KJ/m^3, 33.47 KJ/m^3, and 8.33 KJ/m^3, respectively. When the CPB height ratio increased to 0.7, the three energies decreased to 31.57 KJ/m^3, 30.09 KJ/m^3, and 1.48 KJ/m^3, respectively. With the increase in the CPB height ratio, the proportion of CPB in the RCF samples increased. However, the energy storage capacity of the CPB was smaller than that of the rock. Therefore, the three energies gradually decreased with the increase in the CPB height ratio. Figure 12b shows the relationship between the energy ratio and the PCB height ratio. Overall, with the increase in the CPB height ratio, the ratio of the elastic energy to the total energy (U_E–U_T) increased, the ratio of the dissipated energy to the total energy (U_D–U_T) decreased, and the ratio of the elastic energy to the dissipated energy (U_E–U_D) increased. When the CPB height ratio was 0.2, the U_E–U_T, the U_D–U_T, and the U_E–U_D were 0.77, 0.23, and 3.32, respectively. When the CPB height ratio increased to 0.7, the U_E–U_T and the U_E–U_D increased to 0.90 and 9.43, respectively, and the U_D–U_T decreased to 0.10.

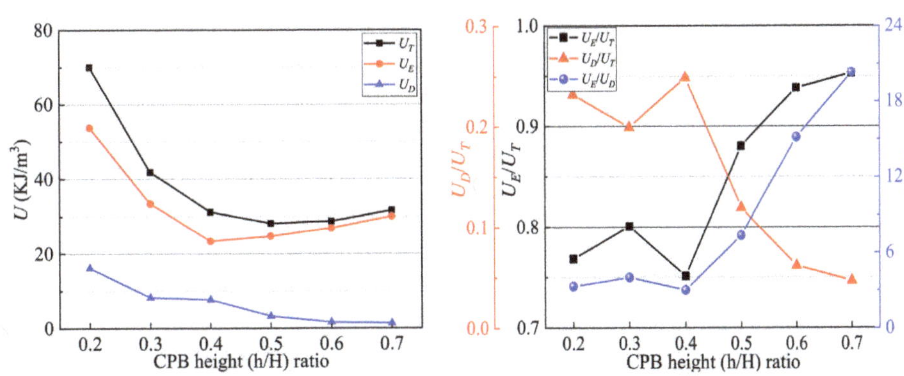

Figure 12. Energy dissipation and energy dissipation ratio of RCF sample at peak point: (**a**) energy dissipation; (**b**) energy dissipation ratio.

3.4. Damage Constitutive Model of RCF Materials

3.4.1. Model Construction

In order to study the damage evolution mechanism of RCF composite materials during uniaxial compression, the damage model that is based on AE ring count [53,54] shows that: assuming that the section area A is destroyed, the cumulative AE ring count of the RCF composite material is N. Then, when the damaged area of the section reaches A_d, the AE cumulative ring count is:

$$N_f = \frac{A_d}{A} N \tag{6}$$

It can be considered that the damage variable D_A in RCF combination materials is:

$$D_A = (1 - \frac{\sigma_r}{\sigma_p}) \frac{N_f}{N} \tag{7}$$

where D_A is the damage variable of the RCF material based on the AE ring count, σ_r is the residual stress of the RCF material, σ_p is the peak stress of the RCF material, N_f is the AE cumulative ring count, and N is the total AE ring count.

Then, the damage constitutive model of RCF composite materials, based on the AE ring count, can be expressed as:

$$\sigma = E\varepsilon(1 - D_A) = E\varepsilon[1 - (1 - \frac{\sigma_r}{\sigma_p}) \frac{N_f}{N}] \tag{8}$$

In addition, according to the research of Ma et al. [50,54], the damage variable of RCF composite materials can be characterized by the energy dissipation relationship:

$$D_E = (1 - \frac{\sigma_r}{\sigma_p}) \times \frac{U_D}{U_{D\max}} \times (1 + \frac{U_{Dp}}{U_{Tp}}) \tag{9}$$

where D_E is the damage variable of the RCF material based on the energy dissipation, U_D is the dissipation energy, $U_{D\max}$ is the cumulative dissipated energy, U_{Dp} is dissipation energy at the peak point, and U_{Tp} is total energy that is input into the RCF material at the peak point.

The damage constitutive model of RCF material, based on the energy dissipation under uniaxial compressive:

$$\sigma = (1 - D_E)E\varepsilon = \left[1 - (1 - \frac{\sigma_r}{\sigma_p}) \times \frac{U_D}{U_{D\max}} \times (1 + \frac{U_{Dp}}{U_{Tp}})\right] E\varepsilon \tag{10}$$

3.4.2. Model Verification

Based on the findings of other scholars [50,53–55], this paper constructs two damage constitutive models of RCF material under uniaxial compression, based on acoustic emission and energy dissipation; however, the rationality and reliability of these models still need to be fully verified. Figure 13 shows the comparison results between the two model curves that were constructed in this paper and the test curve.

The damage constitutive models that are based on acoustic emission and energy dissipation highly coincide with the test curve, indicating that the two damage constitutive models that were constructed in this paper are applicable and can provide reliable theoretical support for subsequent damage evolution analysis.

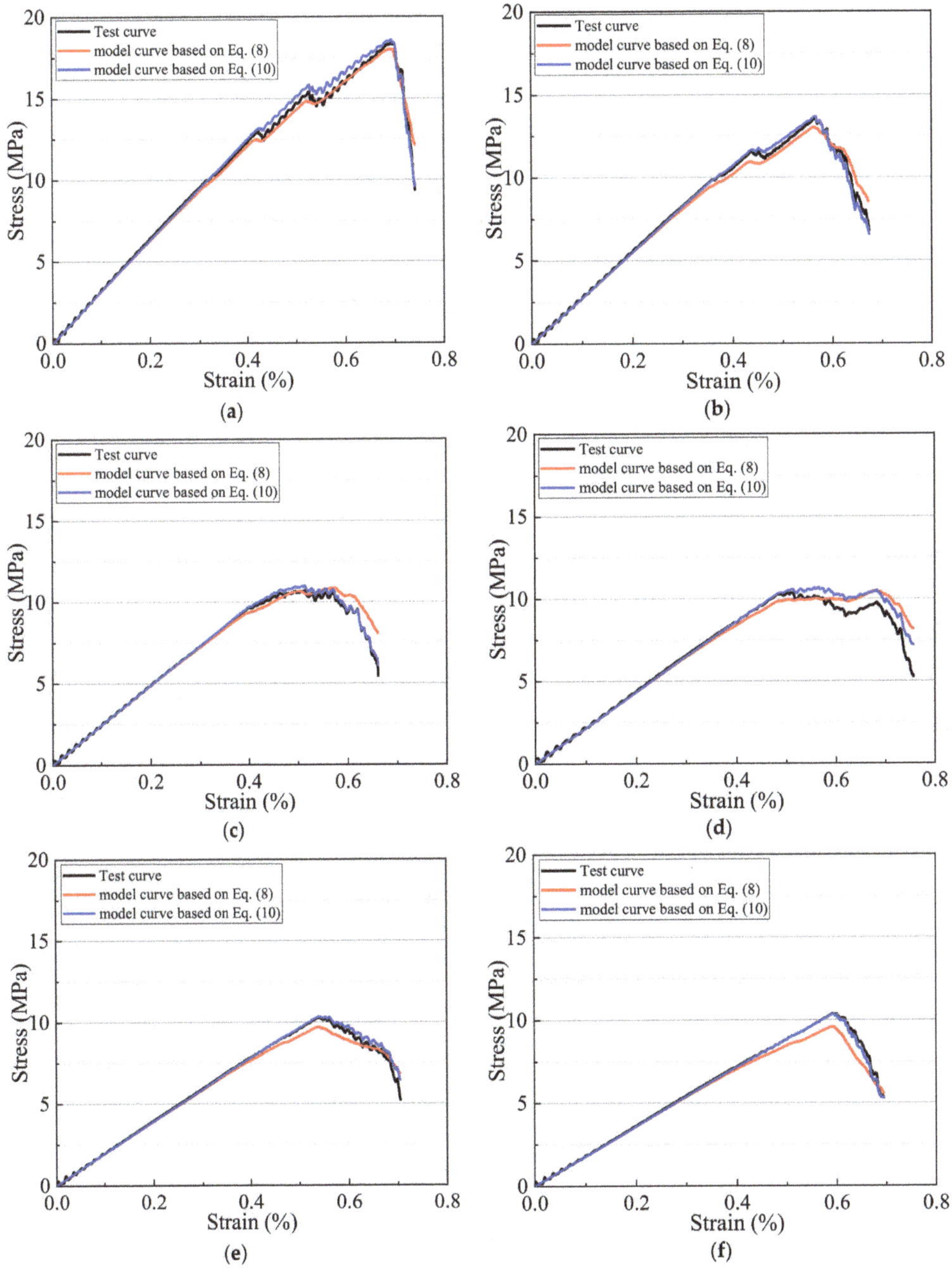

Figure 13. Damage constitutive model curves and tests curves: (**a**–**f**): RCF2-RCF7.

3.4.3. Damage Evolution

According to the established damage evolution Equations (8) and (10), the damage evolution curve of RCF materials can be calculated, and the results of these calculations are shown in Figure 14. The damage evolution law that is based on acoustic emission and energy dissipation is basically the same. In the whole loading process, the damage evolution process of RCF materials can be divided into three stages:

(1) Stage 1: At this stage, the original pore structure in the RCF sample had been compacted and the layered structural plane had been closed. The original pores in the sample began to expand, while new microcracks gradually formed, the sample's damage began to accumulate slowly, and the damage curve began to grow slowly. The damage evolution curve of RCF material has an obvious correlation with the CPB height ratio. It can be seen from the figure that the smaller the CPB height ratio, the faster the damage evolution rate. In the initial compression stage, the sample with a larger CPB height ratio could bear greater deformation, but it was not easy to damage. On the contrary, a smaller CPB height ratio allowed the deformation to reach the upper limit quickly and damage occurred in this case first.

(2) Stage 2: This stage is the stage of damage-stable evolution. Under the load, the internal primary pores and new cracks developed steadily, and the damage accumulated stably. At this time, it could be inferred that the damage accumulation of the central CPB was close to its upper limit, the rocks at both ends were compressed, and a small number of fractures expanded in the rocks at both ends. The CPB height ratio had little effect on the damage evolution process and the damage evolution curves of the different RCF samples had no obvious difference.

(3) Stage 3: This stage is the stage of rapid damage accumulation. As the load that was on the RCF material gradually exceeded its peak load, the overall stable structure of the RCF sample had been destroyed, the internal crack and pore structure expanded rapidly and formed macro cracks, and the internal damage to the sample accumulated rapidly. At this stage, there was no significant difference in the damage evolution curves of the different RCF samples.

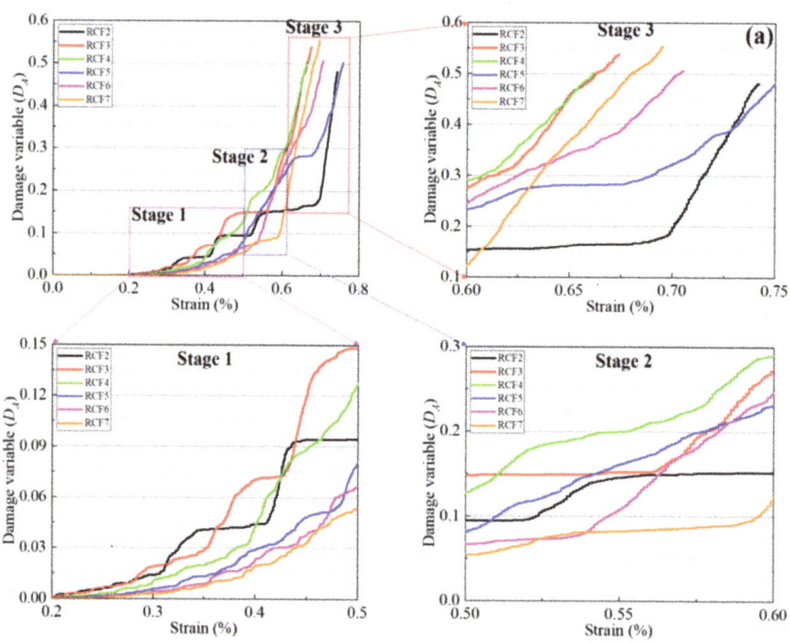

Figure 14. *Cont.*

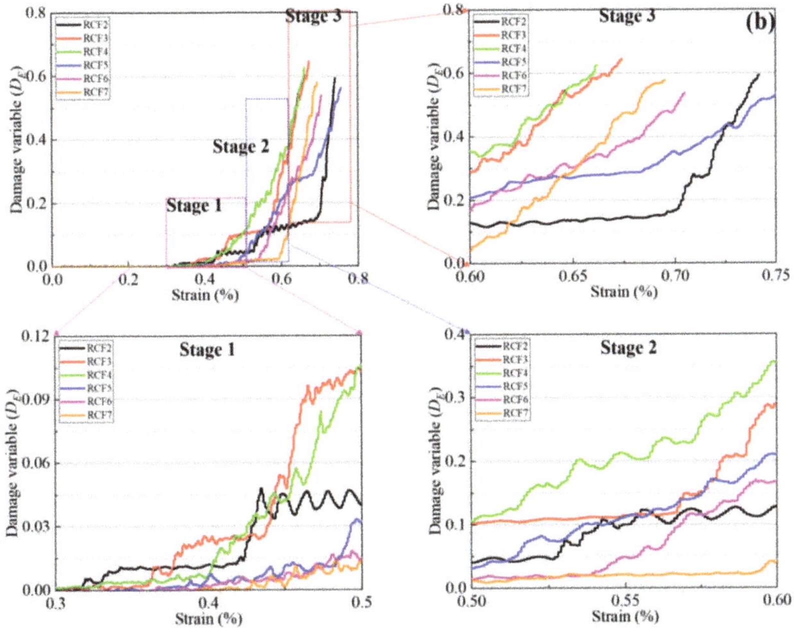

Figure 14. Damage evolution characteristics: (**a**) Damage variable D_A; (**b**) Damage variable D_E.

4. Discussion

According to the analyses in Section 3.1, the CPB height ratio in the RCF layered composite material has an important impact on its mechanical properties. When the proportion of the CPB height increases, the (*UCS* and elastic modulus of the RCF material continues to decrease, while the peak strain of the RCF material first decreases and then increases. These results are shown in Figure 15.

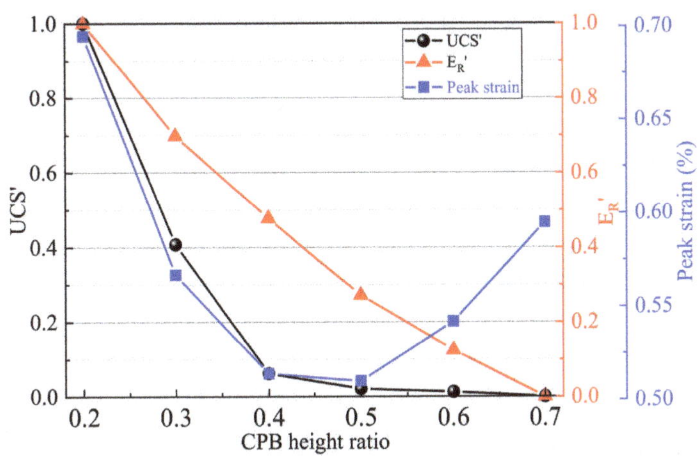

Figure 15. Composite diagram of CPB height ratio, (*UCS*′, E_R′, and peak strain.

In analyzing the deep-seated reasons for these results, it was found that the increase in the CPB height is equivalent to increasing the proportion of the weak medium material in RCF composite material. Under a load, the greater the proportion of weak medium material, the greater the probability of damage. Furthermore, the more damage that is present in a weak medium and the easier the damage in that weak medium evolves into a strong medium, the greater the likelihood of the overall failure of the RCF layered composite material. Therefore, the strength of the RCF material in question is lesser.

In addition, as discovered through the analyses in Section 3.1.2, the stiffness of the CPB material was found to be much less than that of the rock material. When the two media are non-uniformly combined, the stiffness of the weak medium material determines the lower limit of the stiffness of the RCF layered composite material, while the stiffness of the strong medium material determines the upper limit of the stiffness of the RCF material. Obviously, when the proportion of the weak medium is larger, the lower limit of the stiffness of the RCF material is smaller. On the contrary, when the proportion of the weak medium is smaller, that is, the proportion of strong medium is larger, the upper limit of the stiffness of the RCF composite material is larger.

However, there are some differences between the peak strain of an RCF material and its (*UCS* and elastic modulus. As we all know, the deformation of CPB material is greater than that of rock material. When the proportion of the CPB material increases, the overall deformation capacity of the RCF material increases and its peak strain also increases. When the proportion of the CPB material exceeds a certain degree, although the overall deformation capacity of the RCF material is improved its overall strength is also significantly reduced. Under the action of a load, the deformation of the central CPB material is far from reaching its peak deformation and the RCF material reaches its peak strength, resulting in a decreasing trend in the peak strain of the RCF material.

Combined with the analyses in Sections 3.1.1 and 3.1.2. The (*UCS*′ and E_R′ of RCF layered composite materials have a common variation law, both of which decrease with the increase in the proportion of a weak medium, and there is a certain internal relationship between the strength and stiffness characteristics of the material itself. In order to deeply analyze the internal correlation mechanism between (*UCS*′ and E_R′, the two are discussed together. The results are shown in Figure 16.

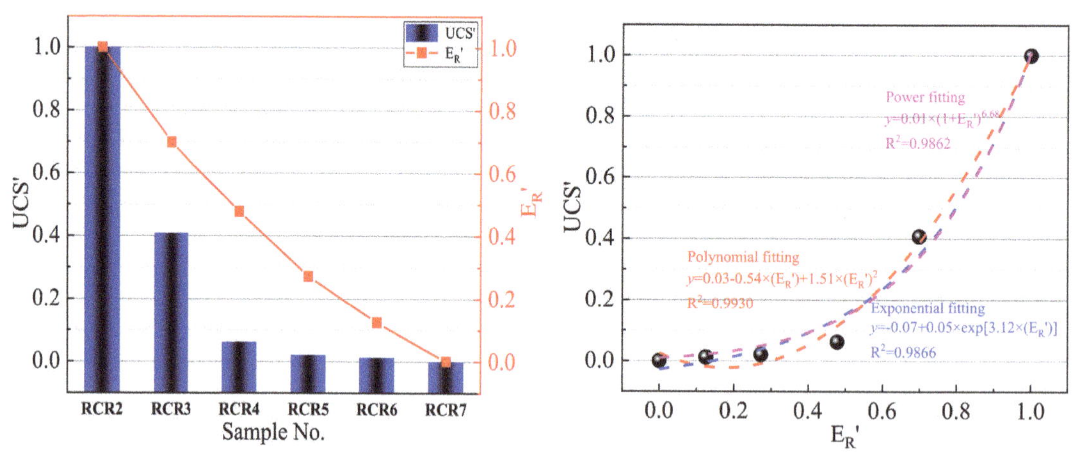

(a) (b)

Figure 16. Relationship between (*UCS*′ and E_R′ of RCF materials.

It can be seen from Figure 16a that the (*UCS*′ and E_R′ of the RCF materials have the same change law and both decrease with the increase in the CPB height ratio (RCF2–RCF7). It can be seen from Figure 16b that the (*UCS*′ of the RCF material increased with the increase in E_R′. When the E_R′ of the RCF materials was 0, the (*UCS*′ of the RCF material was 0;

when the $E_R{}'$ increased to 0.13, 0.27, 0.48, 0.70, and 1.0, the (UCS' increased to 0.01, 0.02, 0.06, 0.41, and 1.0, respectively. Three functions were used to fit the relationship between (UCS' and $E_R{}'$ of the RCF material. The results show that the polynomial function can better characterize the functional relationship between them:

$$(UCS' = 0.03 - 0.54 \times (E_R{}') + 1.51 \times (E_R{}')^2 \tag{11}$$

Under the action of a load, the material continuously absorbs energy from the outside and stores it. When the absorbed energy exceeds the energy storage limit of the material, the material will be damaged and release energy to the outside; the process of releasing energy usually produces acoustic emission signals. Therefore, there is a close relationship between the energy dissipation, damage evolution, and acoustic emission of materials. Figure 17 shows the stress–damage–AE–energy dissipation composite diagram of the RCF4 and RCF5 samples.

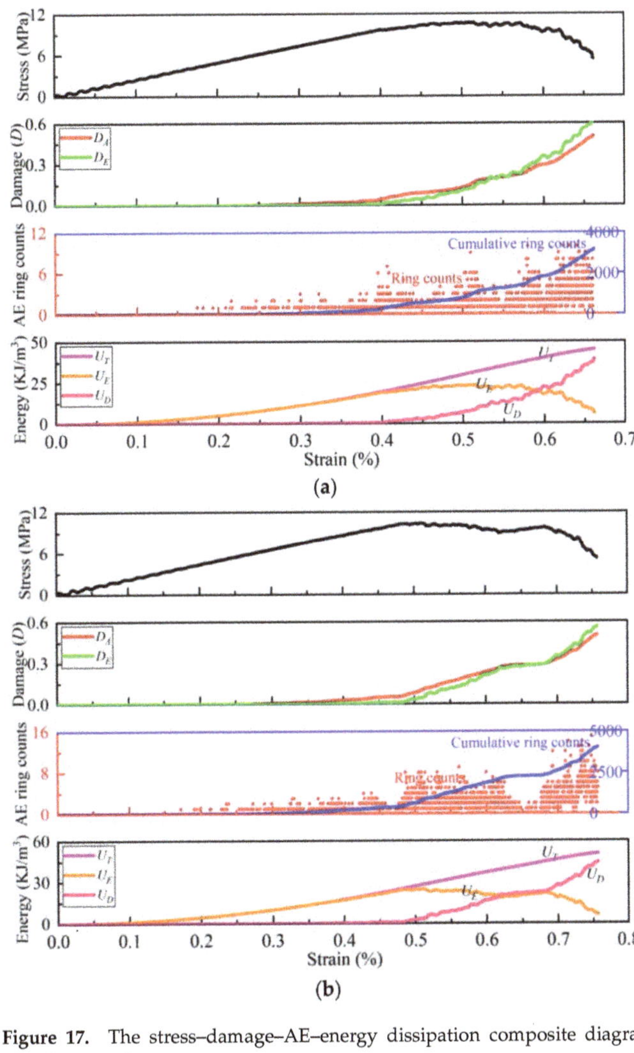

Figure 17. The stress–damage–AE–energy dissipation composite diagram: (**a**) RCF4 sample; (**b**) RCF5 sample.

Taking the RCF4 sample (Figure 17a) as an example, in the initial loading stage (in which the strain increases from 0 to 0.2%), the internal primary pores of the sample were compacted and the layered structural plane was closed. Although the sample continuously absorbed the energy (the U_E increased from 0 to 4.98 KJ/m^3), the sample did not produce damage at this stage (D = 0); however, sporadic AE signals were generated during the compaction of the internal primary pores. When the load was continuously applied, the RCF4 sample entered the linear elastic stage (as the strain increased from 0.2% to 0.4%), the elastic energy continued to accumulate (the U_E increased from 4.98 to 19.6 KJ/m^3), but the dissipated energy did not increase, the sample was not damaged, and a small amount of AE signals were still being monitored (the cumulative ring count was 314). Subsequently, the energy that was accumulated in the RCF4 sample gradually reached its energy storage limit. The RCF4 sample went through the yield stage (in which the strain increased from 0.4% to 0.5%). At this time, the RCF4 sample began to dissipate energy (the U_D increased from 0 to 6.7 KJ/m^3) and the internal damage began to accumulate slowly (the damage value D increased from 0 to 0.13). At the same time, a large number of AE signals were monitored (the cumulative ring count reached 860). Finally, the RCF4 sample entered the post-peak stage, the elastic energy inside the sample began to decrease gradually, the dissipated energy increased rapidly, and the sample entered the stage of rapid damage accumulation. At the same time, the sample began to release a large number of AE signals to the outside world.

5. Conclusions

In this paper, the mechanical properties, AE ring count characteristics, and energy dissipation mechanism of RCF layered composite materials were studied by uniaxial compression AE tests. Based on the AE ring counts and energy dissipation, two damage constitutive models of RCF materials have been established and the damage evolution of RCF materials was deeply analyzed. The main conclusions are as follows:

(1) The (*UCS* and E_R of the RCF samples were lower than those of the RR and FR, but higher than those of the CPB. With the increase in the CPB height ratio, the (*UCS* and E_R of the RCF samples showed a decreasing trend. The CPB height ratio had an exponential function relationship with the (*UCS* and a polynomial function relationship with E_R. The (*UCS* increased as a polynomial function with the increase in the E_R.

(2) The change law of the AE ring counts of the different RCF samples was basically similar. The AE ring count first grew slowly, then increased rapidly, and finally maintained a high-speed increase. The AE cumulative ring count at the peak point of the RCF sample decrease with the increase in the CPB height ratio.

(3) At first, the elastic energy U_E of the RCF sample began to accumulate slowly, then the dissipated energy U_D began to increase. Finally, the U_E began to decrease and the U_D increased almost linearly. The U_T, U_E, and U_D at the peak point showed a decreasing trend with the increase in the CPB height ratio. With the increase in the CPB height ratio, the U_E–U_T ratio and the U_D–U_T ratio decreased and the U_E–U_D ratio increased.

(4) Two damage constitutive models were established based on the AE ring counts and energy principle. The verification results show that the model that was constructed in this paper is reasonable and reliable. The damage evolution process of RCF materials can be divided into three stages: the slow damage accumulation stage, stable damage growth stage, and rapid damage accumulation stage.

(5) The AE ring counts and energy dissipation are closely related to internal damage evolution. First, the AE ring count accumulated slowly, the U_E increased slowly, and there was almost no damage evolution. Subsequently, the AE ring count accumulated rapidly, U_D increased slowly, and damage slowly evolved. Finally, the AE ring count maintained high-speed accumulation, U_E began to decrease, U_D increased rapidly, and damage rapidly evolved.

Author Contributions: J.W. and C.Z. analyzed the experimental data and initiated the writing of the paper; J.F. and W.S. modified the manuscript; Y.Z. corrected the English writing. All authors have read and agreed to the published version of the manuscript.

Funding: This research was supported the National Natural Science Foundation of China (Grant No. 51974012), and the China Postdoctoral Science Foundation (2021M690361).

Data Availability Statement: No new data were created or analyzed in this study. Data sharing is not applicable to this article.

Acknowledgments: The experimental works that were described in this study were conducted at the Key Laboratory of High-Efficient Mining and Safety of Metal Mines of the Ministry of Education in the University of Science and Technology, Beijing. The authors gratefully acknowledge the staff and students at the laboratory for their technical help during testing.

Conflicts of Interest: The authors declare no conflict of interest.

References

1. Xue, G.L.; Yilmaz, E.; Song, W.D.; Cao, S. Mechanical, flexural and microstructural properties of cement-tailings matrix composites: Effects of fiber type and dosage. *Compos. Part B Eng.* **2019**, *172*, 131–142. [CrossRef]
2. Wang, J.; Fu, J.X.; Song, W.D.; Zhang, Y.F. Viscosity and Strength Properties of Cemented Tailings Backfill with Fly Ash and Its Strength Predicted. *Minerals* **2021**, *11*, 78. [CrossRef]
3. Li, J.J.; Yilmaz, E.; Cao, S. Influence of industrial solid waste as filling material on mechanical and microstructural characteristics of cementitious backfills. *Constr. Build. Mater.* **2021**, *299*, 124288. [CrossRef]
4. Xue, G.L.; Yilmaz, E.; Song, W.D.; Cao, S. Analysis of internal structure behavior of fiber reinforced cement-tailings matrix composites through X-ray computed tomography. *Compos. Part B Eng.* **2019**, *175*, 107091. [CrossRef]
5. Zhao, K.; Huang, M.; Yan, Y.J.; Wan, W.L.; Ning, F.J.; Zhou, Y.; He, Z.W. Mechanical properties and synergistic deformation characteristics of tailings cemented filling assembled material body with different cement-tailings ratios. *Chin. J. Rock Mech. Eng.* **2021**, *40* (Suppl. 1), 2781–2789.
6. Cao, S.; Xue, G.L.; Yilmaz, E.; Yin, Z.Y.; Yang, F.D. Utilizing concrete pillars as an environmental mining practice in underground mines. *J. Clean. Prod.* **2021**, *278*, 123433. [CrossRef]
7. Tosun-Felekoglu, K.; Felekoglu, B. Effects of fiber–matrix interaction on multiple cracking performance of polymeric fiber reinforced cementitious composites. *Compos. Part B Eng.* **2013**, *52*, 62–71. [CrossRef]
8. Fang, K.; Fall, M. Shear Behaviour of Rock-Tailings Backfill Interface: Effect of Cementation, Rock Type, and Rock Surface Roughness. *Geotech. Geol. Eng.* **2021**, *39*, 1753–1770. [CrossRef]
9. Xiu, Z.G.; Wang, S.H.; Wang, F.L.; Ren, F.Y.; Vantuan, N. Comparative experimental study on the shear behavior of cemented paste backfill and surrounding rock-backfill interface. *Chin. J. Rock Mech. Eng.* **2021**, *40*, 1628–1642.
10. Cheng, A.P.; Shu, P.F.; Zhang, Y.S.; Wang, P.; Wang, M.X. Acoustic emission characteristics and damage constitution of backfill-surrounding rock combination. *J. Min. Saf. Eng.* **2020**, *37*, 1238–1245.
11. Sarfarazi, V.; Abharian, S.; Ghorbani, A. Physical test and PFC modelling of rock pillar failure containing two neighboring joints and one hole. *Smart Struct. Syst.* **2021**, *27*, 123–137.
12. Sarfarazi, V.; Abharian, S.; Ghalam, E.Z. Physical test and PFC2D simulation of the failure mechanism of echelon joint under uniaxial compression. *Comput. Concr.* **2021**, *27*, 99–109.
13. Wang, Y.R.; Lu, H.J.; Wu, J. Experimental investigation on strength and failure characteristics of cemented paste backfill-rock composite under uniaxial compression. *Constr. Build. Mater.* **2021**, *304*, 124629. [CrossRef]
14. He, Z.W.; Zhao, K.; Yan, Y.J.; Ning, F.J.; Zhou, Y.; Song, Y.F. Mechanical response and acoustic emission characteristics of cement paste backfill and rock combination. *Constr. Build. Mater.* **2021**, *288*, 123119. [CrossRef]
15. Wu, W.L.; Xu, W.B.; Zuo, J.P. Effect of inclined interface angle on shear strength and deformation response of cemented paste backfill-rock under triaxial compression. *Constr. Build. Mater.* **2021**, *279*, 122478.
16. Fang, K.; Cui, L.; Fall, M. A coupled chemo-elastic cohesive zone model for backfill-rock interface. *Comput. Geotech.* **2020**, *125*, 103666. [CrossRef]
17. Fang, K.; Fall, M.; Cui, L. Thermo-chemo-mechanical cohesive zone model for cemented paste backfill-rock interface. *Eng. Fract. Mech.* **2021**, *244*, 107546. [CrossRef]
18. Yu, X.; Kemeny, J.; Tan, Y.Y.; Song, W.D.; Huang, K. Mechanical properties and fracturing of rock-backfill composite specimens under triaxial compression. *Constr. Build. Mater.* **2021**, *304*, 124577. [CrossRef]
19. Zhao, L.D. Numerical investigation on the mechanical behaviour of combined backfill-rock structure with KCC model. *Constr. Build. Mater.* **2021**, *283*, 122782. [CrossRef]
20. Zhao, K.; He, Z.W.; Ning, F.J.; Zhou, Y.; Yan, Y.J.; Wang, Y.L.; Wang, J.Q. Acoustic Emission Characteristics of Cementitious Material with Different Cement-Tailing Ratio. *J. Chin. Ceram. Soc.* **2021**, *49*, 2462–2469.
21. Liu, W.; Ma, L.Q.; Sun, H.; Khan, N.M. An experimental study on infrared radiation and acoustic emission characteristics during crack evolution process of loading rock. *Infrared Phys. Technol.* **2021**, *118*, 103864. [CrossRef]

22. Liu, X.L.; Liu, Z.; Li, X.B.; Gong, F.Q.; Du, K. Experimental study on the effect of strain rate on rock acoustic emission characteristics. *Int. J. Rock Mech. Min. Sci.* **2020**, *133*, 104420. [CrossRef]
23. Liu, J.P.; Wang, R.; Lei, G.; Si, Y.T.; Xu, S.D.; Li, Y.H. Studies of stress and displacement distribution and the evolution law during rock failure process based on acoustic emission and microseismic monitoring. *Int. J. Rock Mech. Min. Sci.* **2020**, *132*, 104384. [CrossRef]
24. Lin, Q.B.; Cao, P.; Cao, R.H.; Lin, H.; Meng, J.J. Mechanical behavior around double circular openings in a jointed rock mass under uniaxial compression. *Arch. Civ. Mech. Eng.* **2020**, *20*, 19. [CrossRef]
25. Wang, J.; Fu, J.X.; Song, W.D.; Zhang, Y.F.; Wu, S. Acoustic emission characteristics and damage evolution process of layered cemented tailings backfill under uniaxial compression. *Constr. Build. Mater.* **2021**, *295*, 123663. [CrossRef]
26. Cheng, A.P.; Shu, P.F.; Deng, D.Q.; Zhou, C.S.; Huang, S.B.; Ye, Z.Y. Microscopic acoustic emission simulation and fracture mechanism of cemented tailings backfill based on moment tensor theory. *Constr. Build. Mater.* **2021**, *308*, 125069. [CrossRef]
27. Zhou, Y.; Yan, Y.J.; Zhao, K.; Yu, X.; Song, Y.F.; Wang, J.Q.; Suo, T.Y. Study of the effect of loading modes on the acoustic emission fractal and damage characteristics of cemented paste backfill. *Constr. Build. Mater.* **2021**, *277*, 122311. [CrossRef]
28. Zhao, K.; Yu, X.; Zhu, S.T.; Yan, Y.J.; Zhou, Y.; He, Z.W.; Song, Y.F.; Huang, M. Acoustic emission fractal characteristics and mechanical damage mechanism of cemented paste backfill prepared with tantalum niobium mine tailings. *Constr. Build. Mater.* **2020**, *258*, 119720. [CrossRef]
29. Wang, Y.Y.; Deng, H.C.; Deng, Y.; Chen, K.P.; He, J.H. Study on crack dynamic evolution and damage-fracture mechanism of rock with pre-existing cracks based on acoustic emission location. *J. Pet. Sci. Eng.* **2021**, *201*, 108420. [CrossRef]
30. Behera, S.K.; Mishra, D.P.; Singh, P.; Mishra, K.; Mandal, S.K.; Ghosh, C.N.; Kumar, R.; Mandal, P.K. Utilization of mill tailings, fly ash and slag as mine paste backfill material: Review and future perspective. *Constr. Build. Mater.* **2021**, *309*, 125120. [CrossRef]
31. Souyang, H.; Huang, Y.; Zhou, N.; Li, J.; Gao, H.; Guo, Y. Experiment on hydration exothermic characteristics and hydration mechanism of sand-based cemented paste backfill materials. *Constr. Build. Mater.* **2021**, *318*, 125870.
32. Cavusoglu, I.; Yilmaz, E.; Yilmaz, A.O. Sodium silicate effect on setting properties, strength behavior and microstructure of cemented coal fly ash backfill. *Powder Technol.* **2021**, *384*, 17–28. [CrossRef]
33. Libos, I.L.S.; Cui, L.; Liu, X. Effect of curing temperature on time-dependent shear behavior and properties of polypropylene fiber-reinforced cemented paste backfill. *Constr. Build. Mater.* **2021**, *311*, 125302. [CrossRef]
34. Ran, H.; Guo, Y.; Feng, G.; Qi, T.; Du, X. Creep properties and resistivity-ultrasonic-AE responses of cemented gangue backfill column under high-stress area. *Int. J. Min. Sci. Technol.* **2021**, *31*, 401–412. [CrossRef]
35. Hou, Y.; Yin, S.; Chen, X.; Zhang, M.; Yang, S. Study on characteristic stress and energy damage evolution mechanism of cemented tailings backfill under uniaxial compression. *Constr. Build. Mater.* **2021**, *301*, 124333. [CrossRef]
36. Yin, S.; Hou, Y.; Chen, X.; Zhang, M. Mechanical, flowing and microstructural properties of cemented sulfur tailings backfill: Effects of fiber lengths and dosage. *Constr. Build. Mater.* **2021**, *309*, 125058. [CrossRef]
37. Qiu, H.; Zhang, F.; Liu, L.; Huan, C.; Hou, D.; Kang, W. Experimental study on acoustic emission characteristics of cemented rock-tailings backfill. *Constr. Build. Mater.* **2021**, *315*, 125278. [CrossRef]
38. Xin, L. Meso-scale modeling of the influence of waste rock content on mechanical behavior of cemented tailings backfill. *Constr. Build. Mater.* **2021**, *307*, 124473. [CrossRef]
39. Zhao, K.; Huang, M.; Zhou, Y.; Yan, Y.; Wan, W.; Ning, F.; He, Z.; Wang, J. Synergistic deformation in a combination of cemented paste backfill and rocks. *Constr. Build. Mater.* **2022**, *317*, 125943. [CrossRef]
40. Liu, H.F.; Zhang, J.X.; Li, B.Y.; Zhou, N.; Li, D.Q.; Zhang, L.B.; Xiao, X. Long term leaching behavior of arsenic from cemented paste backfill made of construction and demolition waste: Experimental and numerical simulation studies. *J. Hazard. Mater.* **2021**, *416*, 125813. [PubMed]
41. Liu, L.; Fang, Z.Y.; Wu, Y.P.; Lai, X.P.; Wang, P.; Song, K.-I. Experimental investigation of solid-liquid two-phase flow in cemented rock-tailings backfill using Electrical Resistance Tomography. *Constr. Build. Mater.* **2018**, *175*, 267–276. [CrossRef]
42. Guicheng, H.; Shenglong, L.; Bingxiang, H.; Zhijun, Z.; Dexin, D. Reasonable matching for cemented waste rock backfill and sand shale. *J. Min. Saf. Eng.* **2017**, *34*, 371–376.
43. Yuye, T.; Xin, Y.; Weidong, S.; Haiping, W.; Shuai, C. Experimental study on combined pressure-bearing mechanism of filling body and surrounding rock. *J. Min. Saf. Eng.* **2018**, *35*, 1071–1076.
44. Wang, J.; Fu, J.X.; Song, W.D.; Zhang, Y.F.; Wang, Y. Mechanical behavior, acoustic emission properties and damage evolution of cemented paste backfill considering structural feature. *Constr. Build. Mater.* **2020**, *261*, 119958. [CrossRef]
45. Wang, J.; Fu, J.X.; Song, W.D.; Zhang, Y.F. Mechanical properties, damage evolution, and constitutive model of rock-encased backfill under uniaxial compression. *Constr. Build. Mater.* **2021**, *285*, 122898. [CrossRef]
46. Fu, J.X.; Wang, J.; Song, W.D. Damage constitutive model and strength criterion of cemented paste backfill based on layered effect considerations. *J. Mater. Res. Technol.* **2020**, *9*, 6073–6084. [CrossRef]
47. Liu, W.Z.; Chen, J.T.; Guo, Z.P.; Yang, H.Z.; Xie, W.W.; Zhang, Y.D. Mechanical properties and damage evolution of cemented coal gangue-fly ash backfill under uniaxial compression: Effects of different curing temperatures. *Constr. Build. Mater.* **2021**, *305*, 124820. [CrossRef]
48. Hou, J.F.; Guo, Z.P.; Li, J.B.; Zhao, L.J. Study on triaxial creep test and theoretical model of cemented gangue fly ash backfill under seepage-stress coupling. *Constr. Build. Mater.* **2021**, *273*, 121722. [CrossRef]

49. Lin, Q.B.; Cao, P.; Mao, S.Y.; Ou, C.J.; Cao, R.H. Fatigue behaviour and constitutive model of yellow sandstone containing pre-existing surface crack under uniaxial cyclic loading. *Theor. Appl. Fract. Mech.* **2020**, *109*, 102776. [CrossRef]
50. Ma, Q.; Tan, Y.L.; Liu, X.S.; Gu, Q.H.; Li, X.B. Effect of coal thicknesses on energy evolution characteristics of roof rock-coal-floor rock sandwich composite structure and its damage constitutive model. *Compos. Part B Eng.* **2020**, *198*, 108086. [CrossRef]
51. Hou, Z.K.; Gutierrez, M.; Ma, S.Q.; Almrabat, A.; Yang, C.H. Mechanical Behavior of Shale at Different Strain Rates. *Rock Mech. Rock Eng.* **2019**, *52*, 3531–3544. [CrossRef]
52. Jia, Z.Q.; Li, C.B.; Zhang, R.; Wang, M.; Gao, M.Z.; Zhang, Z.T.; Zhang, Z.P.; Ren, L.; Xie, J. Energy Evolution of Coal at Different Depths Under Unloading Conditions. *Rock Mech. Rock Eng.* **2019**, *52*, 4637–4649. [CrossRef]
53. Liu, B.; Huang, J.; Wang, Z.; Liu, L. Study on damage evolution and acoustic emission character of coal-rock under uniaxial compression. *Chin. J. Rock Mech. Eng.* **2009**, *28* (Suppl. 1), 3234–3238.
54. Barile, C.; Casavola, C.; Pappalettera, G.; Vimalathithan, P.K. Damage characterization in composite materials using acoustic emission signal-based and parameter-based data. *Compos. Part B Eng.* **2019**, *178*, 107469. [CrossRef]
55. Liu, X.S.; Tan, Y.L.; Ning, J.G.; Lu, Y.W.; Gu, Q.H. Mechanical properties and damage constitutive model of coal in coal-rock combined body. *Int. J. Rock Mech. Min. Sci.* **2018**, *110*, 140–150. [CrossRef]

minerals

Review

Review on Comprehensive Utilization of Magnesium Slag and Development Prospect of Preparing Backfilling Materials

Xiaobing Yang [1,2,*], Fusong Dong [1,2], Xizhi Zhang [1,2,3], Chenzhuo Li [1,2] and Qian Gao [1,2]

[1] School of Civil and Resources Engineering, University of Science and Technology Beijing, Beijing 100083, China
[2] Key Laboratory of High-Efficient Mining and Safety of Metal Mines, Ministry of Education, University of Science and Technology Beijing, Beijing 100083, China
[3] Mining Engineering Branch of Jinchuan Group Engineering Construction Co., Ltd., Jinchang 737102, China
* Correspondence: yangxiaobing@ustb.edu.cn; Tel.: +86-189-1027-1725

Abstract: China is the largest producer of the metal magnesium, which is the third largest metal after steel and aluminum, and magnesium slag (MS) discharged from magnesium production cannot be treated effectively at present. Large amounts of MS occupy the land, making the land salinized and polluting the groundwater, which will threaten the sustainable development of the inland areas of China. To realize the large-scale utilization of MS, this paper reviews the research on the comprehensive utilization of MS and proposes using MS to prepare backfill materials. Firstly, the source and physical and chemical properties of MS are introduced, and the hazards caused by MS are also pointed out. Then, the several utilization methods of MS are summarized, such as cement admixture, chemically activated cementitious materials, clinker sintering, etc. Thirdly, the effect of MS on the properties of cementitious materials and concrete is summarized, including condensation time, workability, mechanical properties, durability, etc. Finally, based on the cemented backfill mining method, MS replaces Portland cement and blast furnace slag (BFS) to prepare cementitious materials. The mechanical properties and fluidity of backfilling slurry composed of MS, gypsum, BFS, and tailings can meet the requirements of backfilling mining.

Keywords: magnesium slag; resource utilization; physicochemical properties; backfilling mining; magnesium slag-based cementitious materials

Citation: Yang, X.; Dong, F.; Zhang, X.; Li, C.; Gao, Q. Review on Comprehensive Utilization of Magnesium Slag and Development Prospect of Preparing Backfilling Materials. *Minerals* **2022**, *12*, 1415. https://doi.org/10.3390/min12111415

Academic Editor: Abbas Taheri

Received: 29 September 2022
Accepted: 4 November 2022
Published: 8 November 2022

1. Introduction

Magnesium is the third most common metal after steel and aluminum. China is the largest producer of magnesium in the world, producing about 823,000 tons of metal magnesium in 2021, accounting for more than 85% of the global output. The Pidgeon process is the main method to produce magnesium in China [1]. However, MS is produced at a rate of 6~8 tons per ton of metallic magnesium. As a result, millions of tons of MS are produced in China every year, which cannot be effectively treated and is mainly stored and buried. Most of the magnesium smelting enterprises in China are positioned in the middle and upper reaches of the Yellow River. The arid climate in this inland area has resulted in a fragile ecological balance, and the threshold and resistance of the ecosystem are low. Large amounts of MS occupy land, make the land salinized, and pollute the groundwater. Those environmental issues have been threatening the sustainable development of the inland areas. In addition, the northwest district is the most centralized area for mineral resources in China, and large-scale mining action also destroys the local ecological balance.

This paper reviewed the research achievements on resource utilization of MS and pointed out the existing problems. Based on the cemented backfilling mining method, this paper outlines an approach to achieving large-scale utilization of MS. MS replaced the Portland cement and BFS to cement the solid waste, such as waste rock and tailings, and then the mixture stirred in the filling station on the surface was transported to the stope through the pipeline. The backfilling body can maintain the stability of the stope and avoid the environmental issues caused by the mining process.

2. Sources and Physicochemical Properties of MS

2.1. Main Sources

The Pidgeon process is the main method to produce magnesium in China, as shown in Figure 1. Dolomite, as raw material, is calcined in the rotary furnace, the product is mixed with reducing agent silicon iron powder and catalyst fluorite, then ground and pressed into pellets, and the pellets are put into the reduction furnace. Under the condition of temperature 1200 °C and internal pressure of at least 13.3 Pa, magnesium appears in the form of steam, passes through the condenser, forms crystals, and is finally refined by solvent to produce magnesium [2]. The remaining material in the reduction furnace is the MS.

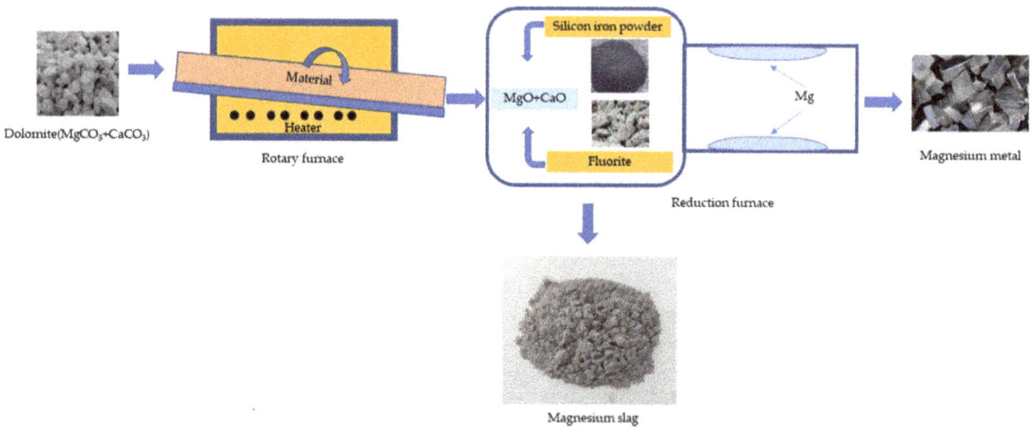

Figure 1. Flow chart of magnesium smelting by Pidgeon process.

MS is derived from the preparation process of magnesium metal, and the main chemical components of dolomite are MgO, SiO_2, Al_2O_3, Fe_2O_3, CaO, etc. Combined with the above analysis, the following chemical reaction occurred in the Pidgeon process:

$$MgCO_3 + CaCO_3 \rightarrow CaO + MgO + 2CO_2 \quad (1)$$

$$CaO{\cdot}MgO + Si(Fe) \rightarrow Ca_2SiO_4 + Mg + Fe \quad (2)$$

The briquettes, which discharged from the retorts after the MgO reduction, contain the following two solid products: dicalcium silicate and Fe(s). During the cooling process of MS, Ca_2SiO_4, as the main product within the pellet, will undergo several phase changes. The change from β-phase to γ-phase, which occurs mainly between 400 and 500 °C. This change will occur in volume expansion, with an expansion rate of about 12% [3–7], and the solid pellet will disintegrate after expansion, producing MS particles [6,7].

2.2. Physical Property

MS is a gray substance, similar in color to cement, odorless, with a particle size between 0 and 10 mm, and slightly sticky to the touch. After experimental analysis, it was learned that [8], using the MS from a large magnesium smelting enterprise in Shanxi, the loose density was measured to be 1.15 g/cm^3, and the average true density was 2.89 g/cm^3. The magnesium residue is sieved, and the particle size distribution is shown in Figure 2 [9–12].

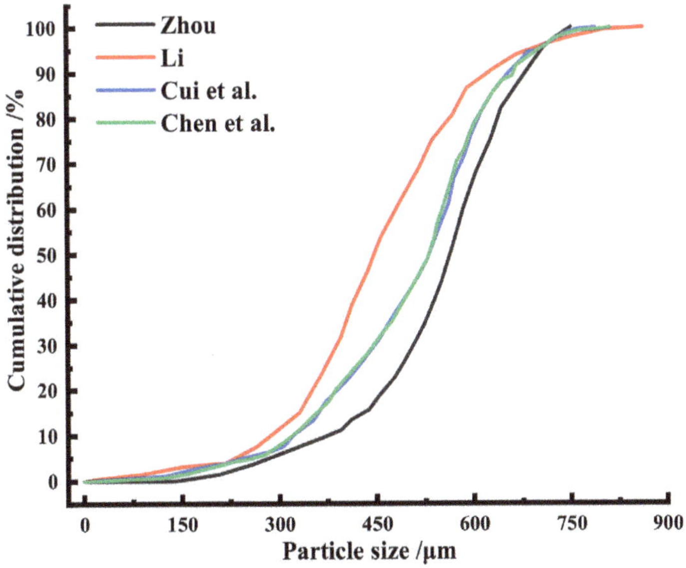

Figure 2. Particle size distribution of MS [9–12].

2.3. Chemical Composition

A total of 20 MS samples were collected from four plants producing metallic magnesium in Shanxi Province [13]. The main chemical components in MS are CaO, SiO_2, MgO, Al_2O_3, and Fe_2O_3. Djokic et al. [14] investigated the chemical compositions of twenty slag samples from the deposit of the Mg-Serbian Company. Han et al. measured the content ratio of CaO to SiO_2, i.e., the basicity, for the slag it is 1.58.

The quality evaluation parameters such as the alkalinity coefficient M_0, the quality coefficient K, and the activity index M_a of MS were calculated according to the results of the chemical composition of MS, and the results were obtained in Table 1.

Table 1. Activity evaluation index of MS based on chemical component.

Source	M_0	K	M_a
Li, Y. et al. [13]	2.143	2.310	0.055
Djokic et al. [14]	1.667	2.997	0.499
Han, F.L. et al. [15]	1.773	1.795	0.023
Tengxiang Magnesium Products Co., Ltd., Hami City, Xinjiang Province, China	2.02	2.098	0.027

Among them, the alkalinity coefficients were all greater than 1.0, which means those MS all belong to alkaline slag powder. It was seen that those MS were classified as medium-quality slag powder as the quality coefficients were all greater than 1.6, and the MS has the potential to be prepared cementitious material. Although the activity indexes were all poor, the content of CaO, one of the main active ingredients, is high, 52.57%, 52.22%, and 42.35%, respectively, so it has potential activity and also can be used as an alkali-activator for cementitious material development.

The XRD pattern of the MS samples is shown in Figure 3. It is seen that γ-Ca_2SiO_4 (γ-C_2S) is the main phase that causes the decomposition of MS, in which *f*-MgO is the secondary phase, and β-Ca_2SiO_4 (β-C_2S) is a trace phase, which indicates that a small amount of β-C_2S can occur in the cooling slag particles.

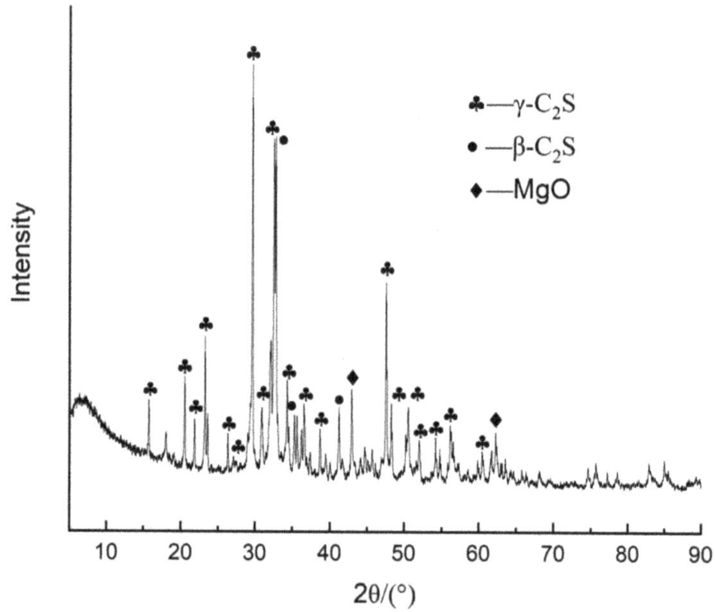

Figure 3. XRD pattern of MS.

The equilibrium phase distribution during the cooling process was obtained with the thermodynamic calculation using FactSage 6.2 [15]. At higher temperature conditions, $Ca_2SiO_4(s_2)$, such as α′-Ca_2SiO_4 polycrystalline, may only be 29% at 1000–1300 °C. $MgOCa_3O_3SiO_4(s)$ as the main phase, such as merwinite formed by MgO, CaO, and SiO_2, has a content of 47% at 650–1300 °C, but it will decompose in large quantities at around 500 °C, becoming the second phase in amount when the temperature is close to the room temperature phase, where the content is 17.5%. Therefore, $Ca_2SiO_4(s)$, such as γ-Ca_2SiO_4 polycrystalline, will become the dominant phase when the temperature is close to room temperature, with a content of 56%. Any other slag with a similar content of γ-Ca_2SiO_4 will disintegrate into fine powder or dust.

2.4. Activity Excitation and Hydration Properties

MS has a chemical composition similar to cement, BFS, and other cementitious materials that contain γ-C_2S or β-C_2S and other mineral compositions with hydration properties; that is, MS has potential activity. Studies have found that [16–18], MS has certain gelling properties; the *f*-CaO and *f*-MgO in MS hydrate in the first. However, because the content of the early hydration mineral C_3S is lower, and the reaction rate of β-C_2S, the main gelling mineral in MS, is lower than that of C_3S, the weak hydration capacity of MS in the early-stage results in the low strength of cementitious materials.

2.5. Carbonization

MS contains a large amount of CaO and MgO, and calcium-containing components are mainly calcium silicate, lime, and silicate. Through CO_2 activation, Ca^{2+} and Mg^{2+} leach from Ca- and Mg-containing minerals can react with ${CO_3}^{2-}$ and form calcium carbonate or magnesium carbonate, respectively. This dissolution-precipitation process is the carbonation of MS. After carbonation, part of the calcium carbonate is converted into calcite and sphalerite [19].

3. Hazards of MS

Magnesium production via the Pidgeon process is energy- and natural resource-intensive. There are several studies of environmental issues relating to magnesium production, including emissions of air pollutants and global warming impacts [20,21]. The global warming impact of magnesium production in China has been estimated near to 42 kg CO_2 eq/kg Mg ingot [21].

MS with fine particles is difficult to settle in the air and easy to cause dust pollution in the natural environment; long-term inhalation will lead to respiratory diseases. In addition, MS is the industrial solid waste from the smelting of metal magnesium; most of it will be directly dumped into the wasteland or landfill storage [22]. The harmful elements and heavy metal elements in the MS will penetrate the soil with surface precipitation, resulting in water pollution.

3.1. *Harmful Components*

During the smelting of magnesium metal, some harmful elements will remain in the MS and be enriched in the MS, such as arsenic (As), chromium (Cr), cadmium (Cd), mercury (Hg), copper (Cu), nickel (Ni), fluorooctane (F), and chlorine (Cl) [23].

The data on pollutants in MS in some regions of China were carried out by Wang et al. [23]. All the contents of F, Cd, Hg, Ni, Cu, and Cr in MS exceeded the values of China's soil environmental background and exceeded 18, 7, 6, 5, 4, and 2 times, respectively.

The hazardous substances contained in the MS are a threat to the safety of the water environment, and Wang et al. [23] determined the leachable concentrations of some heavy metals in the MS, including Hg, Cu, and Cr, whose extraction solutions exceeded the limit values (Class V) of the Groundwater Quality Standard (GB/T 14848-2017 2017), where the leachable concentration of F exceeded the limit values (Class IV–V). It shows that MS and MS-based fertilizers contain a variety of pollutant elements, which can leach into the ground by rainwater and surface water, thus posing a potential safety threat to soil quality and the water environment.

3.2. *Dust Pollution*

Studies have found that under stable weather conditions, MS dust particles will disperse around the reservoir for 1400–2000 m, and with climate change, their impact range will become wider [14].

MS is dust, accounting for about 60% of dust particles with a diameter of <150 μm. Because MS has a difficulty of settling out of the air, MS easily generates dust pollution in the natural environment and causes human respiratory diseases [23].

By adding an optimizing agent, one can effectively stabilize β-C_2S in MS and stop its transformation into γ-C_2S in the cooling process. Wu et al. [24] used several chemical stabilizers to treat the dusting MS at 1200 °C, including borates, phosphates, and rare earth oxides, and obtained volume-stable slag aggregates that were beneficial for environmental protection and recycling of the MS. After being taken out of the furnace, the sample of slag without stabilizers disintegrated immediately. The samples with stabilizer showed no dust at all, even after several months of removal. It was shown in X-ray diffraction patterns that the γ-C_2S is a major phase in the original MS sample, but its peaks are rather weak in the sample with chemical stabilizers, and the β-C_2S becomes a dominant phase. The research shows that the dusting of MS could be prevented effectively by additions of boric acid, calcium dihydrogen phosphate, and used cerium rouge [24].

4. Application of MS

As a common industrial solid waste, MS is widely used in cement admixtures, chemically activated cementitious materials, fertilizers, calcined porcelain, inorganic fibers, desulfurization, etc. in industrial production.

4.1. Cement Admixture

The poor hydration reactivity and safety risks of MS due to high levels of MgO inhibit its widespread use in the construction industry. However, MS has similar physical, chemical, and mineral properties as cement, so MS has potential uses as a mineral admixture in cement. Peng et al. prepared Portland cement by replacing clinker with two kinds of MS and investigated the properties of the cement, such as its water requirement for normal consistency, setting time, and strength. It was found that the addition of MS prolongs the setting time of cement. The cement meets the requirement of the P.C 42.5R standard when the additive content of MS ranges from 10% to 30%. While the additive content increases to 35%–40%, the cement can only reach the P.C 32.5R standard [25]. It was found that the activity of MS is slightly lower than that of BFS when they are mixed with clinker and gypsum to prepare composite Portland cement [26]. The flexural and compressive strengths of cement decrease as the content of MS increase when the content of MS and BFS fix at 50%. The properties of cement can reach the P.C 42.5 standard when the content of MS and BFS is 20% and 30%, respectively.

Studies have shown that MS has a greater impact on the physical and mechanical properties of the concrete, as the MS content increases the fluidity of concrete, and the fineness of MS also has a slight impact on fluidity and strength. Because MS is not conducive to the progress of the hydration reaction, the early strength will be affected but beneficial to the later strength. At the same time, the incorporation of MS optimizes the long-term performance of concrete, reduces dry shrinkage, improves sulfate, and increases freeze resistance [27,28]. The optimum content of MS as a mineral admixture in concrete can be up to 30% [29].

Soil stabilization is one of the methods of soil improvement in geotechnical engineering, of which cement and lime stabilization are the most commonly used methods in recent years and can reliably improve and increase soil strength parameters [30–34]. Through the analysis of the physicochemical characteristics and microstructure of MS, it is known that the addition of MS in the production process of cement can significantly reduce the production temperature, not only reduce the use of water, reduce energy consumption in the production process, shorten the condensation time, and, to a certain extent, make the cement have higher strength [28,29,35]. Through experimental tests, it was learned that as the content of MS and cement increased, the soil pH increased, there was a certain improvement in soil conditions, soil particles were granulated, and soil compressive strength increased [36].

The slaked magnesium slag (SMS) can be used as a blend that is suitable for the production of ordinary Portland cements. Because the MS is slaked, the SMS has no problem with soundness [28].

4.2. Chemically Activated Cementitious Material

Chemically activated cementitious materials in terms of either alkali-activated or acid-activated materials are of growing interest owing to their advantages over non-cement clinker, utilization of industrial wastes, low carbon emission, durable properties, or other special engineering performance [37,38].

MS containing high contents of CaO and MgO can be activated with CO_2 to prepare a calcium carbonate binder. The compressive strength of the MS paste reached up to 119.5 MPa after 14 d of CO_2 curing. This is mainly attributed to the formation of calcium carbonates, which leads to an enhanced product matrix with a dense microstructure in terms of reduced pore diameter and total porosity. Carbonation improves the volume stability of MS pastes [19].

It is known that alkali additions can activate pozzolanic materials to react and obtain strength. However, the strength of the mixture of MS and little clinker could not meet the engineering demand due to the lack of active ingredients. BFS was added to the cementitious system, and several activators were also used to improve the hydration degree. The properties of MS cementitious material achieved the standards of composite cement of 32.5 MPa, and its hydration products are Aft, C-S-H, and $Ca(OH)_2$ [39].

4.3. Clinker Sintering

Most MS contains more than 50% (in mass) CaO, so it can be used as an alternative to lime-based materials for cement clinker production [40]. MS favorably improves clinker burnability with the introduction of MgO and CaF_2. An appropriate amount of MS shortens the setting times of the cement and improves the compressive strength of the cement, allowing more C_3S to be produced in the clinker [40]. Heavy metals in clinkers can be immobilized or stabilized by hydrated matrices.

MS can be used to generate cement clinker, thus replacing part of the limestone, sand, and gravel at 11% [35]. The use of MS reduced the standard coal consumption and grinding time by 14% and 5%, respectively, and the results indicate that the use of MS can significantly improve the combustion properties and grindability and thus reduce energy consumption. The 28-d compressive strength of MS-based cement reached 58.9 MPa, and there was a significant increase in early compressive strength, but the setting time was relatively long, which may be related to the different crystalline states, and the physical properties of the cement using MS clinker met the requirements of the 52.5 strength class specified in GB 175-2007. The disposal of the MS via the production of Portland cement with the substitution of the raw material by MS is also environmentally safe.

The SMS used for the prevention of dust pollution was reactive. As a raw material, SMS can be used for calcining clinker of good quality [28].

4.4. Fertilizer

MS is rich in a large number of Ca, Mg, and other nutrients, which have a favorable impact on improving soil fertility and crop quality [41], so the further use of MS as fertilizer or improver in agriculture can not only effectively treat industrial solid waste but also achieve the comprehensive utilization of resources.

Related studies have shown that by mixing the MS with K_2CO_3 and heat treating it at 1300 °C for one hour, a novel slow-release potassium fertilizer was synthesized (SPF) [42], which could minimize fertilizer nutrient loss and Pidgeon magnesium slag (PMS) disposal problems. The active silicon content of SPF was 2.09 times as much as that of MS.

The silicon–potash fertilizer prepared from MS has great potential for application to Northern China's calcareous and nutrient-limited soil [43]. MS contains a large number of alkaline oxides (CaO and MgO), which would increase the soil pH and improve available K and Si absorption. The alkaline oxide in MS would react with K_2O and form semi-soluble aggregates during the preparation of Si-K fertilizer. The mobility of K^+ was decreased by incorporating it into the Si-O structure, and the solubility of Si was improved due to the depolymerization of Si-O tetrahedrons in the preparation of Si-K fertilizer.

MS, an alkaline byproduct during the refining of magnesium and magnesium alloys, contains some heavy metals (HMs) with a low content. Aiming to prevent soil HMs pollution in the application of MS-based fertilizer, the potential risk in the soil–plant system has been evaluated [44]. The results showed that the Cr, Cu, Pb, and Zn contents of two typical Chinese soil samples met the secondary standards of the Soil Environment Standard (GB 15618-2009), so the application of MS-based fertilizers to the two Chinese soils does not pose a risk of contamination to maize plants under current conditions. For more objective results, long-term field trials are needed to be conducted to verify.

4.5. Calcined Porcelain

With MS as a sintering aid, the sintering temperature of the proppant in the past was usually between 1300 °C and 1600 °C, which would produce huge energy consumption and bring high production costs [45–47]. By sintering MS with solid waste coal gangue at a sintering temperature below 1300 °C to prepare a low-density bauxite-based ceramic proppant, not only greatly reduced energy consumption but reduced production costs at the same time. The studies showed that [48,49], at high temperatures in the sintering process, the MS reacts with the oxides in the raw material to form a liquid phase. Combining with the XRD patterns and SEM images, the main component of the proppant is granular corundum (Al_2O_3) and rod-shaped mullite ($Al_6Si_2O_{13}$). When the temperature increases, the intensity of the mullite diffraction peak increases significantly, and the corundum diffraction peak decreases significantly, which, due to the fact that the formation of the glass phase at high temperature provides favorable conditions for the good growth of the mullite's shaped nucleus and crystal nuclei, the proppant prepared by MS and gangue can form a fully glazed surface and a more compact structure at lower sintering temperatures.

4.6. Inorganic Fiber

At present, the market demand for thermal insulation materials is huge, and in order to effectively recycle a large amount of industrial solid waste, the preparation of slag mineral wool products has become an effective choice. However, as a short fiber, mineral wool cannot perform all the functions of fibrous material such as continuous basalt fibers. It is certainly the best choice that the continuous fibers are produced entirely from industrial solid wastes, both from the perspective of cost and resource utilization. Two inorganic continuous fibers were successfully produced by fly ash and MS with or without feldspar added, through melting at high temperature, following spinning into continuous fibers [50]. Their compressive strengths show a trend of first rising and then falling, with better thermal stability, and arrive up to 903 MPa and 539 MPa at room temperature, respectively. Two fibers also have excellent thermal resistance and stability.

It is known that any inorganic solid waste can be melted at high enough temperatures, but the preparation of the slag wool fibers needs to consider the cost and raw material source factors. One advantage of slag continuous fibers is the ability to artificial match various local slags and manage solid waste resources. It should be noted that the preparation of slag fiber must be considered to avoid secondary pollution.

4.7. Desulfurization

In MS, Ca exists not only in the form of CaO, but also in the mineral form of γ-C_2S or β-C_2S, and it also contains a small amount of MgO. All this means MS may be used in the desulfurization process of coal-fired flue gas. However, the conversion rate of calcium used to evaluate the desulfurization performance of MS is very low when it is directly used for desulfurization.

The steam activation and quenching treatment methods are used to modify MS desulfurization activity, but the current performance is far from enough to meet the needs of industrial applications [51–53]. Compared with the naturally cooled MS, the particles of quenched samples are cracked and pulverized. The relatively rough surface made hydration easier by promoting reactant diffusion and dissolution. The degree of hydration increased with the increase in quenching temperature and hydration temperature. The calcium conversion rate of MS treated with continuous hydration is 30.3%, and the ones treated with non-continuous hydration are 13.3%.

(1) Through four modification methods of MS hydration, MS/fly ash hydration, MS quenching hydration, and MS/additive hydration to improve MSs desulfurization ability, their desulfurization performance is characterized by the calcium conversion rate. At the optimal parameters, the performance of the sample composed of quenching MS and fly ash hydration was the best. On the other hand, the additives such as $H_2C_2Q_4$ and NaOH also enhance the MSs desulfurization performance, and the calcium conversion achieved was 73.7% and 55.54%, respectively.

(2) Based on the modified method of quenching hydration, additives such as NaCl, $CaCl_2$, Na_2SO_4, $CaSO_4$ Na_2CO_3, K_2CO_3, acetic acid, and adipic acid were researched for improving quenched MS hydrate desulfurization performance. The 2% Na_2SO_4-modified MS obtained the best result, and the calcium conversion rate reached 40.16%.

To sum up, combining various modification ways and several additives, the desulfurization performance of MS still cannot meet the requirements of industrial applications, and more relevant research is needed.

5. Effect of MS on the Properties of Cementitious Materials and Concrete

5.1. Condensation Time and Workability

The slump and slump flow of concrete increased as the content of MS increased from 0 to 40%. The improvement in workability was probably due to the filling effect and the relatively lower reactivity of MS. The MS filled in the voids between cement particles to displace the water and increase the amount of free water in the mixture. Besides, a lower content of water would be consumed by the cementitious materials incorporating MS during the early stages of hydration. All this would result in an increased amount of free water to improve the fluidity of the concrete containing MS [27].

As the additive amount of MS in the Portland cement increases, the amount of water needed for cement to retain its normal consistency increases, and both the initial and final setting times are prolonged. All of those are due to the high fineness of the powder and the strong water-absorbing properties of MS [54]. The addition of MS reduces the fluidity of cement without the water reducer, and the fluidity of cement increases with the increasing content of MS when the content of the water reducer is 2% of the cementitious material mass [55].

5.2. Mechanical Property

The compressive strength and the flexural strength of the concrete decreased as the amount of MS increased to 40% at all curing times. It can be known that the MS was not beneficial for the development of the early strength of concrete, but the MS gradually participated in the reaction to form calcium silicate hydration gel and improved the later-age strength [27].

The compressive strength of cement mixed with MS decreased rapidly with the increasing amount of MS. The negative influence of MS on early-period strength is greater than on later-period strength [54]. The 28-d and 60-d compressive strengths of cement paste added 10% MS reached 64.0 MPa and 76.4 MPa and increased by 10.54% and 14.37% compared with pure cement paste, respectively. It was indicated that MS granules dissolved and hydrated in the alkaline environment formed by cement hydration, pores between cement hydration products were filled along with the hydration of MS, which made the structure of cement paste more compact and obtained a higher compressive strength than pure cement paste [55].

5.3. Durability

The sulphate resistance properties of concrete can be improved by increasing MS content. After 150 cycles of sulphate attack, the sulphate-resistant coefficient of the reference sample (without MS) jumped down to 68%. The coefficient of concrete containing MS increased as the MS content increased [27]. This is due to the low reactivity of MS and the comparatively low Al_2O_3 content of MS. The number of hydration products that can react with SO_4^{2-} to form expanded products reduced after the MS replaced the cement [56]. In this case, the internal stress caused by expansion was reduced, and the sulphate resistance of the concrete was greatly improved.

5.4. Volume Stability

The incorporation of MS can effectively improve the resistance to drying shrinkage of concrete. The early drying shrinkage of the concrete increased with the increasing MS content in the early days. This is mainly ascribed that the MS reduced the amount of water involved in the hydration processing and increased the amount of free water in the mixture. The free water evaporation promoted the early shrinkage rate of concrete. At later ages, MS gradually participated in the hydration reaction, and the hydration products filled the pores of the hardened concrete. As a result, the drying shrinkage of the concrete at later ages was significantly reduced by incorporating MS [27].

There are some percent of free CaO and MgO [6,7] that will hydrate in MS; the production of hydration can cause long-term volumetric instability for the slag [57,58]. The special and negative characteristics have made it rather difficult to use MS powder as building materials or aggregates for road construction [15]. The research shows that, by adding the borate agent and sintering at 1200 °C for several hours, the MS powder will be volume stable and can be used in building applications to replace valuable natural aggregates.

5.5. Hydration Heat and Degree

The total hydration heat of the blended cement paste decreased with increasing MS content. The appearance of the peak in the hydration rate curves was delayed by 4 h and 9 h for the blended cement paste containing 10% and 30% MS [27]. The reason was that the amount of C_3A and C_3S producing huge amounts of hydration heat at an early age was reduced with the replacement of MS.

Sun et al. found that the dissolution exothermic peak (the first exothermic peak) was reduced by the incorporation of MS, but the time of peak had not been affected. The addition of MS reduced the value of the hydration exothermic peak and delayed the time of peak, which positively correlated with the content of MS [55]. This study also showed that the hydration degree of cement added with MS decreased with the increasing of MS content; adding 10% MS did not inhibit the cement hydration, but the hydration rate and degree of cement with high content (50%) MS were seriously reduced.

6. Summary and Prospect

6.1. Large-Scale Utilization of MS

The recycling and utilization of MS has achieved great progress in the past years. However, the utilization amount of MS in practice is so limited that a large-scale utilization approach is urgent. The most potent way is to use MS to replace the Portland cement in backfilling mining.

Cemented backfilling mining method is the first option for mineral exploitation due to the recent stricter environmental protection policy in China. The expensive cost of backfill mining has put pressure on the state of enterprise operations, and the cost of cement is 70% of the cost of backfill materials. In addition, the tailings became finer along with the lower ore grade and the development of mineral processing technology. Portland cement has poor performance when ultra-fine tailings are used as aggregate, and its content needs to increase to achieve the same strength. So, it is urgent to exploit new cementitious materials adapted to the ultrafine tailings.

The properties of backfill materials had huge disparities with civil engineering materials, such as a higher water–cement rate, and lower strength request. Further, the shortage of cementitious materials prepared with MS cannot be accepted in civil engineering, but those can be allowed in the backfill mining and the required amounts of backfill materials are huge as more and more mines adopt the backfilling mining method. It is obvious that using MS to prepare backfill materials is the best way to realize large-scale utilization of MS.

Liu et al. modified magnesium slag with a chemical stabilizer and obtained a kind of modified magnesium slag (MMS) with high stability and good activity. Further, MMS and fly ash were mixed as cementitious materials with the aeolian sand as aggregate, and a new paste backfill material that has good mechanical properties and fluidity was prepared [59].

This paper utilized existing MS and industrial solid wastes to prepare cementitious materials for backfilling mining. The properties of cementitious materials composed of MS, gypsum, and BFS (MSGB) were researched. The MS used in this experiment was taken from Tengxiang Magnesium Products Co., Ltd., and it was ground to a certain fineness after drying to obtain higher activity. The particle size distribution of the MS was shown in Figure 4, the particle size characteristic parameters were listed in Table 2, and the chemical composition of the MS was listed in Table 1 in the form of three evaluation parameters.

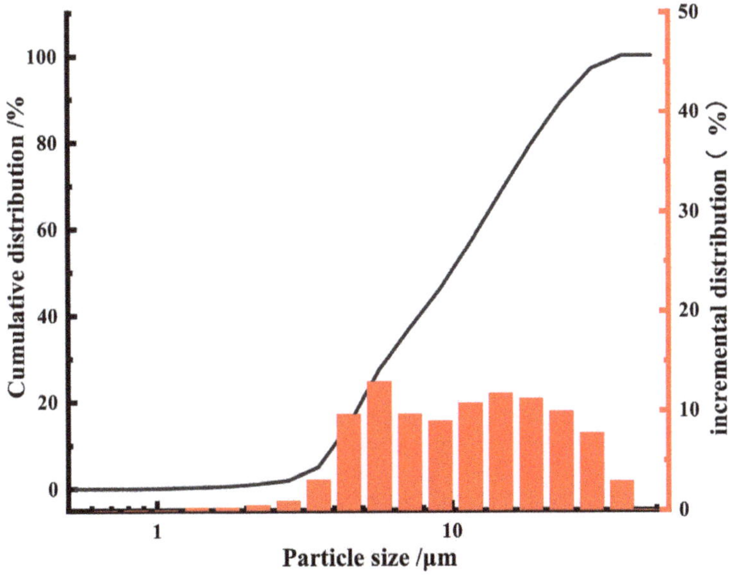

Figure 4. Particle size distribution of MS used in the experiment.

Table 2. Particle size characteristic parameters of MS.

d_{10}/mm	d_{30}/mm	d_{60}/mm	d_{90}/mm	Specific Surface Area m^2/kg
3.99	6.11	12.57	24.07	340

The level of MS was set at 25% and 30%; the level of gypsum was set at 4%, 6%, 8%, 10%, 12%, 14%, and 16%; the resting ingredient was BFS. Tailings taken from a mine in Xinjiang province were used as aggregate; the cement-sand ratio was 1:4, and the solid concentration was 72%. Portland cement was used as a reference. The strength of all samples was shown in Table 3.

Table 3. Strength of samples prepared with MSGB.

Number	Content of Cementitious Materials Ingredient/%			Strength/MPa	
	MS	GS	BFS	7 d	28 d
Y1	25	4	71	1.38	6.08
Y2	25	6	69	1.59	6.07
Y3	25	8	67	1.91	6.37
Y4	25	10	65	2.01	7.14
Y5	25	12	63	2.04	7.34
Y6	25	14	61	1.78	7.09
Y7	25	16	59	1.30	6.92
Z1	30	6	64	1.35	5.09
Z2	30	8	62	2.11	6.73
Z3	30	10	60	2.22	6.75
Z4	30	12	58	2.52	7.01
Z5	30	14	56	1.64	6.87
Z6	30	16	54	2.00	6.74
Z7	30	18	52	1.93	6.15
C1		P.O 42.5 100		1.97	5.29

It was shown in Table 3 that the 28-d strength of most samples prepared with MSGB was higher than P.O 42.5. The 7-d strength was equivalent to P.O 42.5, as the content of GS is 8%, 10%, and 12%. Cement has a fast hydration rate and superior activity, and the sample can obtain better early strength. However, the abundant early hydration production will fill the pores of the sample and inhibit the diffusion of the hydration reaction, which is not conducive to the growth of later strength. In contrast, MSGB has poor activity and a low hydration rate, and the early strength of the sample is low. However, the hydration process of MSGB is long, and the late hydration rate can maintain a relevantly high level, which makes the sample have a higher late strength. Based on the above, it was feasible that MS blend can replace Portland and be used as cementitious materials in backfilling mining.

XRD is used to analyze the hydration products of MSGB (MS at 30% and gypsum at 12%) compared with cement, as shown in Figure 5. It can be seen that the hydration products with a diffraction peak in MSGB were ettringite, gypsum, unhydrated C_3S, C_2S, and MgO. The type of hydration products of MSGB is less than cement which hydration products contain $Ca(OH)_2$ and $CaCO_3$. The XRD patterns of MSGB and cement show a convex hull between 25 and 35°, which is the amorphous C-S-H gels. It is shown in Figure 5 that the diffraction peak of $Ca(OH)_2$ in cement and gypsum in MSGB decreases with the increasing curing age, which means that the $Ca(OH)_2$ and gypsum were constantly consumed in the hydration process and formed more hydration products. By that, the diffraction peak of ettringite in MSGB increased at 28 d curing age.

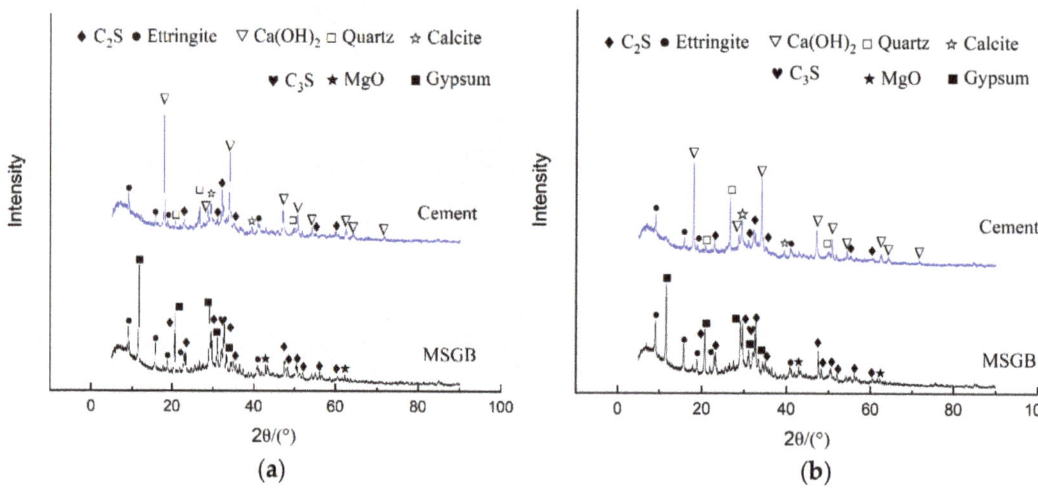

Figure 5. XRD patterns of MSGB and cement. (**a**) 7 d. (**b**) 28 d.

Figures 6 and 7 show the microstructure of the sample prepared with MSGB and cement under different magnifications at 7 d and 28 d curing ages. It can be seen in Figure 6a,c that the holes in MSGB are fewer than in cement, and that means MSGB has a more compact structure. Figure 6b shows that ettringite, which existed in the shape of a needle and rod interspersed in C-S-H gels, formed a dense network of structures. The structure of the sample prepared with MSGB was rather dense and hard to find holes in the surface at 28 d curing age, and the ones with cement still have many holes. It can be seen that the ettringite in the cement is still slim and short, and there are not enough C-S-H gels to connect the ettringite and form a compact structure. In contrast, it can be seen in Figure 7c that the surface of the sample with MSGB is completely covered with C-S-H gels, and ettringite is wrapped in that. The above can be used to explain at the microlevel why the strength of the sample with MGSB is higher than the ones with cement.

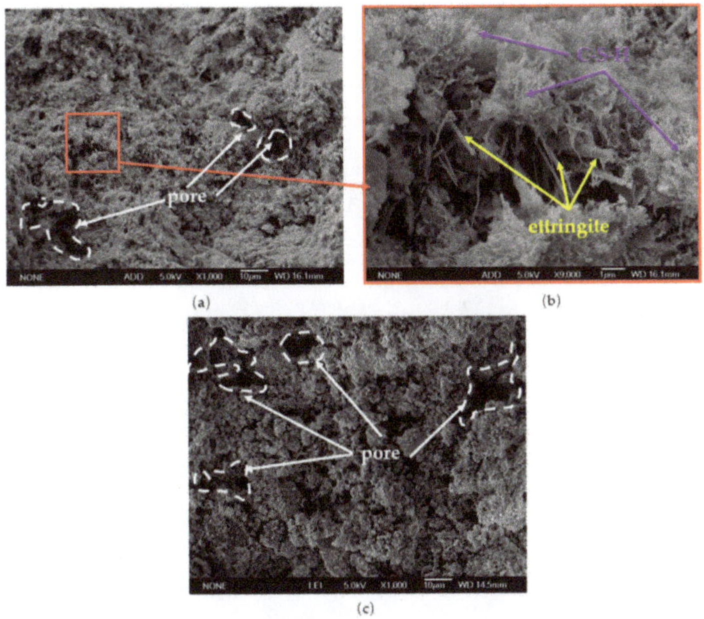

Figure 6. SEM images of MSGB and cement at 7 d curing age. (**a**) MSGB-X1000. (**b**) MSGB-X9000. (**c**) cement-X1000.

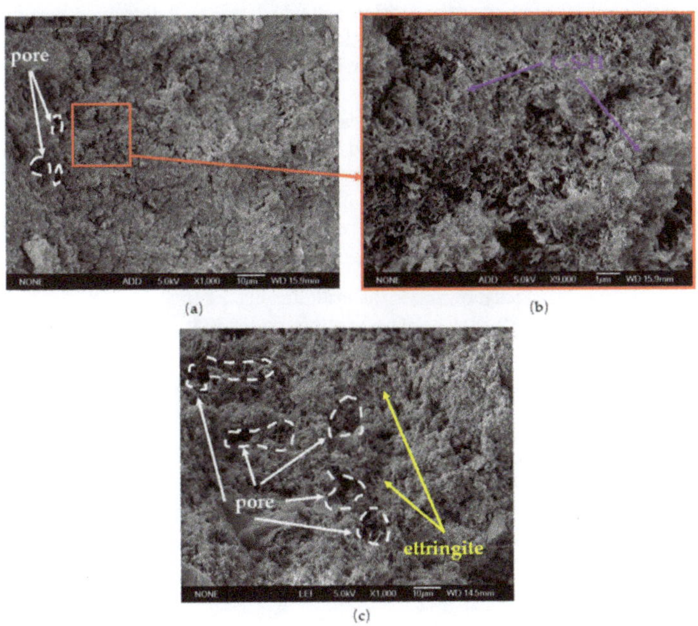

Figure 7. SEM images of MSGB and cement at 28 d curing age. (**a**) MSGB-X1000. (**b**) MSGB-X9000. (**c**) cement-X1000.

Strength and fluidity were the two critical factors to evaluate the performance of backfill materials. So, the rheological experiment of a slurry prepared with MSGB was conducted with an R/S + SST rheometer at different conditions to evaluate its fluidity. The rheological experiment took the controlled shear rate test method (CSR), and the loading method was chosen as ramp loading, with the shear rate rising from 0 s^{-1} to 120 s^{-1} and then decreasing from 120 s^{-1} to 0 s^{-1}, the loading time of 240 s.

The rheological parameters were obtained through regressing test data based on the Bingham model and shown in Table 4. What can be clearly seen in this table is the decrease in slurry yield stress along with the decrease in the cement–sand ratio. Table 4 also reveals that there has been a sharp increase in the viscosity coefficient with the increasing of solid concentration.

Table 4. Rheological parameters of the slurry prepared with MSGB.

Solid Concentration/%	Cement-Sand Ratio	Yield Stress/Pa	Viscosity Coefficient/Pa·s	Correlation Coefficient
68	1:4	18.268	0.094	0.946
	1:6	15.070	0.107	0.959
	1:8	14.881	0.093	0.969
	1:10	13.401	0.091	0.946
70	1:4	32.430	0.177	0.966
	1:6	25.764	0.150	0.963
	1:8	25.681	0.149	0.964
	1:10	24.099	0.154	0.964
72	1:4	44.299	0.328	0.995
	1:6	43.032	0.327	0.996
	1:8	41.429	0.255	0.985
	1:10	41.060	0.276	0.989

In practice, pressure drop was widely used to estimate the pipeline transportation property of backfill slurry, which can be determined through a theoretical formula, L tube experiment, and pipe loop test. The pressure drop is related to the properties of the slurry (yield stress, viscosity, bulk density), the flow state, and engineering conditions [60]. The theoretical calculation is a more convenient and cost-effective method to predict the slurry pressure drop [61]. Before choosing the appropriate model, the Reynolds number *Re* shown in Equation (3) should be calculated to evaluate the flow state of the slurry.

$$Re = \frac{\rho v D}{\mu} \tag{3}$$

where μ is the plastic viscosity of the slurry, Pa·s; D is the pipe diameter, m; v is the flow velocity, m/s; ρ is the bulk density, kg/m^3.

It was assumed in this paper that the pipe diameter is 100 mm, the flow velocity is 1.0 m/s, and all those parameters can be adjusted to match the backfilling technical conditions of specific mines in a practical application. The calculated *Re* ranged from 573 to 1960, as shown in Table 5, which means the slurry fell in the laminar flow regime ($Re < 2000$), and the Swamee–Aggarwal correlation as shown in Equation (4) was used to calculate the friction factor. Further, the friction head loss can be calculated by the Darcy–Weisbach equation and converted to a pressure drop i_m, and the calculation process has been introduced in reference [61]. The calculated pressure drop of the slurry is also shown in Table 5.

$$f = \frac{64}{Re} + \frac{10.67 + 0.1414\left(\frac{He}{Re}\right)^{1.143}}{\left[1 + 0.0149\left(\frac{He}{Re}\right)^{1.16}\right]Re}\left(\frac{He}{Re}\right) \tag{4}$$

$$He = \frac{\tau \rho D^2}{\mu^2} \tag{5}$$

where f is the friction factor, τ is the yield stress; He is the Hedstrom number.

Table 5. R_e and pressure drop of the slurry at a pipe diameter of 100 mm, velocity of 1.0 m/s.

Solid Concentration/%	Cement–Sand Ratio	Yield Stress/Pa	Viscosity Coefficient/Pa·s	*Re*	Pressure Drop /(Pa·m^{-1})
68	1:4	18.268	0.094	1908.51	1227.62
	1:6	15.070	0.107	1671.03	1117.18
	1:8	14.881	0.093	1919.35	1058.91
	1:10	13.401	0.091	1960.44	979.02
70	1:4	32.430	0.177	1038.98	2216.07
	1:6	25.764	0.150	1220.00	1794.23
	1:8	25.681	0.149	1228.19	1786.61
	1:10	24.099	0.154	1187.66	1726.80
72	1:4	44.299	0.328	573.17	3330.55
	1:6	43.032	0.327	573.39	3264.15
	1:8	41.429	0.255	734.12	2934.19
	1:10	41.060	0.276	677.54	2989.84

Rang analysis was employed to compare the influence of solid concentration and cement-sand ratio on pressure drop, and the result is shown in Figure 8. It can be seen that the higher the solid concentration and cement–sand ratio, the higher the pressure drop, and the solid concentration has a greater impact on the pressure drop than the cement–sand ratio. The calculated results of pressure drop could be a reference for the design of the backfilling system, especially in the choice of pipeline transportation form (gravity flow or pumping).

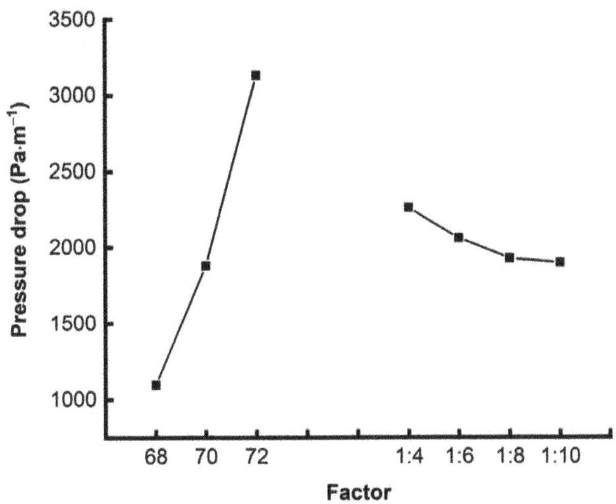

Figure 8. Influence of solid concentration and cement–sand ratio on pressure drop.

6.2. Conclusions and Prospect

This paper introduced the sources, physicochemical properties, and environmental impact of MS, and especially summarized its application approach and effect on the properties of cement and concrete. The conclusion was obtained as follows:

(1) Based on the three quality evaluation parameters of MS, it is known that the MS is alkaline slag of medium- to high-quality and has poor activity. MS has the potential activity to prepare cementitious materials, and can also be used as an alkaline activator.

(2) The properties of cement can reach the P.C 42.5R standard and P.C 32.5R standard as the content of MS in cement ranges from 10%–30% to 30%–45%, respectively. However, clinker is still the necessary substance of composite Portland cement, whose content is 45% at least, and the application method that MS uses as cement admixture is still not

economically and environmentally friendly due to the clinker production process being energy intensive.

(3) Those application methods of MS, such as clinker sintering, fertilizer, calcined porcelain, inorganic fiber, and desulfurization, which all require high-temperature calcination and are also not economically and environmentally.

(4) The addition of MS will prolong the initial and final setting times of cement and increase the slump and slump flow of concrete due to the filling effect and the relatively lower reactivity of MS. The strength of cement and concrete decreased as the content of MS increased, and the negative influence of MS on early-period strength is greater than that on later-period strength. The hydration production of free CaO and MgO in MS can cause long-term volumetric instability, which limits the utilization of MS in building materials or aggregates for road construction.

(5) The special properties of backfilling materials, such as their higher water–cement rate and lower strength request, make it possible to realize the large-scale utilization of MS. The strength of the sample prepared with MSGB, composed of MS, gypsum, and BFS, was equivalent to or higher than the ones with cement. The pressure drops of a slurry prepared with MSGB are calculated through a theoretical formula, and the results can be used as a reference for the design of the backfilling system. The main hydration products of MSGB were ettringite, C-S-H, unhydrated C_3S, C_2S, and MgO.

In conclusion, MS blend can match the requirements of backfill materials both in strength and fluidity and has the capacity to replace Portland cement. The cost of MS blend, where the content of MS is 30% and that of GS ranges from 8% to 12% (the rest is BFS), is about 115–120 RMB/m^3, including raw material cost, transport charges, and processing charges for grinding, and it is 68% cheaper than the cost of cement. Furthermore, the MS blend, as an alternative to cement, can vastly reduce the cost of backfill mining. Nevertheless, the following several issues should be researched in the subsequent work:

(1) The long-term mechanical property of the magnesium blend needs to be verified. Although the higher water–cement ratio of backfill can partly mitigate the negative effect of *f*-MgO in MS in the early curing time. The slow hydration rate and long hydration time of *f*-MgO would be harmful to the long-term strength of backfill, which has had a fixed internal structure. Furthermore, the gypsum as the activator and the hydration products ettringite all have expansibility. This effect will generate interstress on the pore, and affect the volume stability and strength of the backfill body.

(2) In addition to having activity, MS that contain *f*-CaO and *f*-MgO also can be used as alkali-activators to improve the hydration degree of pozzolanic materials. It is critical to find several low-cost activators that can be applied to actual industrial production to improve the activity of MS. Further, the interaction mechanism among the activator, MS, and pozzolanic materials needs to be researched.

(3) As previously mentioned, some harmful elements remain in MS in the process of smelting. The leaching property of cementitious materials prepared with MS must be researched, and guarantee the leaching rate of toxic components is lower than the requirement of relevant norms and standards.

Author Contributions: Conceptualization, X.Y. and F.D.; writing-original draft preparation, X.Y. and F.D.; methodology, F.D., C.L. and X.Y.; validation, Q.G.; investigation, F.D. and X.Z.; supervision, X.Y. and Q.G.; funding acquisition, X.Y. All authors have read and agreed to the published version of the manuscript.

Funding: This research was funded by China Postdoctoral Science Foundation (2021M690363); the Fundamental Research Funds for the Central Universities (FRF-TP-20-039A1); Open Foundation of State Key Laboratory of Mineral Processing (BGRIMM-KJSKL-2021-18).

Data Availability Statement: Data are contained within the article.

Acknowledgments: Thanks for the great efforts of editors and reviewers.

Conflicts of Interest: The authors declare no conflict of interest.

References

1. You, J.; Wang, Y. Reduction Mechanism of Pidgeon Process of Magnesium Metal. *Chin. J. Process Eng.* **2019**, *19*, 560–566. [CrossRef]
2. Pidgeon Method for Magnesium Smelting. Available online: http://www.chinamagnesium.org/index.php?v=listing&cid=102 (accessed on 28 September 2022).
3. Yang, Q.; Engström, F.; Tossavainen, M.; He, M.; Wang, Z. AOD Slag Treatments to Recover Metal and to Prevent Slag Dusting. In Proceedings of the 7th Nordic-Japan Symposium on Science and Technology of Process Metallurgy, Stockholm, Sweden, 15–16 September 2005.
4. Yang, Q.; Nedar, L.; Engström, F.; He, M. Treatments of AOD Slag to Produce Aggregates for Road Construction. In Proceedings of the Iron & Steel Technology Conference, AISTech 2006, Cleveland, OH, USA, 1–4 May 2006; pp. 1–4.
5. Ghosh, S.N.; Rao, P.B.; Paul, A.K.; Raina, K. The Chemistry of Dicalcium Silicate Mineral. *J. Mater. Sci.* **1979**, *14*, 1554–1566. [CrossRef]
6. Durinck, D.; Arnout, S.; Jones, P.; Blanpain, B.; Wollants, P. Global Slag Magazine, Strasbourg, France, 25–30 March 2008. pp. 25–30. Available online: https://www.globalslag.com/magazine (accessed on 28 September 2022).
7. Branca, T.A.; Colla, V.; Valentini, R. A Way to Reduce Environmental Impact of Ladle Furnace Slag. *Ironmak. Steelmak.* **2009**, *36*, 597–602. [CrossRef]
8. Xiang-Bin, X.U.; Cao, H.J. Research of Physicochemical Property of Magnesium Slag. *China Nonferrous Metall.* **2016**, *45*, 73–76.
9. Zhou, P. Study on Preparation and Microstructure Characteristics of Modified Magnesium Slag-Based Backfill Material. Master's Thesis, Xi'an University of Science and Technology, Xi'an, China, 2021.
10. Li, S. The Study of Magnesium Slag on the Drying Shrinkage of Concrete. Master's Thesis, Ningxia University, Yinchuan, China, 2015.
11. Cui, Z.Z.; Chen, D.D.; Zhang, C. Autogenous Shrinkage of Concrete with Magnesium Slag and Fly Ash. *Bull. Chin. Ceram. Soc.* **2017**, *36*, 759–764.
12. Chen, D.; Cui, Z. Effects of Magnesium Slag on the Autogenous Shrinkage of C30 Concrete. *Ningxia Eng. Technol.* **2017**, *16*, 158–160+164.
13. Li, Y.; Fan, Y.; Chen, Z.; Cheng, F.; Guo, Y. Chemical, Mineralogical, and Morphological Characteristics of Pidgeon Magnesium Slag. *Environ. Eng. Sci.* **2016**, *33*, 290–297. [CrossRef]
14. Djokic, J.; Minic, D.; Kamberovic, Z.; Petkovic, D. Impact Analysis of Airborn Pollution Due to Magnesium Slag Deposit and Climatic Changes Condition. *Ecol. Chem. Eng.* **2012**, *19*, 439. [CrossRef]
15. Han, F.L.; Yang, Q.X.; Wu, L.E.; Guo, S.W. Treatments of Magnesium Slag to Recycle Waste from Pidgeon Process. In *Advanced Materials Research*; Trans Tech Publications Ltd.: Wollerau, Switzerland, 2012; Volume 418, pp. 1657–1667.
16. Cui, Z.; Yang, J. Study on Hydration Inertia Mechanism of Magnesium Slag. *New Build. Mater.* **2007**, *158*, 337–339. [CrossRef]
17. Peng, X.; Wang, K.; Li, J.; Yu, Z.; Wang, S. Hydraulic Potential Stimulation and Bricks Preparation of Magnesium Slag. *J. Chongqing Univ.* **2013**, *36*, 48–52.
18. Vadala, M.L.; Rutnakornpituk, M.; Zalich, M.A.; St Pierre, T.G.; Riffle, J.S. Block Copolysiloxanes and Their Complexation with Cobalt Nanoparticles. *Polymer* **2004**, *45*, 7449–7461. [CrossRef]
19. Mo, L.; Hao, Y.; Liu, Y.; Wang, F.; Deng, M. Preparation of Calcium Carbonate Binders via CO_2 Activation of Magnesium Slag. *Cement Concrete Res.* **2019**, *121*, 81–90. [CrossRef]
20. Feng, G.; Nie, Z.; Wang, Z.-H.; Gong, X.-Z.; Zuo, T.-Y. Assessing Environmental Impact of Magnesium Production Using Pidgeon Process in China. *Trans. Nonferrous Metals Soc. China* **2008**, *18*, 749–754.
21. Ramakrishnan, S.; Koltun, P. Global Warming Impact of the Magnesium Produced in China Using the Pidgeon Process. *Resour. Conserv. Recycl.* **2004**, *42*, 49–64. [CrossRef]
22. Mao, J. Using Magnesium Reducing Slag to Produce Ca-Mg-Si Compound Fertilizer and Its Applied Basic Research. Ph.D. Thesis, Taiyuan University of Technology, Taiyuan, China, 2016.
23. Wang, X.; Xiang, Y.A.N.; Li, X. Environmental Risks for Application of Magnesium Slag to Soils in China. *J. Integr. Agric.* **2020**, *19*, 1671–1679. [CrossRef]
24. Wu, L.; Yang, Q.; Han, F.; Du, C. Dusting Control of Magnesium Slag Produced by Pidgeon Process. In *AIP Conference Proceedings*; American Institute of Physics: College Park, MD, USA, 2013; Volume 1542, pp. 1282–1285.
25. Peng, X.; Wang, K.; Gong, M.; Huang, W.; Wang, S. Properties of Portland Cement with Magnesium Slag. *Civ. Constr. Environ. Eng.* **2011**, *33*, 140–144.
26. Deng, J.; Chen, X.; Guo, Y. Effect of Magnesium Slag and Blast Furnace Slag on Properties of Composite Cement. *J. Xi'an Univ. Sci. Technol.* **2008**, *28*, 735–739.
27. Ji, G.; Peng, X.; Wang, S.; Hu, C.; Ran, P.; Sun, K.; Zeng, L. Influence of Magnesium Slag as a Mineral Admixture on the Performance of Concrete. *Constr. Build. Mater.* **2021**, *295*, 123619. [CrossRef]
28. Cai, J.W.; Gao, G.L.; Bai, R.Y.; Lu, F.; Li, L. Research on Slaked Magnesium Slag as a Raw Material and Blend for Portland Cement. In *Advanced Materials Research*; Trans Tech Publications Ltd.: Wollerau, Switzerland, 2011; Volume 335, pp. 1246–1249.
29. Oliveira, C.; Gumieri, A.G.; Gomes, A.M.; Vasconcelos, W.L. Characterization of Magnesium Slag Aiming the Utilization as a Mineral Admixture in Mortar. In Proceedings of the International RILEM Conference on the Use of Recycled Materials in Buildings and Structures, Barcelona, Spain, 8–11 November 2004.
30. Mola-Abasi, H.; Shooshpasha, I. Influence of Zeolite and Cement Additions on Mechanical Behavior of Sandy Soil. *J. Rock Mech. Geotech. Eng.* **2016**, *8*, 746–752. [CrossRef]

31. Jahandari, S.; Saberian, M.; Tao, Z.; Mojtahedi, S.F.; Li, J.; Ghasemi, M.; Rezvani, S.S.; Li, W. Effects of Saturation Degrees, Freezing-Thawing, and Curing on Geotechnical Properties of Lime and Lime-Cement Concretes. *Cold Reg. Sci. Technol.* **2019**, *160*, 242–251. [CrossRef]
32. Saberian, M.; Jahandari, S.; Li, J.; Zivari, F. Effect of Curing, Capillary Action, and Groundwater Level Increment on Geotechnical Properties of Lime Concrete: Experimental and Prediction Studies. *J. Rock Mech. Geotech. Eng.* **2017**, *9*, 638–647. [CrossRef]
33. Jahandari, S.; Li, J.; Saberian, M.; Shahsavarigoughari, M. Experimental Study of the Effects of Geogrids on Elasticity Modulus, Brittleness, Strength, and Stress-Strain Behavior of Lime Stabilized Kaolinitic Clay. *GeoResJ* **2017**, *13*, 49–58. [CrossRef]
34. Jahandari, S.; Saberian, M.; Zivari, F.; Li, J.; Ghasemi, M.; Vali, R. Experimental Study of the Effects of Curing Time on Geotechnical Properties of Stabilized Clay with Lime and Geogrid. *Int. J. Geotech. Eng.* **2019**, *13*, 172–183. [CrossRef]
35. Lu, F.; Bai, R.Y.; Cai, J.W. Study on Clinker Production Using Magnesium Slag on a 4500tpd Line. In *Advanced Materials Research;* Trans Tech Publications Ltd.: Wollerau, Switzerland, 2013; Volume 690, pp. 724–727.
36. Amini, O.; Ghasemi, M. Laboratory Study of the Effects of Using Magnesium Slag on the Geotechnical Properties of Cement Stabilized Soil. *Constr. Build. Mater.* **2019**, *223*, 409–420. [CrossRef]
37. Louati, S.; Baklouti, S.; Samet, B. Geopolymers Based on Phosphoric Acid and Illito-Kaolinitic Clay. *Adv. Mater. Sci. Eng.* **2016**, *2016*, 2359759. [CrossRef]
38. Tchakouté, H.K.; Rüscher, C.H. Mechanical and Microstructural Properties of Metakaolin-Based Geopolymer Cements from Sodium Waterglass and Phosphoric Acid Solution as Hardeners: A Comparative Study. *Appl. Clay Sci.* **2017**, *140*, 81–87. [CrossRef]
39. Xiao, L.G.; Luo, F.; Li, R.B.; Liu, C.Y. The Study of Alkaline-Activated Magnesium Slag Cementitious Material. In *Key Engineering Materials;* Trans Tech Publications Ltd.: Wollerau, Switzerland, 2012; Volume 517, pp. 363–366.
40. Li, H.; Huang, Y.; Yang, X.; Jiang, Z.; Yang, Z. Approach to the Management of Magnesium Slag via the Production of Portland Cement Clinker. *J. Mater. Cycles Waste Manag.* **2018**, *20*, 1701–1709. [CrossRef]
41. Xiao, L.G.; Wang, S.Y.; Luo, F. Status Research and Applications of Magnesium Slag. *J. Jilin Inst. Arch. Civ. Eng.* **2008**, *1*, 1–7.
42. Li, Y.; Cheng, F. Synthesis of a Novel Slow-Release Potassium Fertilizer from Modified Pidgeon Magnesium Slag by Potassium Carbonate. *J. Air Waste Manag. Assoc.* **2016**, *66*, 758–767. [CrossRef]
43. Fan, Y.; Qin, W.; Chen, Z.; Cheng, F. A Silicon–Potash Fertilizer Prepared from Magnesium Slag and How It Can Improve Soil Fertility and Agronomic Performance. *Soil Sci. Plant Nutr.* **2019**, *65*, 274–280. [CrossRef]
44. Fan, Y.; Li, Y.; Li, H.; Cheng, F. Evaluating Heavy Metal Accumulation and Potential Risks in Soil-Plant Systems Applied with Magnesium Slag-Based Fertilizer. *Chemosphere* **2018**, *197*, 382–388. [CrossRef]
45. Ma, X.; Tian, Y.; Zhou, Y.; Wang, K.; Chai, Y.; Li, Z. Sintering Temperature Dependence of Low-Cost, Low-Density Ceramic Proppant with High Breakage Resistance. *Mater. Lett.* **2016**, *180*, 127–129. [CrossRef]
46. Chen, P.; Liu, K.; Liu, Y. Research on the Low Density and High Strength of the Proppant. *J. Shanxi Univ. Sci. Technol.* **2014**, *32*, 63–67.
47. Wu, T.; Zhou, J.; Wu, B. Effect of TiO_2 Content on the Acid Resistance of a Ceramic Proppant. *Corros. Sci.* **2015**, *98*, 716–724. [CrossRef]
48. Hao, J.; Ma, H.; Feng, X.; Gao, Y.; Wang, K.; Tian, Y. Low-Temperature Sintering of Ceramic Proppants by Adding Solid Wastes. *Int. J. Appl. Ceram. Technol.* **2018**, *15*, 563–568. [CrossRef]
49. Hao, J.; Ma, H.; Feng, X.; Gao, Y.; Wang, K.; Tian, Y.; Chai, Y. Microstructure and Fracture Mechanism of Low Density Ceramic Proppants. *Mater. Lett.* **2018**, *213*, 92–94. [CrossRef]
50. Zhang, J.; Wen, X.; Cheng, F. Preparation, Thermal Stability and Mechanical Properties of Inorganic Continuous Fibers Produced from Fly Ash and Magnesium Slag. *Waste Manag.* **2021**, *120*, 156–163. [CrossRef]
51. Jia, L.; Fan, B.; Huo, R.; Li, B.; Yao, Y.; Han, F.; Qiao, X.; Jin, Y. Study on Quenching Hydration Reaction Kinetics and Desulfurization Characteristics of Magnesium Slag. *J. Clean. Prod.* **2018**, *190*, 12–23. [CrossRef]
52. Fan, B.; Jia, L.; Li, B.; Huo, R.; Yao, Y.; Han, F.; Qiao, X.; Jin, Y. Study on Desulfurization Performances of Magnesium Slag with Different Hydration Modification. *J. Mater. Cycles Waste Manag.* **2018**, *20*, 1771–1780. [CrossRef]
53. Fan, B.; Jia, L.; Han, F.; Huo, R.; Yao, Y.; Qiao, X.; Zhao, C.; Jin, Y. Study on Magnesium Slag Desulfurizer Modified by Additives in Quenching Hydration. *J. Mater. Cycles Waste Manag.* **2019**, *21*, 1211–1223. [CrossRef]
54. Deng, J.P.; Wang, X.L.; Guo, Y.P. Research on the Hydration Mechanism of Portland Cement with Magnesium Slag. In *Applied Mechanics and Materials;* Trans Tech Publications Ltd.: Wollerau, Switzerland, 2014; Volume 576, pp. 57–62.
55. Sun, R.; Wu, Z.; Wang, D.; Ding, Y.; Fang, K. Properties and Hydration Mechanism of Ultrafine Magnesium Slag Powder-Cement Composites Cementitious Materials. *Mater. Rep.* **2023**, 1–20. [CrossRef]
56. Mehta, P.K. Mechanism of Sulfate Attack on Portland Cement Concrete—Another Look. *Cement Concrete Res.* **1983**, *13*, 401–406. [CrossRef]
57. Cui, Z.; Ni, X.; Meng, X. Study on the Expansibility of Magnesium Slag. *Fly Ash Compr. Util.* **2006**, 8–11.
58. Zhao, A. Utilize Magnesia Slag to Produce New Wall Materials. *Shanxi Archit.* **2003**, *29*, 48–49.

59. Liu, L.; Ruan, S.; Fang, Z.; Hou, D.; Sun, W. Modification of Magnesium Slag and Its Application in the Field of Mine Filling. *J. China Coal Soc.* **2021**, *46*, 3833–3845. [CrossRef]
60. Yang, X.; Xiao, B.; Gao, Q. Validating the Use of Slag Binder with 91 Percent Blast Furnace Slag for Mine Backfilling. *Adv. Mater. Sci. Eng.* **2020**, *2020*, 2525831. [CrossRef]
61. Yang, X.; Xiao, B.; Gao, Q. Prediction of Friction Pressure for Non-Newtonian Backfilling Pipe-Flow Slurry. *J. Hunan Univ. (Nat. Sci.)* **2020**, *47*, 125–131. [CrossRef]

minerals

MDPI

Article

Evaluation of Stope Stability and Displacement in a Subsidence Area Using 3Dmine–Rhino^{3D}–FLAC^{3D} Coupling

Leiming Wang [1,2,3,4,5], Xingquan Zhang [2], Shenghua Yin [1,2,*], Xuelan Zhang [6], Yifan Jia [2] and Helin Kong [2]

1 Key Laboratory of Ministry of Education for High-Efficient Mining and Safety of Metal, University of Science and Technology Beijing, Beijing 100083, China
2 School of Civil and Environment Engineering, University of Science and Technology Beijing, Beijing 100083, China
3 State Key Laboratory of Coal Mine Resources and Safety Mining, China University of Mining and Technology, Xuzhou 221116, China
4 State Key Laboratory Safety and Health for Metal Mines, Sinosteel Maanshan General Institute of Mining Research Co., Ltd., Maanshan 243000, China
5 Key Laboratory of Green Chemical Engineering Process of Ministry of Education, Wuhan Technological University, Wuhan 430205, China
6 School of Mathematics and Physics, University of Science and Technology Beijing, Beijing 100083, China
* Correspondence: ustb_ysh@163.com or csuysh@126.com; Tel.: +86-138-1166-8481

Abstract: Underground stope stability has important implications for the safety and efficiency of metal mining, especially when there are subsidence zones above. To analyze the stope stability around the subsidence area of the Hongling ore body, this paper innovatively adopts the 3Dmine–Rhino^{3D}–FLAC^{3D} coupled modeling numerical method to model and predict the excavation process of the stope around the Hongling subsidence area; it monitors and analyzes the displacement and stress-distribution changes in a selected stope roof, and its left-hand side and right-hand side walls. This paper realizes the separation and extraction of the special ore body and surface topographic features of the Hongling ore body; it discusses the stope stability characteristics around a subsidence area based on the real topography of the mining area and the occurrence form of the ore body. The study found that the influence of the upper collapse area on the undesired displacement of the downhole stope is mainly reflected in the roof displacement. As the stope position moves away from the collapse area, the roof displacement decreases significantly, the maximum displacement is at the center of the stope roof, and the maximum principal stress occurs at the corner of the stope.

Keywords: stope stability; subsidence area; stress distribution; displacement deformation; 3Dmine–Rhino^{3D}–FLAC^{3D} coupling; underground mining

Citation: Wang, L.; Zhang, X.; Yin, S.; Zhang, X.; Jia, Y.; Kong, H. Evaluation of Stope Stability and Displacement in a Subsidence Area Using 3Dmine–Rhino^{3D}–FLAC^{3D} Coupling. *Minerals* **2022**, *12*, 1202. https://doi.org/10.3390/min12101202

Academic Editor: Mamadou Fall

Received: 11 August 2022
Accepted: 19 September 2022
Published: 24 September 2022

1. Introduction

In a mine subsidence area, there are undesirable displacements and potential stress concentrations in stope mining, which can easily lead to huge safety hazards and potential losses, both economic and of personnel [1–3]. As a result, the detection, simulation and visualization of stope stability and its displacement are very important issues in metal-mining operations [4–7]. For this reason, domestic and foreign scholars undertook research on stope stability in mine subsidence areas using various methods, such as laboratory experiments, physical/mathematical models and numerical simulations, which led to breakthroughs.

In engineering practice, due to the high experimental costs involved, it is generally challenging to experimentally solve various difficult problems. Currently, numerical simulation is required to predict and analyze programs using FLAC^{3D} software (such as Flac^{3D} 6.0 version, Itasca Consulting Group, Inc., Minneapolis, MN, USA); this established the seepage–damage–fracture coupling mathematical model of fractured rock mass. When it is in an unstable state, the overlying rock mass in an excavated stope is prone to collapse. Existing research mainly focused on the collapse and spalling phenomena of the roof and surrounding rock, as well as the evaluation and prediction of stope stability, which is of great significance for guiding industrial mining safety. Due to the large range and large span of stopes, the requirements of conventional indoor research conditions are relatively harsh, so mining experiments on similar materials at a reduced scale are usually used [8–10]. In addition, to compensate for the limitations of rock mechanical experiments, researchers used FlAC^{3D} and other numerical computations to explore the stability of the surrounding rock and slope, the optimization of stope structure parameters and the coupling relationship between roadway and working face in an underground mine [11,12]. Undesirable instability is destructive and can cause the fracturing of the pore structure in porous systems, such as underground stopes and packed beds; however, it can be well understood by simulating predictions [5–7,13,14]. The Barton engineering empirical formula, Pakalnis critical equations and other methods are commonly used to evaluate stope stability [15]. However, since most of the rock masses are set as ideal homogeneous bodies, and the heterogeneity of ore body occurrence and rock mass structure are not usually considered, there are errors in the calculation results [16,17].

In summary, the existing coupling research on FLAC^{3D} and various modeling software was successfully applied in the fields of geotechnical engineering, mining engineering and other fields, and has a good application prospect in realizing numerical and transparent simulations for the mining of underground deposits [18,19]. However, although Flac^{3D} itself is an excellent numerical simulation software, it is not an excellent modeling software. It is difficult to realize the construction of complex models through its use alone. The method of combining 3Dmine–Rhino^{3D} with FLAC^{3D} in this study can better achieve the construction of a heterogeneous model for ore body occurrence and rock mass structure. We used 3Dmine software to build complex models, then used Rhino^{3D} software to draw meshes for the established models; finally, we used FlAC^{3D} software to perform grouping simulations on the software. Previous research still assumes modeling homogenization and idealization; moreover, simulation research considering real topography and ore deposit occurrence forms of mining areas is not sufficiently deep.

In view of this, and based on the stability characteristics of the three stopes 8121 (Stope A), 8122 (Stope B) and 8123 (Stope C) around the Hongling subsidence area after mining, this paper comprehensively considers the differences in the topography and ore deposit occurrence characteristics of the mining area. The 3DMine–Rhino^{3D}–FLAC^{3D} coupling method is innovatively adopted to conduct research from the variation in displacement stress of the side walls of the stope roof. Displacement and stress monitoring points were set up for the corresponding stopes in the model, which effectively predicted the displacement and stress changes in the three stopes after excavation.

2. Ore Deposit Topography Model Established by 3Dmine–Rhino^{3D}–Flac^{3D} Coupling

2.1. Real Topography Model of Hongling Subsidence Area via 3Dmine

The Hongling Lead–Zinc Mine is located in Balinzuo Banner, north of Chifeng City, Inner Mongolia Autonomous Region, and its administrative division is under the jurisdiction of Ulandaba Sumu. Its ore body and surrounding rock have good stability and the degree of development of joints and fissures is low, but local structural damage may lead to poor stability. The grade of the ore body is evenly distributed, and the boundary of the ore rock is clear. The mining method used in the past is the open-field mining method in the layout stage along the strike.

This mining method created a subsidence area in the vicinity of the mining area. In order to predict whether mining procedures would lead to further large-scale subsidence, numerical simulation analysis of the mining process of the stope was performed. The establishment of a surface model is very important for mining engineering as it can reflect the three-dimensional coordinate position relationship of geological bodies, such as ore bodies, roadways, and factory areas. To establish a surface model, as Figure 1 shows, the topographic mine map was used as the original data, and the information required for modeling, such as terrain contours and exploration lines, and the corresponding labeling values, were retained.

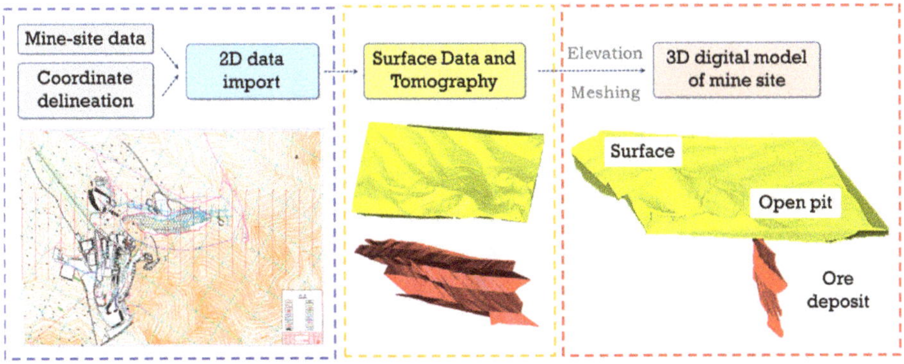

Figure 1. Geological topography and 3D reconstruction of ore bodies in Hongling subsidence area.

This paper analyzes the geological plan of the middle sections at 755 m, 805 m, 855 m, 905 m and 955 m. The ore deposits and surface contour lines were extracted and the CAD data of the target ore deposits were obtained. The mine-site data and coordinate delineations were imported into 3Dmine to visualize the surface data and monography of the Hongling mine-site. These data were then elevated and meshed to obtain the 3D digital model. Following this, optimization and hierarchical management were carried out.

2.2. Ore Deposit Grid Acquisition and Division of Ore Deposit via Rhino^{3D}

To drape the surface, an appropriate threshold range based on the size of the ore deposit was selected. This meshing data file was imported into Rhino, and then the drape was used to reduce its size and make it a poly-surface. Next, a cube with an extent smaller than the surface of the drape was formed in preparation for establishing the mesh. The division operation was performed based on the Boolean operation, and then the drapery surface was divided by the cube, and the redundant surfaces were deleted.

Since there is a subsidence area between the ore deposit and the surface, the range of the subsidence area on the surface must be determined according to the range of the subsidence area of the ore deposit. According to the imported exploration line, a surface was established by pointing several corners to form different surfaces. The part wrapped by this surface was the subsidence area of the ore deposit; the Boolean segmentation operation was then performed to segment the ore deposit.

2.3. Coupling Model and Parameters of Ore Deposit via FlAC^{3D}

To better divide the ore body and surrounding rock, the ore body and surface models were imported into FlAC^{3D} after the initial meshing in Rhino. As Figure 2 shows, the ore beds and surrounding rock meshing were well processed under the Gsurf and GVol comments of Rhino. The model is divided into two parts: external surrounding rock and internal ore body. The former includes ordinary surrounding rock and the subsidence area, and the latter is divided into the excavated stope and planned excavation stope; the subsidence areas in both are connected upward and downward.

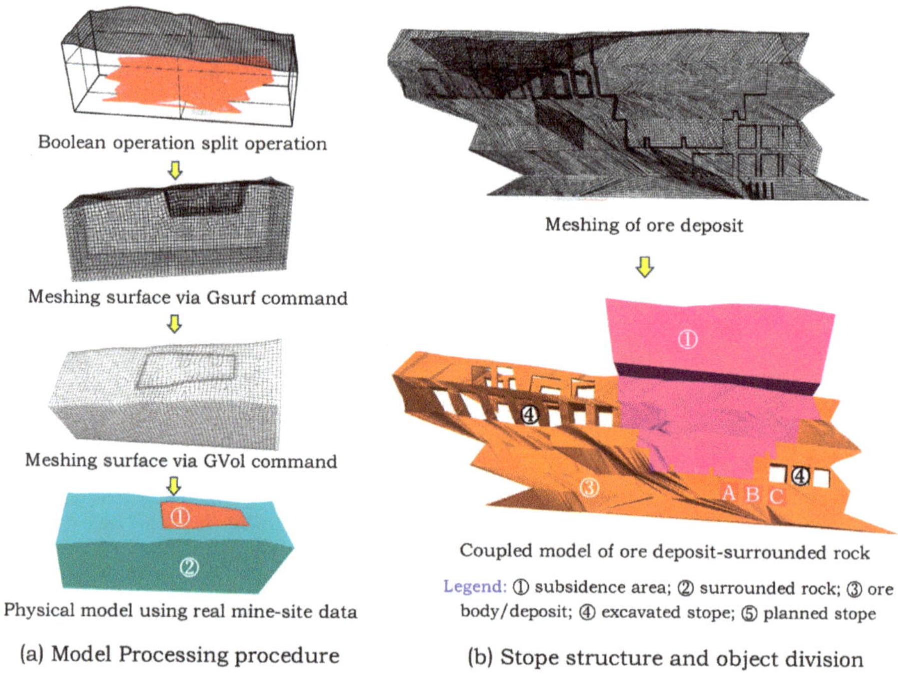

Figure 2. Coupled model processing procedure using Hongling mine-site data, Rhino^{3D} and FlAC^{3D}.

It should be noted that, based on the imported exploration line, a surface was established by pointing several corners to form different surfaces, and the part wrapped by this surface is the subsidence area of the ore body. The Boolean segmentation operation was then performed to segment the ore body with this surface. In this paper, the research objects that we wanted to analyze and predict were the three stopes 8121, 8122 and 8123 that are planned to be excavated: these were named Stope A, Stope B and Stope C, respectively.

Before excavation simulation and monitoring were performed on the model, the constitutive model, material parameters, boundary conditions and initial conditions were assigned to the model. In this paper, the Mohr–Coulomb model was used. The specific material parameters of the ore and rock mass are shown in Table 1 and divided into three parts: the ore body, the surrounding rock and the subsidence area.

Table 1. Key experimental parameters of ore deposit.

Type	Elastic Modulus (GPa)	Poisson Ratio	Tensile Strength (MPa)	Cohesion (MPa)	Internal Friction Angle (°)	Density (g/cm^{-3})
Ore body	21.26	0.19	2	1.8	50	3.56
Rock	18.81	0.21	0.9	1.5	49	2.77
subsidence area	0.4	0.3	0	0	15	2.00

3. Results and Discussion

3.1. The Stress and Ground Displacement of Excavated Stopes in Subsidence Areas

The stress and displacement color map of the three targeted stopes (A/B/C) is shown in Figure 3, and the maximum values of displacement and stress in the stopes (A/B/C) are recorded in Table 2, respectively. Based on Table 2 and Figure 3, the ground displacement phenomenon after the stope excavation in the subsidence area was analyzed: the roof displacement of Stope A in the fall area is the most significant, the maximum vertical displacement of the roofs of Stopes A, B and C being 6.7 cm, 2.6 cm and 1.1 cm, respectively. This indicates that after mining in Stope A, the roof of the roadway is more prone to rupture and deformation, and there is the certain hidden danger of instability.

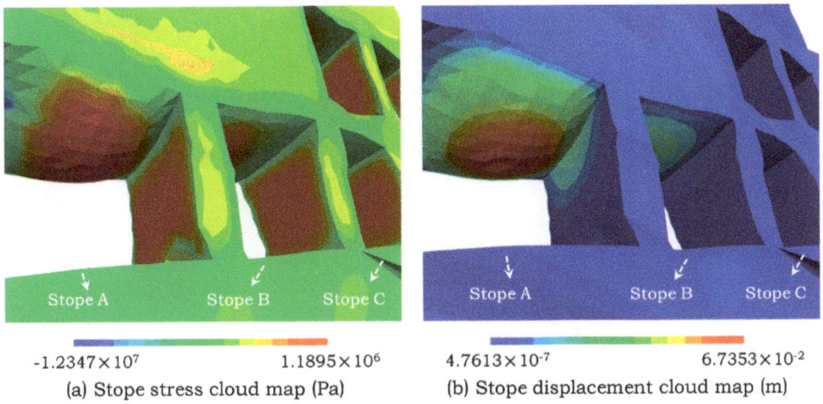

Figure 3. The stress and displacement color map of three targeted stopes (A/B/C).

Table 2. The maximum displacement of each wall after the target stope is excavated.

Stope No.		Stope A	Stope B	Stope C
Maximum Stress/MPa	Roof-Vertical axis	24.4	14.5	9.5
	Left-side wall–Horizontal axis	13.5	7.9	8.3
	Right-side wall–Horizontal axis	10.3	7.4	7.1
Maximum displacement/cm	Roof-Vertical axis	6.7	0.42	−0.31
	Left-side wall–Horizontal axis	2.6	0.86	−0.24
	Right-side wall–Horizontal axis	1.1	0.41	−0.45

Similarly, it can be seen from Figure 3 that, comparing the maximum horizontal displacement of the left- and right-hand sides of the three stopes, the maximum horizontal displacement of the left-hand sides are all positive; that is, they are displaced in the direction close to the center of their respective stope. The two sides are displaced toward the center of the stope: the maximum horizontal displacement of the left-hand side was obtained for Stope B, and was 0.86 cm, and the maximum horizontal displacement of the right-hand side was obtained for Stope C, and was 0.45 cm. Moreover, the horizontal displacements in the three stopes are all less than 1 cm and are also lower than the roof failure tendency. In other words, the roof is more prone to vertical displacement after stope excavation, and is regarded as a roof-falling accident, commonly observed in actual metal deposit mining operations.

3.2. 2D Slice Features of the Excavated Stopes (A/B/C) in the Subsidence Areas

To simulate the actual mining operations, we first excavated each stope separately. After solving the balance, we then excavated all three stopes (A/B/C) simultaneously, and set up monitoring points to monitor the displacement and stress in the roof and side walls after excavation.

The monitor points were set up as follows: first, the initial coupled model was imported and the three stopes (A/B/C) to be excavated were cut using the orthogonal test method. Next, each stope was cut twice along the x-axis and y-axis, and a total of

12 slices were obtained. Each slice included three groups of monitoring points. The monitoring points were then set equidistantly at the edge of each slice for recording. The slice-processing procedure is shown in Figure 4.

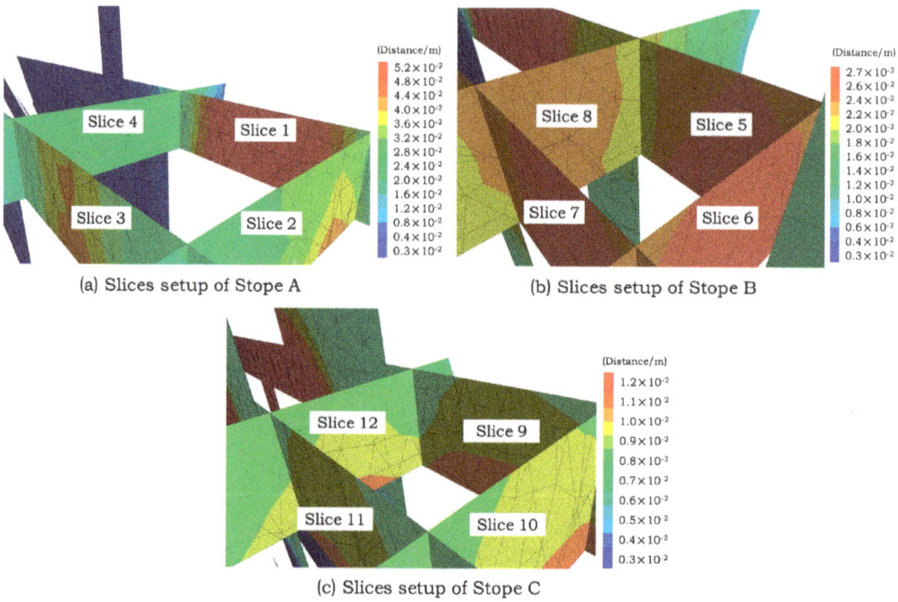

Figure 4. The 2D slices of excavated stopes (A/B/C) in the subsidence areas.

Slice 1, Slice 3, Slice 5, Slice 7, Slice 9 and Slice 11 were cut along the x-axis to monitor the displacement and stress in the z-direction of the top plate, and the displacement and stress in the x-direction of the left- and right-hand side walls, respectively. Eight monitoring points were set.

Slice 2, Slice 4, Slice 6, Slice 8, Slice 10 and Slice 12 were cut along the y-axis to monitor the displacement and stress in the z-direction of the top plate, respectively. Eight monitoring points were set at an equal distance for each group of monitoring points.

After the monitoring points were established, excavation calculations were performed on the three stopes A, B and C in the model, and the monitoring point values and displacement-stress evolution cloud map during the calculation process were output.

Among the three stopes, the left side of Stope A changes greatly, and the stress is concentrated in the upper left side of the stope near the collapse area, where the stress reaches 24.4 MPa, and the collapse area is likely to continue to collapse from this point. The stress changes in the other sides are similar. The maximum stress near the center of the side is less than 1 MPa, and the closer to the edge, the greater the stress; the maximum does not exceed 6 MPa, which is relatively safe and stable.

3.3. *Displacement Features of Monitor Points in a Typical Excavated Stope*

To better detect the displacement features of excavated stopes, we selected the 12 typical slices from the excavated stopes. The vertical displacement changes in the roof of Stope A (Slices 1~4) are shown in Figure 5. It was found that for the roof slices in different directions, the displacement of the monitoring point first increased and then tended to be stable as the number of calculation steps increased. When the step size was 6000 steps, the calculation for each slice tended to be stable overall.

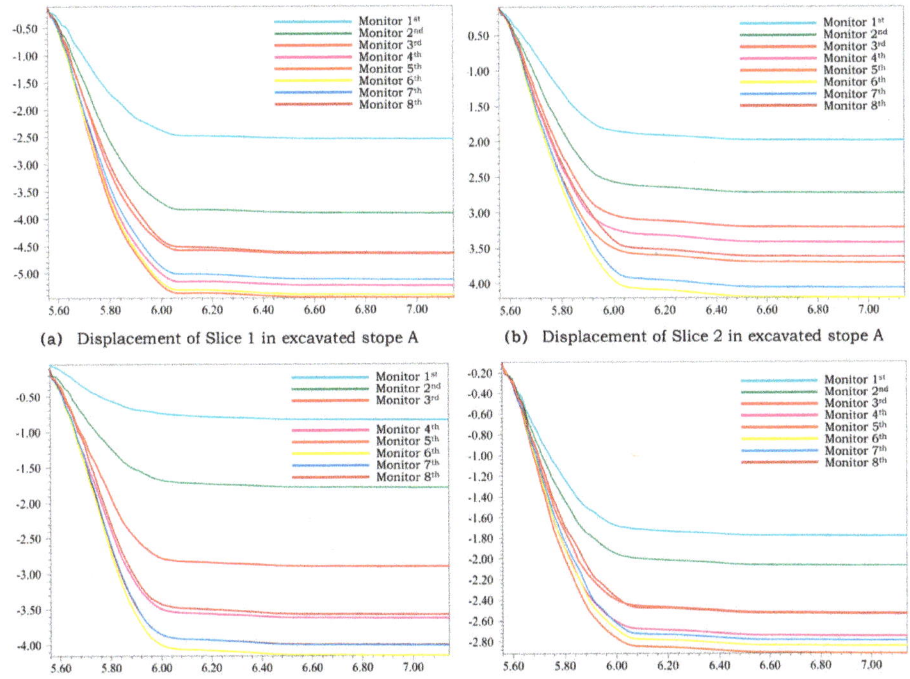

(a) Displacement of Slice 1 in excavated stope A

(b) Displacement of Slice 2 in excavated stope A

(c) Displacement of Slice 3 in excavated stope A

(d) Displacement of Slice 4 in excavated stope A

Figure 5. Displacement features of Slices 1 to 4 in excavated Stope A.

To compare and reveal the vertical displacement variation characteristics in different parts of the roof, the vertical displacement variation in the roof of Stope A was further analyzed, as shown in Figure 6a. It is inferred that the vertical displacement of each slice is larger near the subsidence area. In other words, (1) the roof displacement is negatively correlated with the distance from the center of the roof, and the peak value of the vertical displacement is obtained at the center of the vertical displacement;(2) when a subsidence area exists, the displacement of the slices close to the collapse area is larger. Specifically, the maximum displacement is the fifth monitoring point on Slice 4, with a peak value of 5.4 cm; this slice is the closest to the subsidence area and its overall displacement is the largest relative to the other three curves. Slice 2 and Slice 4 are parallel slices in the same direction; Slice 4 is closer to the subsidence area, and its overall top-plate displacement is significantly larger than that of Slice 2.

Based on Figure 6b, it can be seen that the horizontal displacement changes in the left- and right-hand side walls of Stope A, showing a gradually increasing trend with the unidirectional changes in the monitoring points. Specifically, the displacements of the right-hand side of Slice 1 and the right-hand side of Slice 3 undergo a change from positive to negative; that is, the monitoring points on the right-hand side shifted from top to bottom, first in the positive direction of the x-axis, and then in the negative direction of the x-axis, while the left side is uniformly displaced in the positive direction of the x-axis, and its overall displacement direction is toward the inside of the stope. The peak is 0.4 cm on the left-hand side of Slice 1. In addition, comparing Figure 6a,b, it can be seen that the horizontal displacement in the stope is obviously smaller than the vertical displacement.

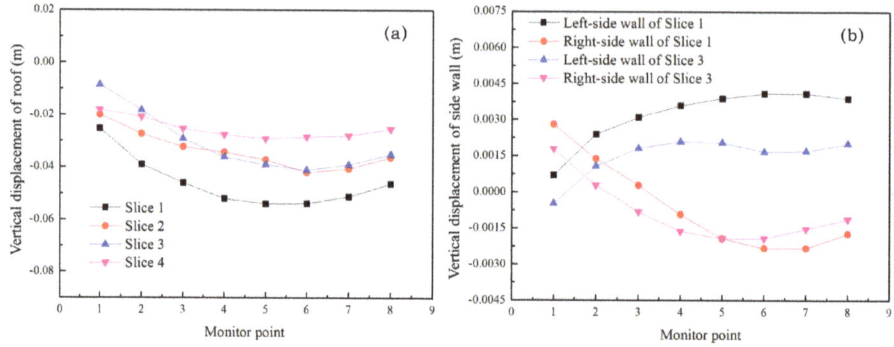

Figure 6. Displacement curve of roof, left- and right-hand side walls of excavated Stope A. (**a**) Vertical displacement of roof; (**b**) Horizontal displacement of side wall.

3.4. Stress Features of Monitor Points in a Typical Excavated Stope

The vertical and horizontal stress distributions of an excavated stope is an important factor affecting mining safety, especially when the excavated stope is located in the subsidence areas [20,21]. To better understand the stress distribution of the roof, left-hand side walls and right-hand side walls, eight monitoring points in the excavated stope were carefully selected and the stress changes at each point were recorded, as shown in Figure 7. This stress fluctuation is very significant in the initial stage. It tends to be stable after a step size of 6000, which is similar to the displacement of each slice. Moreover, the variations are significantly higher in Slice 1 of the excavated Stope A.

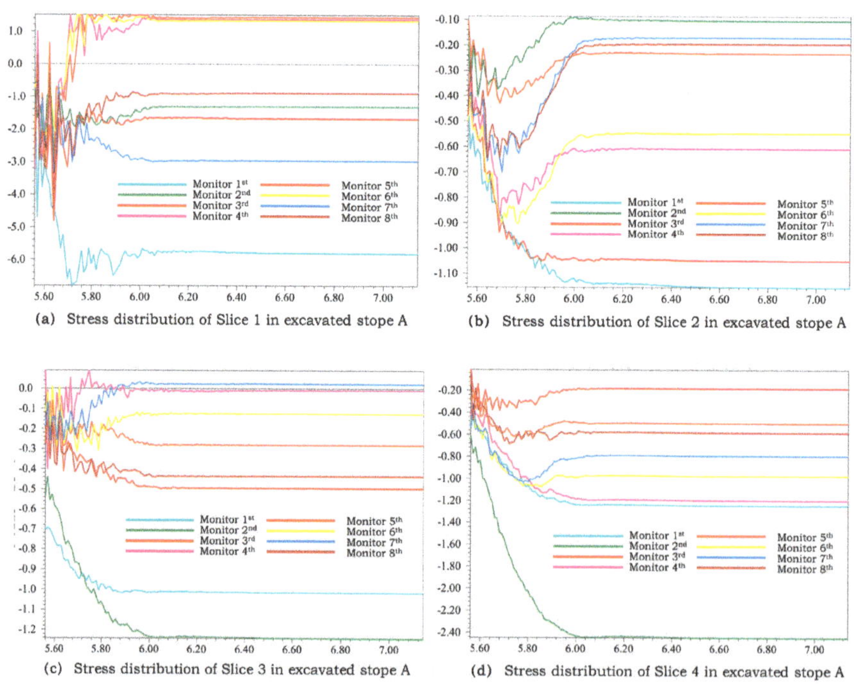

Figure 7. Stress distribution features in Slices 1 to 4 for excavated Stope A.

As Figure 8a shows, the vertical stress of the roof changed greatly when Stope A was excavated. Tensile and compressive stresses co-exist in the stope side walls after excavation, especially in Slice 1. Besides, the maximum stress near the center of the roof was within 5 MPa. The tensile stress appeared near the center of Slice 1, but the maximum stress at the edge reached as high as 24.4 MPa. In addition, there is obvious stress concentration in Slice 4 near the subsidence area, and the analysis shows that the stability of the roof is already poor at this time, thus there is a danger of cracking [22]. Further, it can be inferred that a pressure arch is formed above the roof, and the weight of the rock mass in the arch is transferred to the surrounding rock and pillars nearby, resulting in stress concentration; the pillars are gradually damaged by compression [23,24].

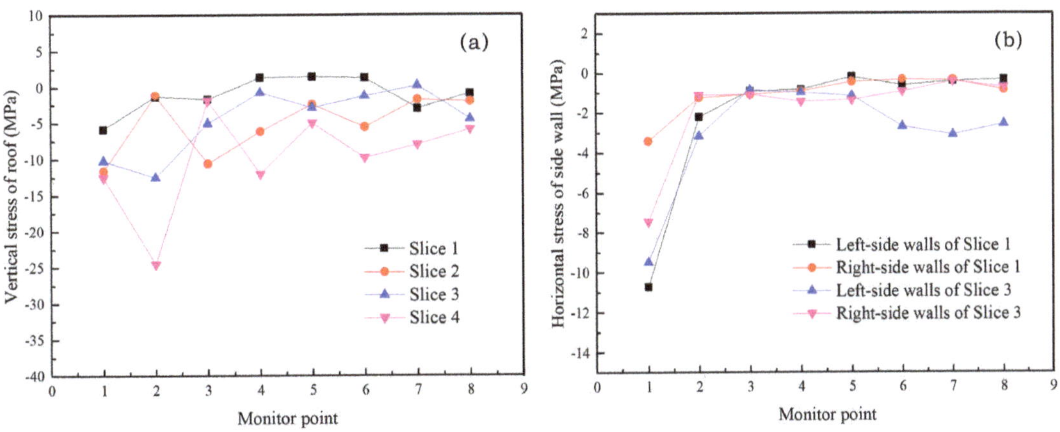

Figure 8. Stress curve for roof, left- and right-hand side walls of excavated Stope A. (**a**) Vertical stress of roof, (**b**) Horizontal stress of side wall.

Different from the serious fracture deformation of the roof, for the horizontal stress of the left- and right-hand side walls of Stope A (Figure 8b), the stress in the horizontal direction of the side walls is mainly concentrated in the upper part of the side walls, where it is close to the roof of the excavated stopes. There is no tensile stress for the horizonal stress of the side walls, and the maximum stress does not exceed 12 MPa, which is much lower than the stress value in the roof. This, therefore, indicates that the rock stability of excavated stopes is much better in the side walls compared with that in the roof.

4. Conclusions

In view of the disadvantages of the homogenization and simplification of many models, this paper comprehensively considers the differences in the topography and features of a mining area and the characteristics of a deposit, and innovatively adopts the 3DMine–Rhino^{3D}–FLAC^{3D} coupling method to realize the simulation of the stress and displacement visualization of the excavated stopes in the subsidence area. The main conclusions are summarized as follows:

1. Based on the numerical prediction and analysis of the displacement and stress after the excavation of Stope A, it is known that the roof displacement of this stope is significantly larger than that of the side wall, and the maximum displacement is 6.7 cm. There is a very obvious stress concentration at the maximum point of 24.4 MPa.
2. The influence of the collapse area on the stope is mainly reflected in the roof displacement. As the stope is far away from the collapse area, the roof displacement decreases significantly, the maximum displacement is at the roof center of the stope, and the maximum stress occurs at the corner of the stope.

3. This method has some limitations. It does not take into account the influence of factors such as stope temperature or rock water permeability, and does not include ventilation pipes, mining quasi-cutting and other projects. In the future, we will include these aspects to make the simulation more accurate and to improve the corresponding expression between the displacement and stress in the roof.

Author Contributions: Conceptualization, L.W., X.Z. (Xuelan Zhang) and Y.J.; methodology, X.Z. (Xingquan Zhang), Y.J. and H.K.; software, X.Z. (Xingquan Zhang), Y.J. and H.K.; validation, L.W. and S.Y.; data curation, L.W., S.Y. and H.K.; writing—original draft preparation, L.W. and X.Z. (Xingquan Zhang); writing—review and editing, L.W., S.Y. and H.K.; visualization, L.W.; supervision, S.Y. and L.W.; project administration, L.W. and S.Y.; funding acquisition, L.W. and S.Y. All authors have read and agreed to the published version of the manuscript.

Funding: This research was funded by the China National Postdoctoral Program for Innovative Talents (BX20220036); the Program of the National Natural Science Foundation of China (52204124); the China Postdoctoral Science Foundation (2022M710356); an Open Project at the State Key Laboratory of Coal Mine Resources and Safe Mining (SKLCRSM22KF006); an Open Project at the State Key Laboratory of Safety and Health for Metal Mines (2021-JSKSSYS-01); an Open Project at the Key Laboratory of Green Chemical Engineering Process of the Ministry of Education (GCP202108); and the Key Program of National Natural Science Foundation of China (52034001).

Data Availability Statement: The data are not publicly available for reasons of confidentiality.

Acknowledgments: Thanks for the great effort by editors and reviewers.

Conflicts of Interest: The authors declare no conflict of interest.

References

1. Bogert, H.; Jung, S.J.; Lim, H.W. Room and Pillar Stope Design in Highly Fractured Area. *Int. J. Rock Mech. Min. Sci.* **1997**, *34*, 145.e1–145.e14. [CrossRef]
2. Abdellah, W.; Raju, G.D.; Mitri, H.S.; Thibodeau, D. Stability of Underground Mine Development Intersections during the Life of a Mine Plan. *Int. J. Rock Mech. Min. Sci.* **2014**, *72*, 173–181. [CrossRef]
3. Emad, M.Z.; Mitri, H.; Kelly, C. Dynamic Model Validation Using Blast Vibration Monitoring in Mine Backfill. *Int. J. Rock Mech. Min. Sci.* **2018**, *107*, 48–54. [CrossRef]
4. Mukhopadhyay, S.; Liu, H.-H.; Spycher, N.; Kennedy, B.M. Impact of Fluid–Rock Chemical Interactions on Tracer Transport in Fractured Rocks. *J. Contam. Hydrol.* **2013**, *154*, 42–52. [CrossRef]
5. Wang, L.; Yin, S.; Deng, B. Understanding the Effect of Stepwise Irrigation on Liquid Holdup and Hysteresis Behavior of Unsaturated Ore Heap. *Minerals* **2021**, *11*, 1180. [CrossRef]
6. Wang, L.; Yin, S.; Wu, A. Ore Agglomeration Behavior and Its Key Controlling Factors in Heap Leaching of Low-Grade Copper Minerals. *J. Clean. Prod.* **2020**, *279*, 123705. [CrossRef]
7. Wang, L.; Yin, S.; Deng, B.; Wu, A. Copper Sulfides Leaching Assisted by Acidic Seawater-Based Media: Ionic Strength and Mechanism. *Miner. Eng.* **2022**, *175*, 107286. [CrossRef]
8. Huang, Y.; Zolfaghari, N.; Bunger, A.P. Cohesive Element Simulations Capture Size and Confining Stress Dependence of Rock Fracture Toughness Obtained from Burst Experiments. *J. Mech. Phys. Solids* **2022**, *160*, 104799. [CrossRef]
9. Ke-ping, Z.; Chao-Lan, Z.; Huan-Yu, D.; Tao, L. Impact Breakage Behavior of Stope Floor in Mined-Out Space. *Procedia Eng.* **2011**, *26*, 1731–1739. [CrossRef]
10. Yang, Z.; Zhai, S.; Gao, Q.; Li, M. Stability Analysis of Large-Scale Stope Using Stage Subsequent Filling Mining Method in Sijiaying Iron Mine. *J. Rock Mech. Geotech. Eng.* **2015**, *7*, 87–94. [CrossRef]
11. Hakami, H. Rock Characterisation Facility (RCF) Shaft Sinking—Numerical Computations Using FLAC. *Int. J. Rock Mech. Min. Sci.* **2001**, *38*, 59–65. [CrossRef]
12. Bock, S. New Open-Source ANSYS-SolidWorks-FLAC3D Geometry Conversion Programs. *J. Sustain. Min.* **2015**, *14*, 124–132. [CrossRef]
13. Wang, L.; Yin, S.; Wu, A. Visualization of Flow Behavior in Ore Segregated Packed Beds with Fine Interlayers. *Int. J. Miner. Metall. Mater.* **2020**, *27*, 900–909. [CrossRef]
14. Wang, L.; Yin, S.; Wu, A.; Chen, W. Synergetic Bioleaching of Copper Sulfides Using Mixed Microorganisms and Its Community Structure Succession. *J. Clean. Prod.* **2020**, *245*, 118689. [CrossRef]
15. Xiao, P.; Mao, H.; Qian, B.; Li, B.; Yang, X.; Xu, N. Stability Analysis of Surrounding Rock Mass in Underground Powerhouse Considering Damage Effect of Microfractures. *J. Rock Mech. Geotech. Eng.* **2022**, *14*, 1115–1130. [CrossRef]
16. Peng, H.; Cai, Q.; Zhou, W.; Shu, J.; Li, G. Study on Stability of Surface Mine Slope Influenced by Underground Mining below the Endwall Slope. *Procedia Earth Planet. Sci.* **2011**, *2*, 7–13. [CrossRef]

17. GEO-SLOPE International Ltd. *Preferential Flow in Vertically Layered Soil*; GEO-SLOPE International Ltd.: Calgary, AB, Canada, 1997; pp. 1–8.
18. Jia, Q.; Tao, G.; Liu, Y.; Wang, S. Laboratory Study on Three-Dimensional Characteristics of Gravity Flow during Longitudinal Sublevel Caving. *Int. J. Rock Mech. Min. Sci.* **2021**, *144*, 104815. [CrossRef]
19. Wang, R.; Li, C.; Xu, J.; Pan, L. Development and Verification of Large Deformation Model Considering Stiffness Deterioration and Shear Dilation Effect in FLAC3D. *Int. J. Min. Sci. Technol.* **2018**, *28*, 959–967. [CrossRef]
20. Wang, L.; Yin, S.; Wu, A.; Chen, W. Effect of Stratified Stacks on Extraction and Surface Morphology of Copper Sulfides. *Hydrometallurgy* **2020**, *191*, 105226. [CrossRef]
21. Zhai, M.; Wang, D.; Zhang, Z.; Zhang, L.; Yang, F.; Huang, B.; Zhong, A.; Li, L. Numerical Simulation and Multi-Factor Optimization of Hydraulic Fracturing in Deep Naturally Fractured Sandstones Based on Response Surface Method. *Eng. Fract. Mech.* **2022**, *259*, 108110. [CrossRef]
22. Wu, F.; Fan, X.; Li, D.; Yang, H.; Liu, J.; Li, X. Experimental Study on Fracability and Fracture Modes of High Permeability Sandstone. *Eng. Fract. Mech.* **2021**, *259*, 108128. [CrossRef]
23. Dai, J.; Yang, J.; Yao, C.; Hu, Y.; Zhang, X.; Jiang, Q.; Zhou, C. Study on the Mechanism of Displacement Mutation for Jointed Rock Slopes during Blasting Excavation. *Int. J. Rock Mech. Min. Sci.* **2022**, *150*, 105032. [CrossRef]
24. Basahel, H.; Mitri, H. Probabilistic Assessment of Rock Slopes Stability Using the Response Surface Approach—A Case Study. *Int. J. Min. Sci. Technol.* **2019**, *29*, 357–370. [CrossRef]

minerals

Article

Flocculation and Settlement Characteristics of Ultrafine Tailings and Microscopic Characteristics of Flocs

Chongchong Han [1,2], Yuye Tan [1,2,*], Lishen Chu [3], Weidong Song [1,2] and Xin Yu [1,2]

1 State Key Laboratory of High-Efficient Mining and Safety of Metal Mines of Ministry of Education, University of Science and Technology Beijing, Beijing 100083, China; g20198065@xs.ustb.edu.cn (C.H.); songwd@ustb.edu.cn (W.S.); aurora.xinyu@gmail.com (X.Y.)
2 School of Civil and Resource Engineering, University of Science and Technology Beijing, Beijing 10083, China
3 Daye Iron Mine Co., Ltd., Wuhan Iron and Steel Resources Group, Huangshi 435006, China; chulishen123@163.com
* Correspondence: tanyuye@ustb.edu.cn; Tel.: +86-132-600-35766

Abstract: Aiming to solve the problems related to the slow settling speed and the long-term consumption of ultra-fine tailings in mine filling, the effect of flocculant type on the flocculation and settling performance of ultra-fine tailings was studied through static sedimentation experiments on tailings. The microstructure of the flocculation was observed and analyzed using an electron microscope. On this basis, the selection of the optimum flocculant type and dosage parameters was carried out. The results show that the best addition amount of the AZ9020 anionic flocculant was 30 g/t, a solution concentration of 0.3%, and a stirring time of more than 45 min. The floc structure of the full-tailings flocculation solution was formed by the AZ9020 anionic flocculant. Moreover, the size of less than 0.1 μm was still relatively large; thus, the overall size of the structure was small and uniformly dispersed. The floc solution had the smallest porosity, the fractal dimension was the largest, the molecular weight of the floc was the largest, and the floc was the most compact, making it appropriate for the rapid removal of floc structures from water. Sedimentation is also the best flocculant for flocculation and sedimentation. The size of the flocs decreased as the height of the flocculation sediment bed increased during flocculation and sedimentation. The research results provide a microscopic view for the selection of the best flocculant type.

Keywords: ultrafine tailing; flocculating sedimentation; floc microstructure; fractal dimension; gray value

Citation: Han, C.; Tan, Y.; Chu, L.; Song, W.; Yu, X. Flocculation and Settlement Characteristics of Ultrafine Tailings and Microscopic Characteristics of Flocs. *Minerals* **2022**, *12*, 221. https://doi.org/10.3390/min12020221

Academic Editor: Geoffrey S. Simate

Received: 27 December 2021
Accepted: 7 February 2022
Published: 9 February 2022

1. Introduction

As a cement filling, tailings can prevent surface subsidence and can reduce the occupation of ground space by tailing stacking. Thus, the use of tailings as a cement filling is an inevitable choice for the green development of current mines [1–3]. However, ultrafine tailings encounter problems, such as a long natural settlement time and slow speed, due to their high content of fine particles, thus failing to meet the large-scale continuous filling requirements of mines [4–6]. Increasing the sedimentation efficiency by adding flocculants is a common practice in current mines to meet the needs of large-scale continuous filling underground. The sedimentation speed and concentration of tailings particles in the flocculation sedimentation process are affected by many factors, such as the type of flocculant [7,8]. Therefore, selecting a reasonable flocculation type according to the characteristics and properties of tailings is crucial to ensure the flocculation and settlement effect of mine-filling systems.

In recent years, scholars at home and abroad have conducted various studies on the sedimentation laws of ultrafine tailings. Jiao created a tailings sedimentation velocity model through experiments and divided the tailings sedimentation process into the following six stages: turbulent flow affected, accelerated sedimentation, final sedimentation velocity, interference sedimentation, compaction sedimentation, and ultimate sedimentation [9]. Yang conducted static flocculation sedimentation and slurry rheological test experiments by investigating three factors, including the flocculant type, the volume fraction of the feed, and the unit consumption of the flocculant [10]. Li conducted a static flocculation and sedimentation experiment on ultrafine tailings and obtained the optimal flocculant unit consumption as well as other parameters [11]. Xue performed deep-cone dense dynamic sedimentation experiments to analyze the influence of the feeding speed, rake stirring speed, and slurry mass fraction on the settlement behavior of flocs and obtained the optimal feeding and rake stirring speeds [12]. Wu studied various factors influencing the flocculation and sedimentation processes through laboratory experiments and determined the degree of influence [13]. Gheshlaghi M. E. and Ruan Z. E. applied numerical simulation methods to study the flocculation and sedimentation behaviors of the tailings in a deep-cone thickener [14,15]. Hou conducted scanning electron microscopy (SEM) observations of tailings flocculation sedimentation samples, analyzed the particle size distribution of tailings particles in different settlement areas, and explored the sedimentation characteristics of tailings particles of various sizes [16]. Chen used SEM technology to analyze the floc structures, explained the characteristics of the concentration distribution and gel point concentration of the viscous sediment from the perspective of the floc microstructure, and discussed common morphological pore parameters [17]. Yang used SEM to observe the microstructure of the mortar in the compaction zone and found that the flocs had three different forms of sedimentation, compaction, and stability, which were accompanied by changes in the size of the flocs [18]. Hu analyzed the pore structure of the tailing fillings, quantified the fractal dimension of the pores [19], and studied the relationship between the grading index, pore structure, and strength.

The aforementioned studies show that laboratory experiments are an effective method through which to examine the rules of tailings flocculation and sedimentation, while nuclear magnetic resonance and micro-electron microscopy provide the possibility of investigating the microstructure characteristics of tailings flocs during flocculation and sedimentation. In this paper, laboratory experiments are used to study the flocculation and sedimentation characteristics of ultrafine tailings in the Daye Iron Mine, and the microstructure characteristics and floc parameters at different bed heights during flocculation and sedimentation are analyzed through nuclear magnetic resonance (NMR) and micro-electron microscope observations, respectively, thus providing a scientific reference for selecting the best type of flocculant. This paper takes ultrafine iron ore tailings as the research object to conduct experiments, and the research conclusions can provide a certain degree of reference and a basis for other similar iron ore mines with the same grain size and chemical composition.

2. Materials and Methods

2.1. Physical and Chemical Properties of Ultrafine Tailings

Ultrafine tailings from the Daye Iron Mine were selected for the experiment. The physical parameters are shown in Table 1, and the particle size composition is presented in Figure 1. The figure shows that the gradation of ultrafine tailings is not particularly uniform. Several coarse and fine particles are observed, and only a few intermediate particles are found. The average median and surface area volume and average particle sizes of the tailings were relatively small, belonging to the category of ultrafine particle tailings. The tailings sorting coefficient was relatively large, making it suitable as a raw material for downhole filling.

Table 1. Physical properties of ultrafine tailings from the Daye Iron Mine.

Median Size d50/μm	Mean Diameter of Surface Area Volume D(3, 2)/μm	Sorting Coefficient/S_0	Specific Surface Area by Weight $\omega/m^2/kg$
37.925	14.319	5.595	6400

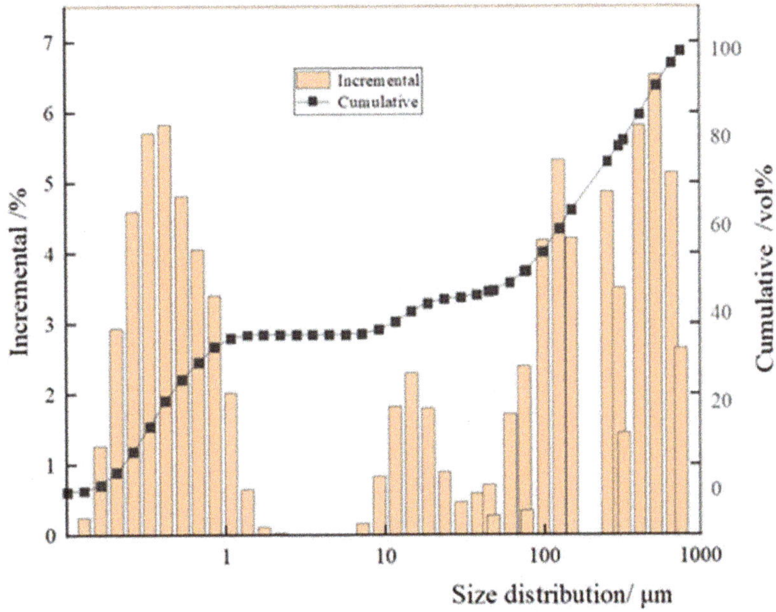

Figure 1. Ultrafine tailings from the Daye Iron Mine: screening size statistics chart.

Table 2 shows that the main mineral components of the tailings are inert oxides SiO_2, TFe and SFe and contain active oxides, such as CaO, Al_2O_3, and MgO. Therefore, ultrafine tailings are inert materials. However, these tailings have corresponding cementing activity and can be stimulated for comprehensive utilization.

Table 2. Main chemical composition of Daye Iron tailings.

Component	SiO_2	TFe	SFe	CaO	FeO	Al_2O_3	MgO
Content (%)	26.30	20.79	20.02	12.45	10.90	6.07	5.55

2.2. Flocculant Parameters

Numerous studies have shown that anionic flocculants have superior flocculation effects on tailings [20,21]. Therefore, the flocculants used in this article, namely AZ358, AZ625, AZ9020, and AZ505, are all anionic. The relevant parameters are shown in Table 3.

Table 3. Technical flocculant index.

Type	Experiment Number	Exterior	Molecular Weight/10,000	Solid Content/%
AZ358 anionic	L2	White particle	1400	≥95.5
AZ625 anionic	L3	White granules or powder	1500–1600	≥95
AZ9020 anionic	L4	White granules or powder	1500–1600	≥95.5
AZ505 anionic	L5	White granules or powder	1600	≥95

2.3. Experimental Program and Process

Regardless of the cross-effect between the ultrafine tailings concentration and the additional amount of flocculant, the type of flocculant, unit consumption, concentration, and mixing time using a fixed ultrafine tailings concentration of 30% were determined through indoor sedimentation experiments via the single-factor analysis method. The research, experimental program, and process are shown in Table 4. The height and time of the solid–liquid separation surface drop during the experiment were observed and recorded, as shown in Figure 2.

Table 4. Experimental scheme and process of ultrafine tailings.

Experimental Steps	Experiment Code	Variable/Experiment Number	1	2	3	4	5	Purpose
1	L	Type	Natural setting	AZ358	AZ625	AZ9020	AZ505	Determine the best flocculant model
2	D	Unit consumption	0 g/t	10 g/t	20 g/t	30 g/t	40 g/t	Determine the best unit consumption of the flocculant
3	N	Concentration	0	0.05%	0.1%	0.2%	0.3%	Determine the optimal flocculant concentration
4	T	Churning time	10 min	15 min	25 min	35 min	45 min	Determine the best flocculant mixing time

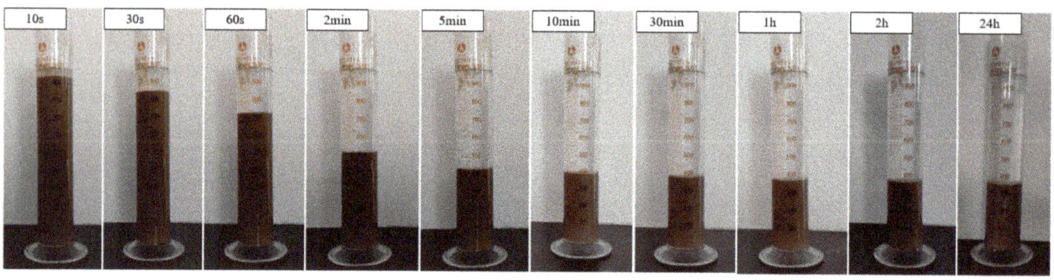

Figure 2. Change process of the liquid surface settlement height of ultrafine total tailings flocculation solution at different times.

2.4. Nuclear Magnetic Resonance (NMR) Analysis Experiment

The experimental system used for the NMR analysis in this experiment (MESOMR23-060H-I, Suzhou, China), which has a hydrogen spectrum, is a kind of nuclear magnetic resonance effect of H^{-1} that can be used for nuclear magnetic resonance spectroscopy and has the following parameters: resonance frequency: 23 MHz, magnet temperature: 25–35 °C, temperature control accuracy: ±0.05 °C. The NMR spectrum is shown in Figure 3a.

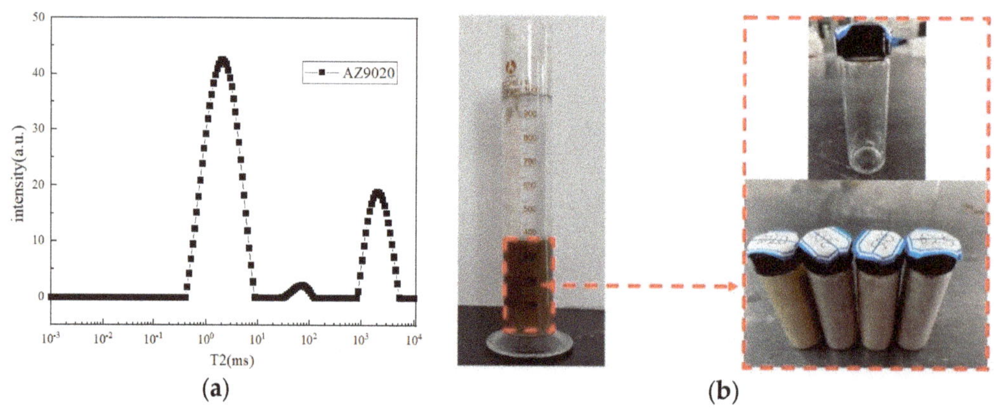

Figure 3. Floc extraction location and NMR analysis specimen (**a**) NMR analysis specimen (**b**) Floc extraction location.

The flocculants AZ358, AZ625, AZ9020, and AZ505 were used to perform the flocculation and sedimentation experiments with a solution concentration of 0.3%, a unit consumption of 30 g/t, and a stirring time of 45 min. The stirrer was taken out, and the solution was allowed to stand, and it was then timed to observe the height of the clarified layer in the cylinder. The flocculation solution was extracted at 200 mL of the graduated cylinder after 30 min and was placed in a sealed glass bottle (Figure 3b). The experimental system used for NMR analysis was also used for observation.

2.5. Micro-Electron Microscope Scanning Experiment (SEM)

For the SEM observation experiment, an SEM EVO 18 tungsten filament from Carl Zeiss (ZEISS) was adopted, along with an image size of 1000 × 750 pixels and an acceleration voltage of 20 kV.

SEM experiment ①: Figure 3 shows the extraction of the four flocculant floc solutions at a volume of 200 mL in a cylinder. This extraction was performed using a long pipette according to the exact scale of the extraction floccules and by gently dropping the cut-out on good filter paper. Liquid nitrogen freezing and fixed, conductive adhesive spraying carbon treatment procedures, the prepared samples were into the micro-electron microscope for scanning observation of the samples, and the electron microscope was operated at a magnification of 2000 times.

SEM experiment ②: The flocculant with the smallest porosity and fractal dimension was selected to conduct the flocculation settlement experiment of the ultrafine whole tailings a second time. The flocculant solutions at different bed positions (positions 1, 2, and 3, located at 300, 200, and 100 mL, respectively, in the measuring cylinder scale), such as those shown in Figure 4, were extracted and prepared for use as SEM observation samples according to the aforementioned method. The sample was observed by an operation electron microscope with magnification of 2000 times.

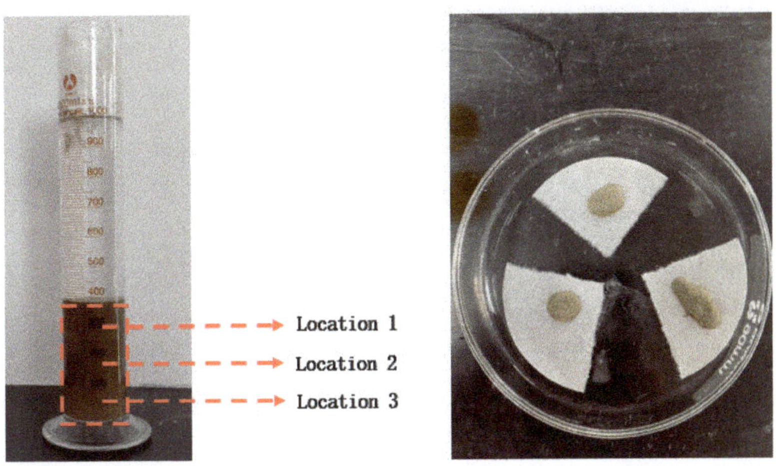

Figure 4. Floc extraction position and the scanning observation test piece under the microscope and electron microscope.

3. Results and Discussion

3.1. Flocculation Sedimentation Characteristics of Ultrafine Tailings

3.1.1. Analysis of the Influence of the Flocculant Type on the Settlement Effect

Figure 5a shows that the addition of a flocculant accelerates the sedimentation rate of the ultrafine tailings solution in the early stages (within 60 min), and the declining height of the clear layer of L4 is always lower than that of the four other groups (L1, L2, L3, and L5). The maximum difference is 80 mm. Figure 5b reveals that the curve of the average settling velocity experienced two peaks: the first peak appeared within the first 3 min, and the second peak appeared between 5 and 15 min. The calculation shows that the maximum settlement velocity of L4 is K4 = 7.80 mm/min, which is higher than that of the other groups. The limit concentrations of experimental groups L1, L2, L3, L4, and L5 after 24 h of sedimentation were 60%, 58.48%, 58.82%, 58.03%, and 58.59%, respectively, which met the mine-filling requirements for underflow concentrations. Experimental group L4 (AZ9020) generally showed the best reduction in the maximum settlement velocity and in the 1 h height of the clarification layer.

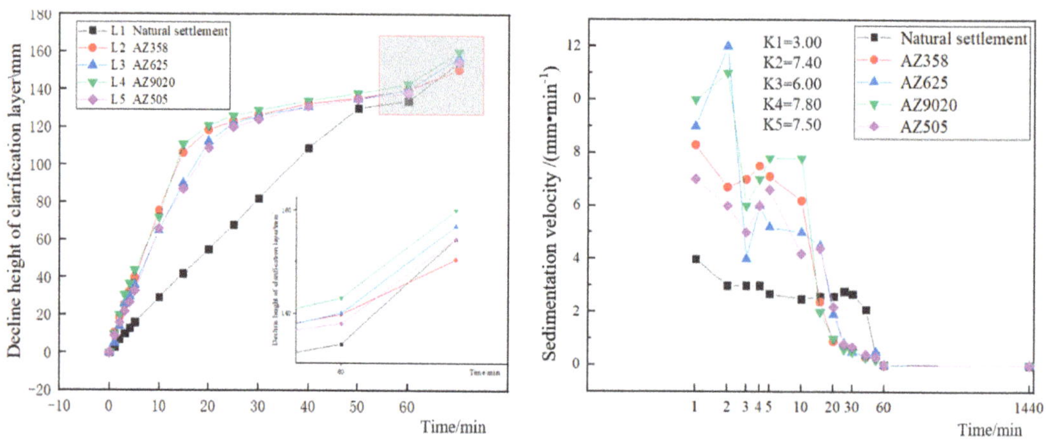

(a) (b)

Figure 5. Change curves of the height and average settling velocity of the ultrafine tailings solution under different types of flocculants (**a**) Drop height of clarification layer (**b**) Sedimentation velocity.

3.1.2. Analysis of the Influence of the Flocculant Unit Consumption on the Settlement Effect

Figure 6a shows that the height of the clarified layer of the tailings solution at first increased and then simultaneously decreased when the unit consumption of the flocculant increased. The unit consumption of experiment D3 was 30 g/t, representing the best consumption that was reached. Figure 6b indicates that the maximum settlement velocity first increased and then decreased as the single consumption of the flocculant increased. The single consumption in experiment D4 was 30 g/t, and the maximum settlement velocity was 8.20 mm/min.

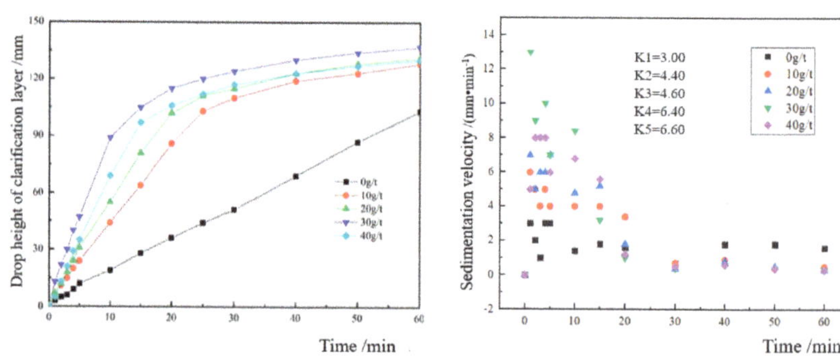

(a) (b)

Figure 6. Change curves of height and sedimentation velocity of clarifier under different flocculant consumptions (AZ9020) (**a**) Drop height of clarification layer (**b**) Sedimentation velocity.

3.1.3. Analysis of the Influence of Flocculant Concentration on the Sedimentation Effect

Figure 7a shows that the clarification layer height increases at the initial sedimentation stage (0−10 min) as the concentration of the flocculant solution rises. The height of the clarification layer is the largest at the initial sedimentation stage, when the experimental N5 concentration is 0.3%, resulting in the best flocculation sedimentation effect. Figure 7b shows that the average sedimentation velocity at each concentration reached its maximum between 5 and 10 min, and the maximum appeared when the experimental N5 concentration was 0.3%, showing a height of 71.00 mm/min.

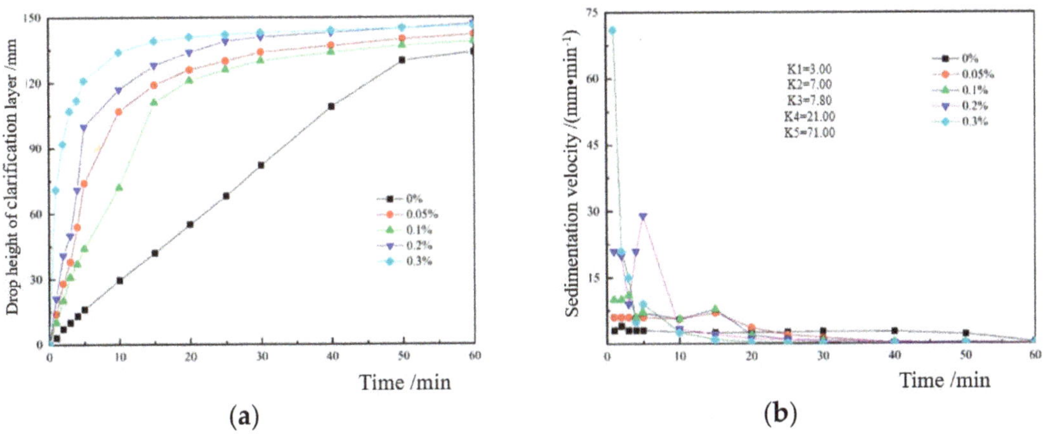

Figure 7. Change curves of height and sedimentation velocity of the clarifier under different flocculant solution concentrations (AZ9020) (**a**) Drop height of clarification layer (**b**) Sedimentation velocity.

3.1.4. Analysis of the Influence of Flocculant Stirring Time on the Settlement Effect of Ultrafine Tailings

Flocculant AZ9020 was chosen under the 0.3% concentration solution and 30 g/t unit consumption conditions (step 4). Figure 8a shows that the height of the clarified layer increased under the same settling time conditions even though the stirring time increased. Experiments T1, T2, and T3 revealed the presence of transparent floccules of undissolved flocculant that were visible to the naked eye, indicating that the flocculant was only partially dissolved in the water at this time and that the concentration did not reach 0.3%. Figure 8b shows that the clarified liquid height curve is close to the curve when the flocculant concentration is 0.05% when the stirring time is 15 min and that the curve is close to the curve when the concentration is 0.20% and at the stirring times of 25 and 35 min. To achieve an improved settling effect, the stirring time should be more than 45 min.

3.2. *Microscopic Characteristics of Floc Structure of Ultrafine Tailings*

In order to study the flocculation effect of different flocculants from the microscopic point of view further, nuclear magnetic resonance analysis was used to observe all of the tailings flocculant solutions formed by the different flocculants and to analyze the porosity and other microscopic parameters of the flocculant solutions.

3.2.1. Analysis of Pore Distribution Characteristics of Ultrafine Tailing Flocculent Solution

The pore size distribution curve of the flocculation solution under the four flocculants is shown in Figure 9a based on the NMR detection and analysis. The spectrum is shown in Figure 9b.

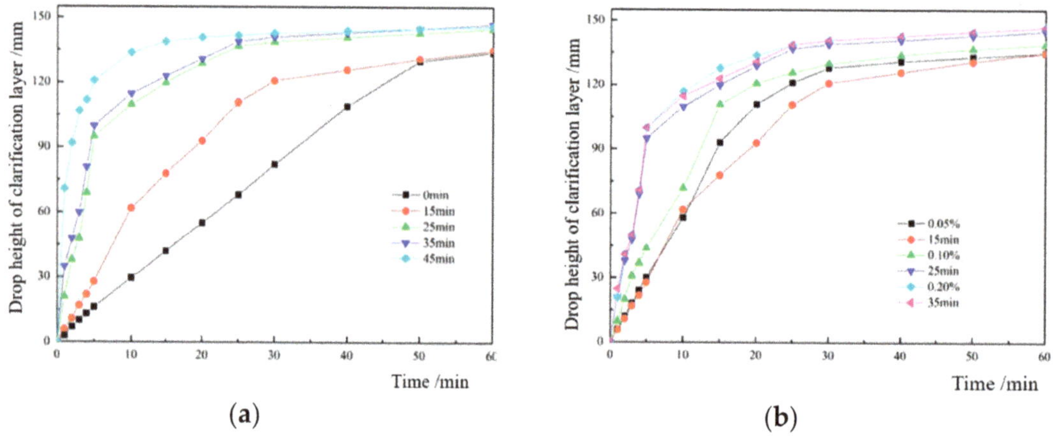

Figure 8. Height changes and transverse comparison curves of the clarifying layer at different stirring times (AZ9020) (**a**) Different mixing time (**b**) Different mixing time and concentration.

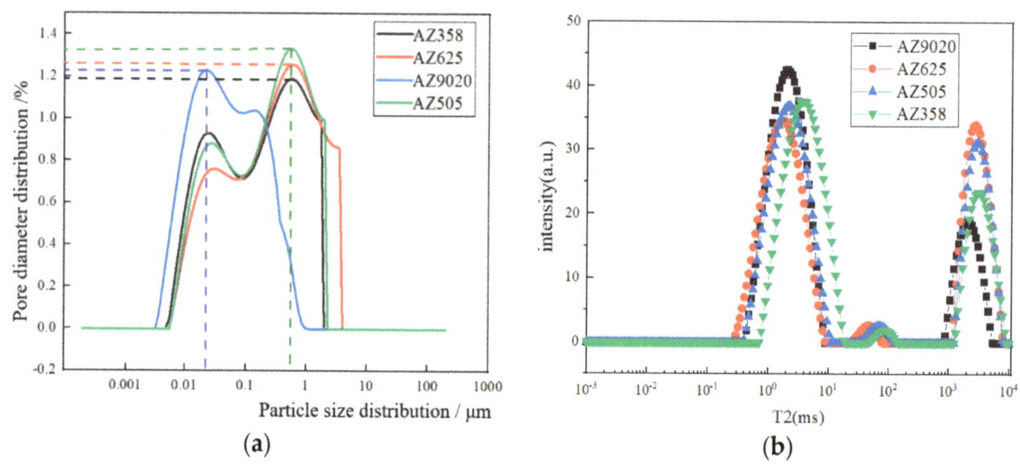

Figure 9. Pore size distribution curve and NMR analysis specimen of the flocculent solution of ultrafine tailings (**a**) Pore size distribution (**b**) NMR analysis specimen.

Figure 9 reveals that the pore size distribution of the floc solution under the AZ359, AZ625, and AZ505 flocculants is only slightly different. The main peaks are at the same position, and the width of the pore size distribution is insignificantly different. The two main peaks appear at 0.01−0.1 μm. The main peak at 0.1−1 μm and the width of the pore size distribution are both large, indicating that the flocs with a pore size larger than 0.1 μm account for a substantial proportion and that the size of the floc structure is also large.

The pore size distribution of the AZ9020 flocculant was significantly different from that of the other flocculants. The two main peaks were located at 0.01 and 0.1 μm. The largest main peak appeared at 0.01 μm, and the pore size distribution between 0.01 and 0.1 μm was also large, indicating that the flocs with a pore size smaller than 0.1 μm accounted for a substantial proportion. The results show that the flocs were small in size and that the whole solution was homodispersed. Solutions that were formed by adding the flocculants had a certain proportion of particles with a small pore size.

Research shows that the tightness of the material structure is negatively related to the porosity [22]; that is, a smaller porosity leads to a tighter material structure. The porosities of the flocculation solution corresponding to flocculants AZ358, AZ625, AZ9020, and AZ505 are 35.96%, 38.84%, 32.65%, and 37.37%, respectively. Among these solutions, the porosity of the flocculation solution corresponding to flocculant AZ9020 is the smallest. Flocculant AZ9020 has the best flocculation and sedimentation effects under the tailings concentration, the amount of additional flocculant, and the concentration.

3.2.2. Fractal Characteristics of Floc Structure of Ultrafine Tailings

The SEM images are shown in Figure 10 and were obtained through the scanning experiment ① using the electron microscope. The dark parts of the figure represent the pores, and the other parts represent the floc structure.

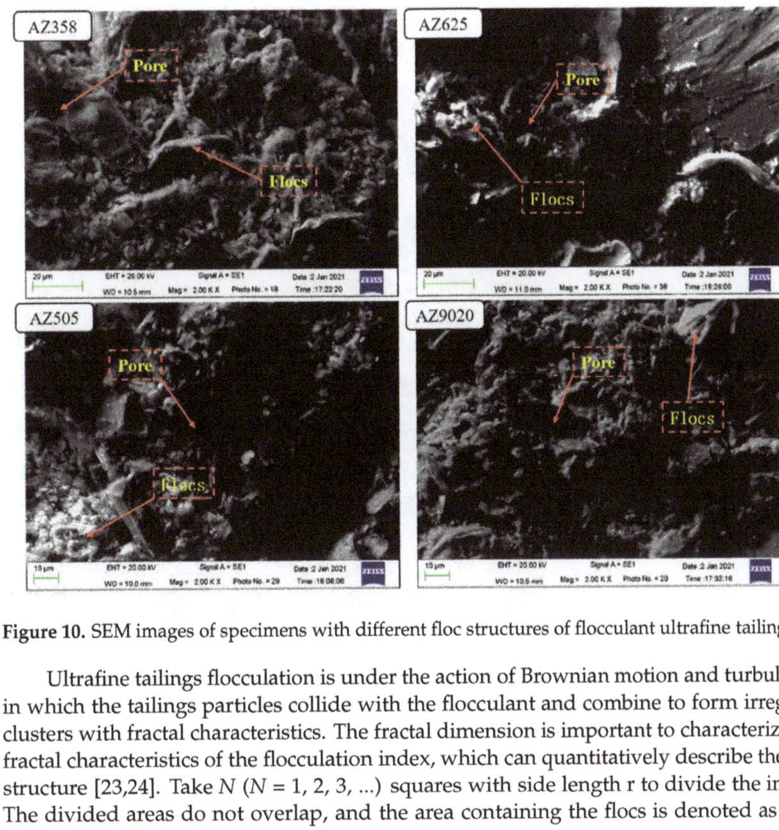

Figure 10. SEM images of specimens with different floc structures of flocculant ultrafine tailings.

Ultrafine tailings flocculation is under the action of Brownian motion and turbulence, in which the tailings particles collide with the flocculant and combine to form irregular clusters with fractal characteristics. The fractal dimension is important to characterize the fractal characteristics of the flocculation index, which can quantitatively describe the floc structure [23,24]. Take N (N = 1, 2, 3, ...) squares with side length r to divide the image. The divided areas do not overlap, and the area containing the flocs is denoted as $N(r)$. Formula (1) is then established as follows:

$$N(r) = 1/r^D \tag{1}$$

Formula (2) is obtained after taking the logarithm:

$$D = \frac{\lg N(r)}{-\lg r} \tag{2}$$

where D represents the fractal dimension.

The quantitative analysis based on the SEM scanning images reveals that image binarization work is the basis for obtaining fractal analysis. Binarization is also called threshold segmentation. Binarization is the process of setting the pixels of an image to 0 or 255 and then presenting the entire image with a clear black and white effect. Thus, binarization is essentially the process of classifying each pixel. Assuming that the size of the SEM picture is M × N, $f(x, y)$ represents the gray value of the pixel in the (x−1)th row and (y−1)th column of the image, where $0 \leq x \leq M$, $0 \leq y \leq N$, and x, y are integers. The principle of the gray-scale image binarization process is then presented as follows:

$$f(x,y) = \begin{cases} 1, f(x,y) \geq T \\ 0, others \end{cases} \tag{3}$$

where T refers to the threshold. All pixels in the overall image are either black or white after the binarization process. The binarization of the image markedly reduces the amount of data in the image, thus making the contour of the target prominent. Figure 11 shows the binarized image obtained after processing the SEM image, and Figure 12 reveals the obtained fractal characteristic curve.

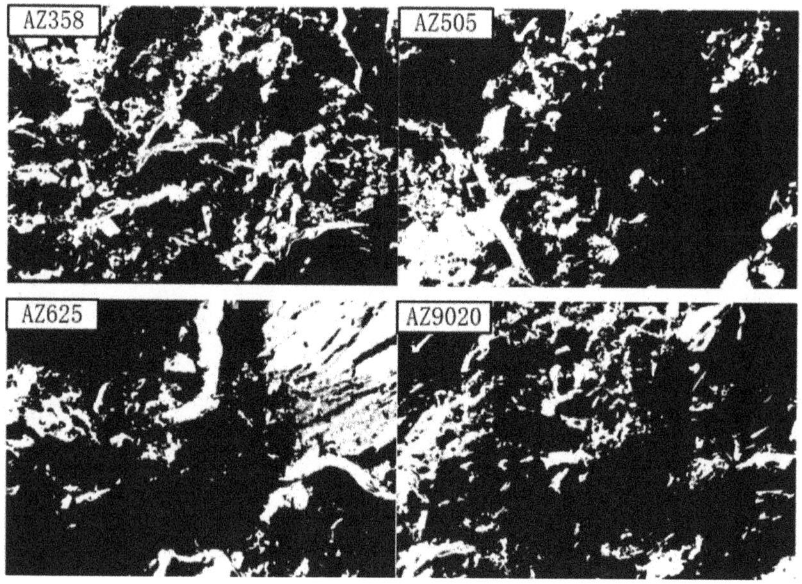

Figure 11. Binary image of tailings flocculation with different flocculants.

The calculated fractal dimensions of the samples under the four flocculants are 1.887, 1.894, 1.903, and 1.914. The largest fractal dimension indicates the compact flocs and the large molecular weights of the flocculants, which are conducive to rapid sedimentation from the water. A small particle spacing inside the flocs results in substantial differences between the densities of the flocs and the liquid, and large sedimentation speeds facilitate improved flocculation effects [25]. The fractal theory indicates that the order of the flocculation effects of the four flocculants is as follows: AZ9020 > AZ625 > AZ505 > AZ358. Flocculant AZ9020 has the best flocculation and sedimentation effects for ultrafine tailings.

3.3. Microscopic Characteristics of Floc Structure of Ultrafine Tailings

In this experiment, scanning electron microscopy was used to observe the different positions of the whole tailings floc solution formed by the same flocculant (AZ9020) to analyze the spatial morphological characteristics and microscopic parameters of the floc structure.

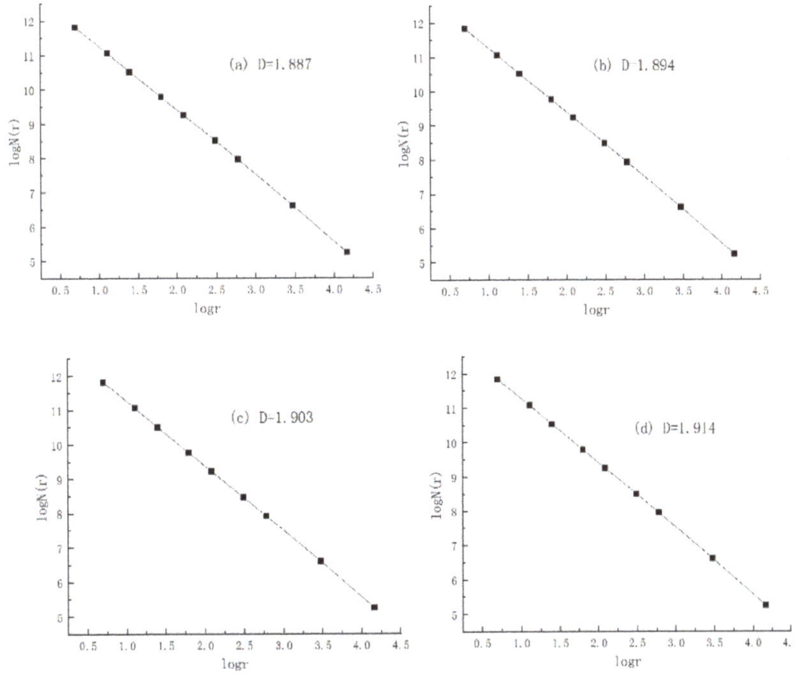

Figure 12. Fractal characteristic curve of floc structure: (**a**) AZ358, (**b**) AZ505, (**c**) AZ625, and (**d**) AZ9020.

3.3.1. Analysis of the Particle and Morphological Characteristics of Tailings Floccules

SEM images of different sedimentation bed heights of flocculant AZ9020 were obtained through the scanning electron microscope experiment ②, and the tailings flocs were acquired through binarization processing, image contour perfection, edge detection, and boundary discrete-point sealing treatment [26–29]. The block area is shown in Figure 13. The white part in the picture is the tailings flocs, and their outlines are clearly demonstrated. The particles of tailings flocs of various bed settlement heights show different characteristics. The upper tailings are dispersed, the size of the middle tailings is obviously larger, and the lower tailings become denser as a whole.

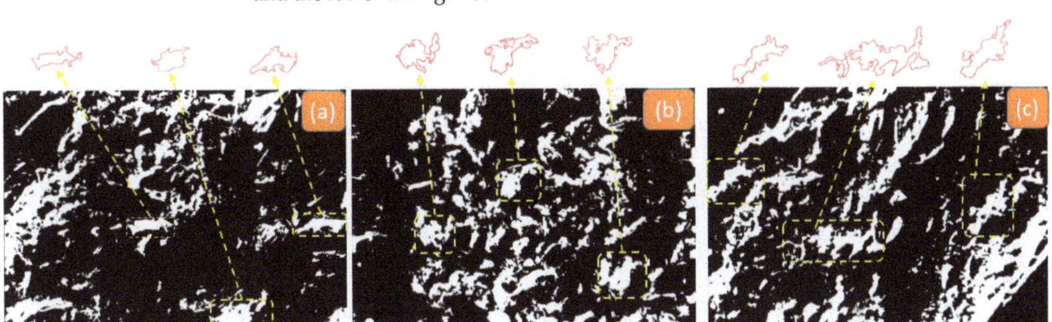

Figure 13. Particle size and morphological characteristics of tailings flocs with different settlement heights under ZYZ-type flocculant: (**a**) location 1; (**b**) location 2; (**c**) location 3.

The equivalent diameter of the floc was obtained in accordance with the enclosed area delineated by the selected floc profile, and the normal distribution curve of the tailings floc size was acquired as shown in Figure 14. The figure reveals the following: the upper part of the small-sized flocs accounted for a large proportion of the solution and were concentrated between 5 and 18 μm; the middle part of the flocs increased in size and were distributed between 10 and 30 μm; the lower part of the large-sized flocs accounted for substantially concentrated distributions between 20 and 60 μm. The average equivalent diameters of the upper, middle, and lower tailings floccules were 10.55, 16.2, and 38.79 μm, respectively. The diameter of a tailings floc is positively correlated with the settlement height.

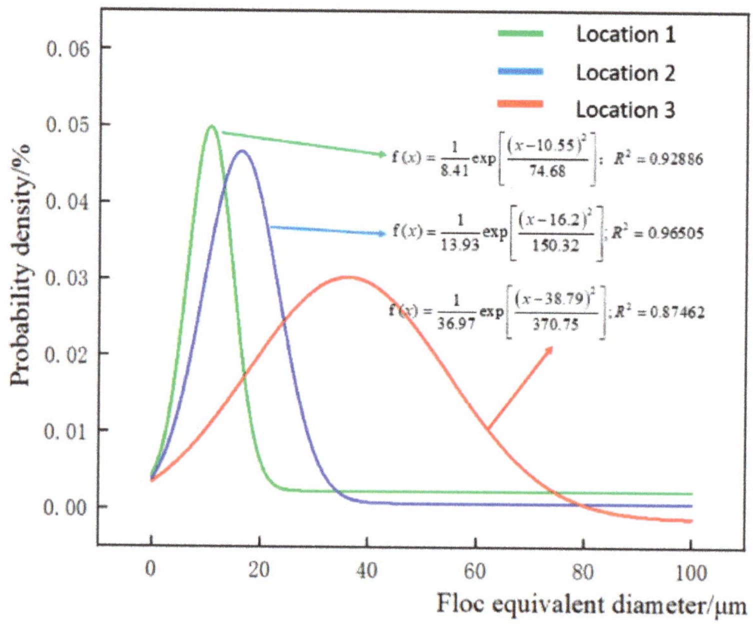

Figure 14. Distribution curves of equivalent diameters of tailings flocs at different heights under AZ9020-type flocculant.

3.3.2. Analysis of Gray-Scale Characteristics of Ultrafine Tailing Floc Structure

The gray value refers to the dark range of the image, where the white value is 255, and the black value is 0, which can intuitively represent the difference between pores and entities [30]. The original SEM images obtained from the scanning electron microscope ② experiment are analyzed on the basis of the following gray-scale characteristics:

Figure 15a shows that the gray value of flocs at the top of the settlement bed is approximately 180, and the gray value of only one pore is below 50, with an average gray value of 112. Figure 15b reveals that the gray value of the flocs in the middle is approximately 110, with an average gray value of 78. Figure 15c shows that the gray value of the flocs at the bottom is lower than 130, and the gray value of most pores is lower than 50, with an average gray value of 52. Thus, the average gray value of flocs decreases with the bed height. This finding indicates that the water content of the upper flocs is high, and the flocs are bright white. The internal water is constantly drained during the floc sedimentation process, the flocs gradually become dark, and the gray value slowly decreases, thus forming a high concentration of ultrafine total tailings floc solution.

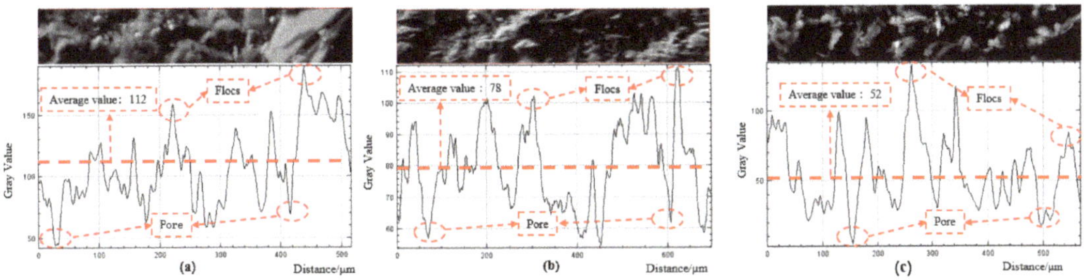

Figure 15. Floc gray-scale characteristic curves: (**a**) location 1; (**b**) location 2; (**c**) location 3.

3.3.3. Analysis of the Spatial Morphology of the Ultrafine Tailings Floc Structure

The different gray values of the original SEM images obtained by SEM experiment (2) were defined as yellow, green, cyan, and blue, and the gray values of some regions were extracted and converted into 3D graphics [31–33], such as those shown in Figure 16. The upper part of the bed settlement (Figure 16a) reveals that the size of flocs is large, most of which are over 20 μm. Moreover, the distribution is concentrated, and macropores that are 10 μm in size are found. The middle of the bed settlement (Figure 16b) shows that the size of flocs ranges from 15 μm to 20 μm, and the pore size is evenly distributed. The lower part of the bed settlement (Figure 16c) demonstrates that the size of most flocs ranges from 3 μm to 10 μm and that there are many small flocs and a relatively increased number of pores distributed around the flocs. The flocs were also evenly distributed. The size of tailings flocs decreased as the settlement height decreased. This finding indicates that large-sized flocs gradually settle, dehydrate, and disperse into small-sized flocs during the flocculation and sedimentation processes, forming a uniform and dense floc pore structure.

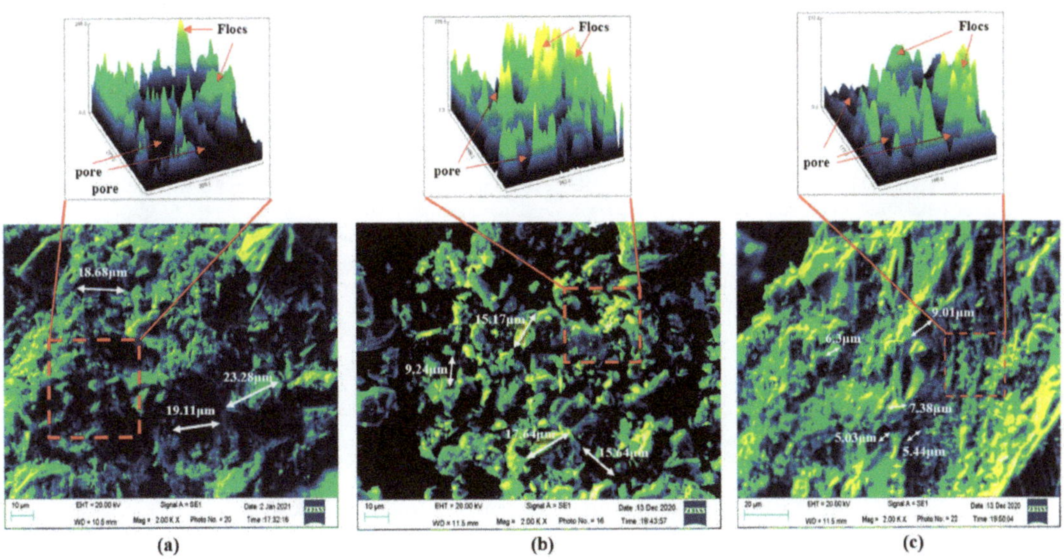

Figure 16. SEM image of tailings floccules of AZ9020 flocculant: (**a**) location 1; (**b**) location 2; (**c**) location 3.

4. Conclusions

The flocculation and sedimentation characteristics of ultrafine tailings were studied in this paper through indoor sedimentation experiments, NMR analysis, and micro-electron microscope scanning observations. A reasonable type of flocculant was selected, and the microscopic characteristics of the floc structure in the flocculation and sedimentation processes were analyzed. The main conclusions are as follows:

(1) The sedimentation characteristics of ultrafine tailings under different types of flocculants were studied by indoor flocculation sedimentation experiments, NMR monitoring and analysis, and scanning electron microscope observations. The results indicate that the optimal additional flocculant amount of ultrafine tailings from the Daye Iron Mine was 30 g/t, and the optimal concentration for the flocculant solution was 0.3%. The flocculant was completely dissolved when the flocculant was stirred more than 45 min, and the flocculation sedimentation effect was optimal under these conditions.

(2) The analysis of the pore distribution and fractal characteristics of the ultrafine tailings flocculation solution revealed that the size of flocs in the ultrafine tailings flocculation solution formed by the AZ9020 flocculant was less than 0.1 μm and that the overall structure size was small and evenly dispersed. The flocculant solution had the smallest porosity, the largest fractal dimensions, and the most compact flocs, indicating that it had the best flocculant sedimentation effect.

(3) The analysis of the spatial morphology and gray-scale characteristics of the ultrafine total tailings flocs showed that the average gray-scale value and size of the flocs decreased as the height of the flocculation sedimentation bed decreased. In the flocculation and sedimentation processes, the large-sized flocs gradually dehydrated and dispersed into small-sized flocs, forming a more uniform and compact pore structure for the flocs.

Author Contributions: Y.T., W.S. and C.H. designed the project; X.Y. and L.C. conducted the original literature reviews; C.H. and Y.T. wrote and organized the paper with a careful discussion and revision by W.S. All authors have read and agreed to the published version of the manuscript.

Funding: National Natural Science Foundation of China Youth Fund (52004019).

Informed Consent Statement: Informed consent was obtained from all subjects involved in the study.

Data Availability Statement: Not applicable.

Acknowledgments: We are very grateful to Daye Iron Mine Co. Ltd., for the experimental raw materials.

Conflicts of Interest: The authors declare no conflict of interest.

References

1. Yao, H.H.; Cai, L.B.; Liu, W.; Tan, W.Q.; Jiao, F.; Yang, R.C. Status quo and development of comprehensive utilization of waste rock from metal mines in my country. *Chin. J. Nonferrous Met.* **2021**, *31*, 1649–1660.
2. Yang, Z.L.; Tan, Y.Y.; Fu, J.X.; Zhen, D.; Wang, Y. Selection of Optimal Flocculant for whole tailings and Floc NMR Analysis. *Min. Res. Dev.* **2021**, *41*, 136–140.
3. Shui, C.; Wang, X.J.; Wang, X.M.; Zen, Q.; Wang, H.; Liu, J. Study on the law and mechanism of APAM unit consumption on the flocculation of particles of different sizes in the flocculation and settlement of whole tailings. *Min. Res. Dev.* **2020**, *40*, 67–73.
4. Du, J.F.; Hou, C.; Zhu, Z.W.; Liu, H.L.; Liu, X.G.; Wang, J.S. Tailings flocculation settlement test and its engineering application. *Met. Mine* **2020**, *1*, 95–100.
5. Wu, A.X.; Wang, Y.; Wang, H.J. Current status and trend of paste filling technology. *Metal Mine* **2016**, *7*, 1–9.
6. Shi, C.X.; Guo, L.J.; Chen, X. Static and dynamic flocculation settlement characteristics of tailings. *Chin. J. Nonferrous Met.* **2021**, *31*, 194–202.
7. Li, X.; Guo, L.J.; Yang, C. Experimental study on flocculation and settlement of whole tailings in a gold mine in anhui. *China Min. Mag.* **2017**, *26*, 326–329.
8. Guo, J.B.; Wang, H.J.; Tian, Z.G.; Du, X.H. The influence of different types of flocculants on the density of ultrafine tailings. *Min. Res. Dev.* **2021**, *41*, 141–145.
9. Jiao, H.Z.; Wu, A.X.; Wang, H.J.; Liu, X.H.; Yang, S.K.; Xiao, Y.T. Experimental study on flocculation and settlement characteristics of whole tailings. *J. Univ. Sci. Technol. Beijing* **2011**, *33*, 1437–1441.
10. Yang, L.H.; Wang, H.J.; Wu, A.X.; Li, H.; Hui, H.; Hui, C.Y. The effect of flocculation settlement on the rheological properties of whole tailings slurry. *J. Cent. South Univ. (Sci. Technol.)* **2016**, *47*, 3523–3529.
11. Li, J.X.; Shun, W.; Zhang, S.Y.; Liu, W.D. Factors Influencing Flocculation and Sedimentation of Tailings. *J. Kunming Univ. Sci. Technol. (Sci. Technol.)* **2021**, *46*, 45–53.

12. Xue, Z.L.; Yan, Z.P.; Jiao, H.Z.; Gan, Q.D.; Zhang, Y.Z.; Liu, Z.Y. Dynamic settlement of flocs in the deep cone thickening process of tailings. *Chin. J. Nonferrous Met.* **2020**, *30*, 2206–2215.
13. Wu, A.X.; Zhou, L.; Yin, S.H. Influencing factors of flocculation and settlement of whole tailings. *Chin. J. Nonferrous Met.* **2016**, *26*, 439–445.
14. Gheshlaghi, M.E.; Goharrizi, A.S.; Shanhrivar, A.A. Modelingindustrial thickener using computational fluid dynamics (CFD), acase study: Tailing thickener in the Sarcheshmeh copper mine. *Int. J. Min. Sci. Technol.* **2013**, *23*, 885–893. [CrossRef]
15. Ruan, Z.E.; Li, C.P.; Shi, C. Numerical simulation of flocculation andsettling behavior of whole-tailings particles in deep-conethickener. *Int. J. Cent. South Univ.* **2016**, *23*, 740–746. [CrossRef]
16. Hou, H.Z.; Li, C.P.; Wang, S.Y. Settlement velocity change and particle settling characteristics of thick and middle mud layer in tailings. *J. Cent. South Univ. (Sci. Technol.)* **2019**, *50*, 1428–1436.
17. Chen, Y.P.; Luo, W.; Xu, C.Y. Microstructure of high-concentration viscous sediment flocs and its effect on settlement characteristics. *J. Sediment Res.* **2020**, *45*, 8–14.
18. Ying, Y.; Wu, A.X.; Wang, H.J. Torque mechanics model and mechanism analysis of rake frame based on mud height. *J. Cent. South Univ. (Sci. Technol.)* **2019**, *50*, 165–171.
19. Hu, J.H.; Ren, Q.F.; Yang, D.J. Cross-scale characteristics of backfill material using NMR and fractal theory. *Trans. Nonferrous Met. Soc. China* **2020**, *30*, 1347–1363. [CrossRef]
20. Chen, D.; Song, W.D.; Wu, S. Experimental study on flocculation and settlement characteristics of tailings slurry. *Non-Ferr. Met. (Min. Part)* **2016**, *68*, 41–44.
21. Dwari, R.K.; Angadi, S.I.; Tripathy, S.K. Studies onflocculation characteristics of chromite's ore process tailing: Effect of flocculants ionicity and molecular mass. *Colloids Surf. A Physicochem. Eng. Asp.* **2018**, *53*, 467–477. [CrossRef]
22. Su, S.L.; Rao, Q.H.; He, Y.H. Effects of porosity on tensile mechanical properties of porous FeAl intermetallics. *Trans. Nonferrous Met. Soc. China* **2020**, *30*, 2757–2763. [CrossRef]
23. Zhao, J.; Fu, X.H.; Wang, J. Analysis of the effect of shear field on ultra-clean coal separation based on fractal dimension. *J. China Coal Soc.* **2016**, *41*, 2078–2085.
24. Zhou, X.; Wu, A.X.; Wang, H.J. The Evolution Law of Flocculation in the Thickening Process of Tailings Based on FBRM. *J. Eng. Sci.* **2021**, *43*, 1425–1432.
25. Zheng, D.; Song, W.D.; Tan, Y.Y. Research on fractal and microscopic quantitative characterization of unclassified tailings flocs. *Int. J. Miner. Metall. Mater.* **2021**, *28*, 1429–1439. [CrossRef]
26. Hu, W.W.; Zhou, R.G.; Fan, P. Quantum Image Edge Detection Based on Canny Algorithm. *J. Zhengzhou Univ. (Nat. Sci. Ed.)* **2020**, *52*, 37–52.
27. Li, J.; Chen, G.F.; Ding, X.Q. Research on Image Edge Detection Method Based on Improved Canny Algorithm. *Comput. Simul.* **2021**, *38*, 371–375.
28. Li, J.; Li, H.Y.; Yao, R.J. Check of MR image edge information of lumber intervertebral disc using an improved Canny algorithm. *J. Jilin Univ. (Eng. Technol. Ed.)* **2016**, *46*, 1704–1709.
29. Zhang, H.H.; Zhao, N.L.; Chao, Q. Influences of CO_2 on the Microstructure in Sheared Olivine Aggregates. *Minerals* **2021**, *11*, 493. [CrossRef]
30. Cai, G.P.; Wang, L.; Luo, X.Y. Multi-threshold Binary Algorithm for Ore Image Based on Block Processing. *Min. Res. Dev.* **2020**, *40*, 153–157.
31. Xu, H.K.; Han, X.; Qu, H.J. Image gray scale transformation combining false edge extraction and histogram analysis. *Opt. Precis. Eng.* **2017**, *25*, 538–544.
32. Sun, F.; Wang, F.Y.; Jiang, H.L. Analysis on the flocculation characteristics of algal organic matters. *J. Environ. Manag.* **2022**, *302*, 114094. [CrossRef] [PubMed]
33. Rosa, M.F.L.; Francisco, D.P.V.F.A. Characterization and concentration by selective flocculation/magnetic separation of iron ore slimes from a dam of Quadrilátero Ferrífero—Brazil. *J. Mater. Res. Technol.* **2020**, *9*, 2021–2027. [CrossRef]

MDPI
St. Alban-Anlage 66
4052 Basel
Switzerland
Tel. +41 61 683 77 34
Fax +41 61 302 89 18
www.mdpi.com

Minerals Editorial Office
E-mail: minerals@mdpi.com
www.mdpi.com/journal/minerals

www.ingramcontent.com/pod-product-compliance
Lightning Source LLC
Chambersburg PA
CBHW040840250926
51850CB00007BA/1362